Endothelial Cytoskeleton

Endothelial Cytoskeleton

Editors

Juan A. Rosado

and

Pedro C. Redondo

Department of Physiology

University of Extremadura

Cáceres

Spain

CRC Press

Taylor & Francis Group

Boca Raton London New York

CRC Press is an imprint of the

Taylor & Francis Group, an **informa** business

A SCIENCE PUBLISHERS BOOK

CRC Press
Taylor & Francis Group
6000 Broken Sound Parkway NW, Suite 300
Boca Raton, FL 33487-2742

First issued in paperback 2019

© 2014 Copyright reserved
CRC Press is an imprint of Taylor & Francis Group, an Informa business

No claim to original U.S. Government works

ISBN-13: 978-1-4665-9035-9 (hbk)
ISBN-13: 978-0-367-37951-3 (pbk)

Library of Congress Cataloging-in-Publication Data

Endothelial cytoskeleton / editors, Juan A. Rosado, Pedro C.
          Redondo.
       p. ; cm.
   Includes bibliographical references and index.
   ISBN 978-1-4665-9035-9 (hbk. : alk. paper)
   I. Rosado, Juan Antonio. II. Redondo, Pedro C.
   [DNLM: 1. Cytoskeleton--physiology. 2. Endothelial Cells--
                cytology. 3. Endothelial Cells--physiology.
                4. Endothelium, Vascular--cytology.   QU 350]

   QH603.C96
   571.6'54--dc23
                                                        2013016221

Visit the Taylor & Francis Web site at
http://www.taylorandfrancis.com

and the CRC Press Web site at
http://www.crcpress.com

Preface

Endothelial cytoskeleton, is intended to present a concise synthesis of the current knowledge and recent advances in the structure, organization and functional role of the cytoskeleton in endothelial cells (ECs).

The cytoskeleton regulates a number of processes within the cell. Thus, several proteins like the Rho superfamily, require association with the cytoskeletal in order to be fully active. Hence, understanding the molecular processes underlying cytoskeleton remodelling and its interaction with other cellular structures might be crucial to expand our knowledge regarding cell physiology. Particular attention is given to the involvement of the cytoskeleton in the regulation of vascular function mediated by the endothelium. In this sense, disruption of the endothelial layer underlies a number of cardiovascular diseases such as heart stroke, thrombosis, thrombotic thrombocytopenic purpura, deep venous thrombosis or pulmonary embolism. Therefore, this book intends to serve as a comprehensive resource for those interested in the fascinating biological processes associated with vascular biology.

The authors who have contributed in various ways to the compilation of this book are well known and are also active researchers in their respective fields. As a result the book contains chapters that are of high standard and present novel approaches to the subject at hand. Some chapters include original figures derived from the authors' current experimental results.

We would like to thank all contributors and reviewers that have shared their knowledge and experience with us to make this book project a reality. We hope that the information contained in this book will be helpful to the scientists working in this field as well as to students that initiate their research in this area.

We dedicate this book to our beloved families: we appreciate their bearing up with us for the time spent by us in the preparation of this book.

March 2013

Juan A. Rosado
Pedro C. Redondo

Contents

Preface   v

1. The Endothelial Cytoskeleton: Multifunctional Role of the   1
   Endothelial Actomyosin Cytoskeleton
   Yuka Shimizu and *Joe G.N. Garcia*

2. Emerging Themes on Pulmonary Microvascular Endothelial   27
   Cell: Microtubules and Microtubule Associated Proteins
   Cristiaan D. Ochoa, Ron Balczon and *Troy Stevens*

3. Endothelial Actin Cytoskeleton and Angiogenesis   53
   Sadiqa K. Quadri

4. Adherens Junctions and Endothelial Cytoskeleton   74
   Younes Smani and *Tarik Smani-Hajami*

5. Mechanical Force Transmission via the Cytoskeleton in   91
   Vascular Endothelial Cells
   Bori Mazzag, Cecile L.M. Gouget, Yongyun Hwang and
   Abdul I. Barakat

6. The Functional Role of the Microtubule/Microfilament   116
   Cytoskeleton in the Regulation of Pulmonary Vascular
   Endothelial Barrier
   Irina B. Alieva and *Alexander D. Verin*

7. Membrane-Cytoskeleton Interactions and Control of   146
   Vesicle Trafficking in Endothelial Cells
   Felicia Antohe

8. Cell-Matrix Adhesion Proteins in the Regulation of   167
   Endothelial Permeability
   Jurjan Aman, Geerten P. van Nieuw Amerongen and
   Victor W.M. van Hinsbergh

9. Endothelial Cells and the Regulation of Platelet Function   200
   P.C. Redondo, N. Dionisio, E. Lopez, A. Berna-Erro and *J.A. Rosado*

Index   231

Color Plate Section   233

1

The Endothelial Cytoskeleton
Multifunctional Role of the Endothelial Actomyosin Cytoskeleton

Yuka Shimizu[1] and Joe G.N. Garcia[2,]*

Cytoskeleton Overview

The mammalian cytoskeleton plays an essential role in providing the mechanical support necessary to maintain cell shape, cell motility and movement and in signaling transducing functions. The key cytoskeletal components are three inter-communicating networking protein filaments: actin microfilaments, microtubules, and intermediate filaments. Endothelial cells comprise the lining of the inner vessel wall and display a highly dynamic regulation of vascular integrity via opening and closing of paracellular gaps in response to infiltrating cells, ischemia or inflammatory processes (Dudek and Garcia 2001). As endothelial cells have direct contact with circulating blood cells and lymphatic fluid, the cytoskeleton plays a critical role in mechanotransduction, angiogenesis and apoptosis and in the vascular response to inflammation, serving as the major regulator of the endothelial "gate-keeper" function via effects on paracellular gap

[1]Institute for Personalized Respiratory Medicine, Department of Medicine, The University of Illinois at Chicago, Chicago, IL 60612, USA.
[2]Earl Bane Professor of Medicine, Pharmacology & Bioengineering, Chicago, IL 60612.
Email: jggarcia@uic.edu
*Corresponding author

regulation, leukocyte diapedesis and barrier regulation (Dudek and Garcia 2001, Garcia et al. 1986, Hirata et al. 1995, Majno and Palade 1961).

It is estimated that ~15–20% of endothelial cell proteins are comprised of cytoskeletal actin and myosin components, portraying the critical importance of actomyosin filament functions (Gottlieb et al. 1991). Actin filaments bind to plasma membrane proteins to control cell-cell and cell-matrix interactions by stabilizing intercellular junctions, tightly regulating the vascular barrier and therefore fluid and nutrition exchange between circulating blood and tissues. The major component of the actin filament is the small globular ATP-binding/hydrolyzing protein or G-actin. Activation of G-actin monomers by reversible ATP hydrolysis stimulates the assembly of filamentous F-actin polymers in coordination with formation and maintenance of actin filaments dynamically regulated by actin-associating proteins (Table 1) such as Arp2/3 complex, profilin, and gelsolin (Dudek et al. 2004, Garcia et al. 2001, Shasby et al. 1982). Actin filaments undergo "treadmilling", a process where the filament length remains approximately constant while actin monomers are added to the plus (+) end and dissociate from the minus (–) end of the polarized filament (DeMali et al. 2002, Gunst 2004). Arp2/3 is a nucleation protein stimulating the formation of new filaments and the branching of existing actin polymers. Arp2/3 complex appears in the leading edge of motile cells along with polymerizing proteins such as profilin that synergistically stimulates the addition of actin monomers at the (+) end. Capping proteins, such as gelsolin, stabilize the newly formed filaments by inhibiting the addition of G-actin to the polymers, and are well appreciated to be involved in cell movement (Fujita et al. 1997, Spinardi and Witke 2007).

The role of actin filaments in cell motility is well recognized, particularly in muscle contraction through association with the motor protein, myosin. In response to contractile stimuli, actin filaments and myosin form membrane-bound, parallel organized units termed "stress fibers", stimulating myosin sliding along the actin filaments that produces an increase in intracellular tension leading to cell contraction (Dudek et al. 2004). The microtubule component of the endothelial cytoskeletal network consists of α- and β-tubulins that are compression-resistant hollow rods that support the cellular structure and facilitate formation of spindles during mitosis (Goode et al. 2000, Klymkowsky 1999). The remaining cytoskeletal components, intermediate filaments, consist of vimentin and other filament proteins that form structurally conserved subunits in coiled coil filaments that serve to bear tension in nuclear envelope and peripheral cell junctions (Helfand et al. 2003, 2004). These cytoskeletal filaments associate with the actomyosin and microtubular elements as well as with neighboring cell and extracellular matrix components through intercellular junction to aid cells in higher levels of structure and function.

Table 1. Actin-binding proteins involved in endothelial cell barrier regulation.

Actin-Interacting Partners	Function	Molecular masskDa
Spectrin	Cross-links F-actin in cell periphery. Stimulates myosin II ATPase activity. Links surface receptor to cortical actin filaments. Maintains plasma membrane integrity (Broderick and Winder 2005, Thomas 2001)	260/240
α-Actinin	Reorganizes actin cytoskeleton in cell movement. Links actin to plasma membrane and integrins. Part of focal adhesion plaque. Attaches actin to a variety of intracellular structures. Binds vinculin, nebulin, clathrin, and β1 integrins (Lum and Malik 1994, 1996)	100
Fimbrin	Links actin cytoskeleton to vimentin network at sites of cell adhesion. Present in F-actin adhesion structures as well as microvilli. Calcium regulated (Dubreuil 1991)	68
Cortactin	Cross-links actin in cell periphery. Involved in actin rearrangement. Substrate for Abl-and Src-modulated tyrosine phosphorylation-associated cytoskeletal rearrangement (Dudek and Garcia 2003, Dudek et al. 2004, Garcia et al. 1997)	80
Filamin	Participates in vascular permeability. Recognizes peripheral actin. Anchors various transmembrane proteins to actin cytoskeleton. Regulated by cAMP-dependent PKA-mediated phosphorylation and Ca^{2+}/CaM kinase (Garcia et al. 1995)	280
FAK	Tyrosine kinase. Early recruitment to adhesion sites during cell migration. Activated by autophosphorylation and Src kinases. FAK-Src mediates disassembly of adhesion complex (Furuta et al. 1995, Ilić et al. 1995)	120
Integrins	Major components of focal adhesion complex. Transmembrane protein that links ECM and actin filaments (Kornberg et al. 1992, Schaller et al. 1995)	90–160
Paxillin	A focal adhesion protein that interacts with vinculin. Links actin to plasma membrane, recruits other adhesion proteins and signaling regulators. It is tyrosine phosphorylated (Shikata et al. 2003)	68
Tensin	Present only in mature (large) adhesion, interact with protein phosphatases (Lo et al. 1994)	
nmMLCK	Ca^{2+}/CaM dependent kinase. Regulated by Ser/Thr/Tyr phosphorylation. Key EC barrier regulatory protein involved in both stress fiber and cortical actin formation, angiogenesis and paracellular gap regulation (Dudek et al. 2004, Garcia et al. 1993, 2000)	210
Catenins (α, β)	Components of cadherin cell adhesion complex, bind to actin and links actin filaments and cadherins (Taveau et al. 2008)	80, 88

Table 1. contd....

Table 1. contd.

Actin-Interacting Partners	Function	Molecular masskDa
Caldesmon	Binds both actin and myosin. Serves as a nuclear switch for regulation of actomyosin contraction. Phosphorylation by ERK p38 MAPK and PKC facilitates actomyosin contraction. Involved in MLCK independent mechanisms of stress fiber formation (Bogatcheva and Verin 2008, Mirzapoiazova et al. 2006)	77
Vinculin	Binds actin at adherens junction site and stabilizes cellular adhesions. Involved in thrombin-induced barrier regulations. Substrate for PKC (Rüdiger 1998, Ziegler et al. 2006, 2008)	130
Coronin	Found in crown shaped, actin dense area on the cell's dorsal surface. Potential G-protein messenger directly coupled to the cytoskeleton. Cells lacking coronin are defective in cytokinesis and cell motility (Uetrecht and Bear 2006)	55
ERM	Family of proteins that serves as anchor proteins for actin and plasma membrane, involved in regulation of actin dynamic and signal transduction (Tsukita and Yonemura 1999, Tsukita et al. 1997)	75–82
c-Abl	Non-receptor tyrosinekinase, contains SH2, SH3, DNA-binding and actin-binding domains. Serves as barrier enhancer. Proto-oncogene (Welch and Wang 1993)	140

Membrane and Junctional Interactions with the Endothelial Cytoskeleton

In combination with membrane-bound junction proteins, the endothelial cytoskeleton regulates: 1) cell shape and mechanical stability, 2) motility and migration, and 3) cell-cell and cell-matrix interactions; interactions which preserve the integrity of the endothelial monolayer via specialized inter-endothelial contacts (Dudek et al. 2004, Mehta and Malik 2006). The actin microfilament system is focally linked to multiple membrane adhesive proteins such as the cadherin, glycocalyx components, functional intercellular proteins of the zonaoccludens (ZO) and zonaadherens, and focal adhesion complex proteins (Dudek et al. 2004). Involved in this actin network are >80 actin-binding proteins that are critical participants in cytoskeletal arrangement and tensile force generation, and serve to provide a high level of fine tuning of cell shape, adhesion, and orchestrated cell migration as well as regulation of endothelial junction stability (Linz-McGillem et al. 2004). Myosin represents the most abundant actin-binding protein and, as noted above, serves as the molecular machinery to generate

Vascular Barrier Enhancement

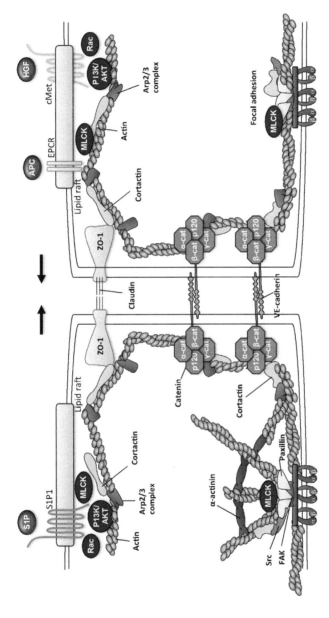

Figure 1. Schematic representation of junction complexes involved in cell-cell and cell-matrix adhesions. Barrier enhancing agonists activate intracellular signaling pathways that lead to peripheral cortical actin formation, resulting in barrier enhancement.

Color image of this figure appears in the color plate section at the end of the book.

tension via an actomyosin motor. Focally distributed changes in tension and relaxation can be accomplished by regulation of the levels of myosin light chain (MLC) phosphorylation and actin stress fiber formation (Dudek et al. 2004). Endothelial specific VE-cadherin is the most prominent protein in the adherens junctional complex, forming a complex with catenin proteins (α, β, and γ) linked to an F-actin-based bridge between the cytoskeleton of adjacent cells. The homodimerized extracelluar domain of VE-cadherin ensures the regulation of intercellular barriers and enforcing the restricted access of vascular components to the interstitial, a process corrupted during high permeability states associated with paracellular gaps (Corada et al. 1999, Dejana et al. 1999). Permeability-enhancing factors transduce signaling to the endothelial cytoskeleton and induce actomyosin crossbridging that increases contractile elements to retract intercellular junctions, creating gaps between cells (Becker et al. 2003, Birukova et al. 2004b, Dudek et al. 2004, Dull and Garcia 2002, Garcia et al. 2001, Moore et al. 1998, Petrache et al. 2001).

Prominent endothelial tight junctions (also called zonaoccludens) create apical and basolateral polarized sub-compartments and a branching network around the parameter of cells that serve as a sealing strand. The tight junction complex is anchored to actin filaments, and the stability of this junction is largely dependent on adherens junction, contributing the vascular permeability regulations. The ezrin, radixin, and moesin (ERM) family of actin-binding proteins act both as linkers between the actin cytoskeleton and plasma membrane proteins as well as signal transducers for agonists that induce cytoskeletal remodeling. Despite structural similarities and reported functional redundancy, ERM proteins differentially modulate lung EC permeability (Adyshev et al. 2011).

Focal adhesions and cell-matrix contact complex also play an important role in regulating the integrity of the vascular endothelium and therefore endothelial cell barrier permeability. The focal adhesion complex links the extra-cellular matrix (ECM) of the inner vessels and abluminal cells with actin cytoskeleton via α and β integrins that interact with various actin-binding proteins such as α-actinin, talin, filamin, vinculin, paxillin, zyxin, and tensin (Mehta and Malik 2006). The membrane-spanning integrins are intimately associated with the tyrosine kinase known as focal adhesion kinase (FAK), which is recruited to the complex by tyrosine phosphorylation signaled by permeability-increasing agonists (Burridge et al. 1987, Romer et al. 1992). FAK is critical for the formation of focal adhesions as well as FA turnover. Actin stress fibers anchored to focal adhesion complexes act as fixed points against tension developed by permeability-inducing signals. FAK is known for its critical role in vascular permeability as FAK knockout mice are embryonic lethal due to vascular defects (Furuta et al. 1995, Ilić et al. 1995).

Cytoskeletal Contractile Activation and Stress Fiber Formation

Majno and Palade observed that lung endothelial cell rounding and paracellular gap formation at the sites of inflammatory edema within the lung vasculature (Majno and Palade 1961) leading to the ultimate development of a barrier regulatory based on paracellular gap opening and closing in processes driven by the endothelial cytoskeleton (Fig. 1). Disruption in the integrity of the EC monolayer is now recognized as a cardinal feature of inflammation, ischemia-reperfusion injury and angiogenesis and occurs in response to a variety of mechanical stress factors, inflammatory mediators, and activated neutrophil products (reactive oxygen species, proteases, cationic peptides). The dramatic cell shape change that results in paracellular gap formation implicates the direct involvement of endothelial structural components comprised of cytoskeletal proteins such as microfilaments and microtubules.

Critical Role of Calcium Transients and Rho Family GTPases

Increases in cytosolic calcium via storage operated channels or by IP3 driven mobilization of intracellular Ca^{2+} stores is critical to the development of contractile tension via transcellular actomyosin stress fiber formation, cortical actin ring disassembly and paracellular gap formation (Kolodney and Wysolmerski 1992, Phillips et al. 1989, Majno and Palade 1961). The Rho family of small GTPases is intimately involved in cytoskeletal rearrangement and the distribution and assembly of intercellular adherens complexes and focal adhesions. Phosphorylation of regulatory myosin light chains (MLC) is primarily catalyzed by Ca^{2+}/calmodulin (CaM)-dependent myosin light-chain kinase (MLCK) in close coordinated activity with the Rho/Rho kinase pathway. Activation of Rho by the guanosine nucleotide exchange factor, p115-RhoGEF, induces phosphorylation of downstream endothelial target MLC phosphatase (MYPT1, Thr^{686}, Thr^{850}), resulting in MYPT1 inactivation and accumulation of dephosphorylated MLC, actin remodeling, and cell contraction (Amano et al. 1996, Birukova et al. 2004b, Kimura et al. 1996, Wettschureck and Offermanns 2002, Metha et al. 2001). In multiple *in vitro* and *in vivo* models of increased endothelial cell permeability, augmented MLCK activity generates centripetal cellular tension via its ability to enhance actomyosin motor activity (Dudek et al. 2004, Garcia et al. 1995) and produces increased MLC phosphorylation within newly formed stress fibers, intracellular tension, cell rounding, paracellular gap formation, and subsequent endothelial barrier disruption (Dudek et al. 2004, Goeckeler and Wysolmerski 1995, Parker 2000).

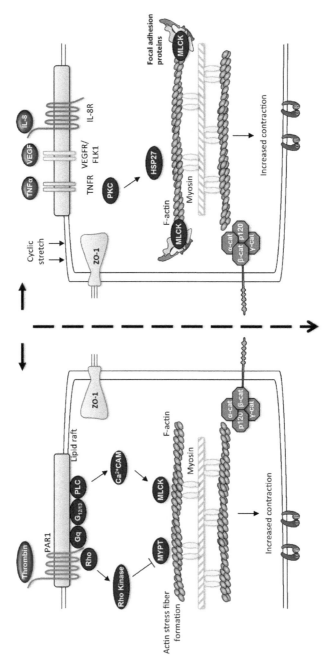

Paracellular Gap Formation

Figure 2. Intracellular signaling pathway activated by contractile agonists. Formation of actin stress fiber results in increased cell contraction and formation of paracellular gap.

Color image of this figure appears in the color plate section at the end of the book.

Non-muscle myosin light chain kinase (MLCK) isoform and alternatively spliced variants

A key actin-binding protein and central regulator of the EC contractile apparatus is the Ca^{2+}/calmodulin-dependent non-muscle isoform of myosin light chain kinase (nmMLCK) which is encoded by the human *mylk* gene located on chromosome 3q21. *mylk* encodes three MLCK proteins (Fig. 3): the nmMLCK isoform, the smooth muscle MLCK isoform (smMLCK, 130–150 kD) and telokin (Garcia et al. 2000, Lazar and Garcia 1999, Potier et al. 1995, Verin et al. 1998). While smMLCK is expressed ubiquitously in different tissues, nmMLCK is the predominant form expressed in endothelial cells (Verin et al. 1998) existing with a high molecular weight of 214kDa (1914 amino acids) compared to smooth muscle smMLCK (130–150kDa) and telokin (17kDa) which retains the C-terminus of nm/smMLCK. Endothelial cells respond to Ca^{2+} mobilizing receptor ligands, such as permeability-enhancing thrombin and vascular endothelial growth factor (VEGF), with increased intracellular Ca^{2+} binding to calmodulin (Becker et al. 2003, Garcia et al. 1993). The activated Ca^{2+}/CaM complex interacts with Ca2+/CaM-binding domain of nmMLCK, resulting in an active conformation and MLCphosphorylation (Thr^{18}, Ser^{19}) which increases actomyosin ATPase activity and shifts the equilibrium between the folded and unfolded myosin forms (Kamisoyama et al. 1994), thus providing the assembling and functioning of the contractile stress fibers.

Smooth muscle smMLCK and nmMLCK are identical in the actin-binding, catalytic, CaM-binding and KRP domains; however, nmMLCK contains a unique 922 amino acid N-terminal domain comprising multiple sites for protein-protein interaction (SH2- and SH3-binding domains) as well as potential regulatory phosphorylation sites (Garcia et al. 2000, Lazar and Garcia 1999, Potier et al. 1995, Verin et al. 1998). For example, tyrosine residues Y^{464} and Y^{471} have been identified as $p60^{src}$ phosphorylation sites which upregulate nmMLCK kinase activity (Birukov et al. 2001, Dudek et al. 2010).

The nmMLCK isoform, like the endothelial cytoskeleton in general, is highly multifunctional and serves as an important effector in numerous endothelial processes. In multiple *in vitro* and *in vivo* models of EC permeability, increased MLCK activity produces increased MLC phosphorylation within newly formed stress fibers, intracellular tension, cell rounding, paracellular gap formation, and subsequent EC barrier disruption. Inhibition of nmMLCK attenuates or prevents vascular leak produced by ischemia/reperfusion, neutrophils, TGFβ, thrombin, and mechanical stress as well as neutrophil influx in response to LTB4 or FMLP. We have described increased MLC phosphorylation in a cortical distribution during EC barrier enhancement suggesting that spatially-defined MLCK

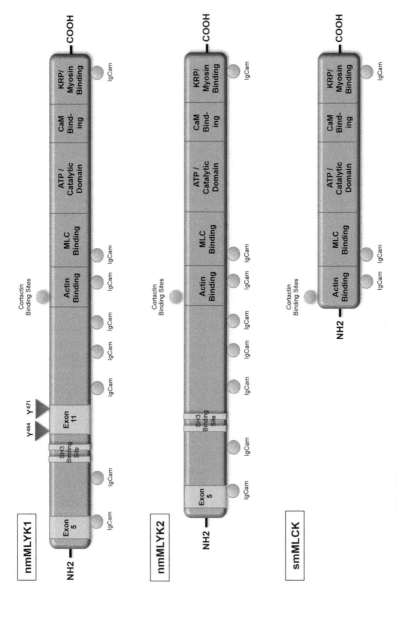

Figure 3. Representation of MLCK isoform/splice-variants.

Color image of this figure appears in the color plate section at the end of the book.

activation can differentially regulate permeability (Garcia et al. 2001). nmMLCK has been implicated in TNF-induced endothelial cell apoptosis, mechanotransduction, and calcium signaling.

mylk splice variants

There are four known *mylk* splice variants of nmMLCK: nmMLCK1, nmMLCK2, nmMLCK3a, nmMLCK3b, with the existence of nmMLCK4 strongly implicated. The MLCK1 and MLCK2 variants differ in the display of a single exon 11 in MLCK1 that is deleted in MLCK2 (nucleotides 1428–1634, 69 amino acids). In endothelial cells, expression of MLCK2 is comparable to MLCK1 whereas MLCK1 accounts for 97% of MLCK in GI epithelium. The 69 amino acid deletion (exon 11) in MLCK2 contains a critical regulatory domain for enzymatic activity via phosphorylation of Tyr^{464} and Tyr^{471} residues. Other splice variants lack various domains of the full-length proteins encoded by exon 30 such as MLCK3a (nucleotides 5081–5233), MLCK3b (nucleotides 1428–1634, 5081–5233, and MLCK4 (nucleotides 4534–4737).

We have shown that nmMLCK knockout mice as well as mice treated with an inhibitory peptide to reduce MLCK activity, are protected against ventilator-induced lung injury (Mirzapoiazova et al. 2006). In addition, we have shown that genetic variants (single nucleotide polymorphisms) in *mylk* confer significant susceptibility to sepsis, and sepsis- and trauma-induced ALI (Gao et al. 2006), as well as contribute to risk of severe asthma, another inflammatory lung disorder, in African Americans (Flores et al. 2007). A key regulatory feature of nmMLCK is the post-translational modification (PTM) by increased levels of nmMLCK tyrosine phosphorylation catalyzed by either p60[src] kinase, c-Abl kinase, or by inhibition of tyrosine phosphatases (vanadate). This PTM serves to increase kinase activity and modulates EC barrier responses (Carbajal and Schaeffer 1998, Dudek et al. 2004, 2010, Garcia et al. 1997, 1999). Diperoxovanadate, a potent tyrosine phosphatase inhibitor, also increases nmMLCK activity, stress fibers, and EC contraction via activation of p60[src] tyrosine kinase (Garcia et al. 1999, 2000). The nmMLCK isoform binds cortactin, an 80/85 kD actin-binding protein involved in barrier regulation (Dudek and Garcia 2001, Dudek et al. 2004) that localizes to numerous cortical structures within cells (Garcia et al. 1997). Cortactin exhibits an amino acid structure which contains an N-terminal acidic region (NTA) that stimulates actin polymerization by binding the Arp2/3 complex (murine AA #1–90), a unique tandem repeat site for the direct binding of F-actin (AA #91–326), a proline- and tyrosine-rich area containing sites for p60[src] phosphorylation (AA #401–495), and a C-terminal SH3 domain (#496–546) (Weed and Parsons 2001) which binds junctional proteins such as ZO-1 as well as key cytoskeletal effectors such as nmMLCK

(Dudek and Garcia 2001, 2003, Garcia et al. 1993) with this interaction, enhancing cortical actin formation and tensile strength. The central region of cortactin binds and cross-links actin filaments, with its C-terminus site for p60^{src} kinase-mediated phosphorylation that reduces cross-linking activity. Tyrosine phosphorylation of cortactin by p60^{src} potentiates and stabilizes actin polymerization, as well as strengthens cortactin-nmMLCK interactions (Pries et al. 2000) and is a key step in a sequence of events that produce cytoskeletal changes, reassembly of adherens junctions, and barrier restoration during lung inflammation.

Actomyosin contraction following permeability-enhancing factors

A variety of agonists, cytokines, growth factors, and mechanical forces alter the endothelial cytoskeleton and therefore influence pulmonary vascular barrier properties in a wide range of pulmonary diseases such as acute lung injury (ALI) or acute respiratory distress syndrome (ARDS), ventilator-induced lung injury (VILI), sepsis, asthma, and ischemia reperfusion (I/R)-induced injury (Becker et al. 2003, Birukova et al. 2004b, Dudek et al. 2004, Dull and Garcia 2002, Garcia et al. 1999, 2001, Petrache et al. 2001, Wave and Matthay 2000). The serine protease, thrombin, is a coagulation factor that is an important mediator in the pathogenesis of ALI as thrombin evokes EC responses that regulate hemostasis and thrombosis (Dudek et al. 2004, Garcia et al. 1986). Thrombin increases EC leakiness to macromolecules by ligating and proteolytically cleaving the extracellular NH$_2$-terminal domain of the thrombin receptor, a member of the family of proteinase-activated receptors (PARs) (Garcia et al. 1993, 1995, Lollar and Owen 1980, Vu et al. 1991). The cleaved NH$_2$-terminus, acting as a tethered ligand, activates the receptor and induces Ca^{2+} flux into the cells that subsequently increase MLC phosphorylation by Rho-mediated MLC phosphatase inhibition and calcium-dependent MLCK activation (Mehta and Malik 2006), causing actomyosin contraction. *In vivo* studies detailed events that followed thrombin infusion into the pulmonary artery of the chronically-instrumented lung lymph sheep model, initiating a cascade of events that culminate in intravascular coagulation, inflammation, and vascular leak (Aschner et al. 1990, Minnear et al. 1989, Vogel et al. 2000).

Naturally occurring agonists, such as the cytokines TNF-α and IL-1β, have a prominent effect early in ALI, causing microthrombosis, and eliciting a cascade of inflammatory signals which result in capillary endothelial production of P-selectin, an adhesion molecule which enhances leukocyte-EC migration (Angelini et al. 2006, Mantovani et al. 1997, Nwariaku et al. 2002) and actin reorganization, and paracellular gap formation (Orfanos et al. 2000). TNF-α also increases tyrosine phosphorylation of VE-cadherin leading to increased paracellular gaps in human lung endothelium (Angelini et al. 2006).

New blood vessel formation, or angiogenesis, is defined by the generation of new capillaries by EC either by sprouting or by splitting from pre-existing vessels. Sprouting angiogenesis involves EC detachment from the basement membrane, migration and subsequent proliferation, tube formation, and finally, functional maturation of the new vessel (Patterson et al. 2000). Vascular endothelial growth factor (VEGF) increases EC permeability and was originally named "vascular permeability factor" for its profound effects on vascular barrier function (Dvorak 2006). VEGF levels are highest in the lungs and plasma and are increased in patients with ALI compared to other groups (Mirzapoiazova et al. 2006). VEGF increases cytosolic calcium and levels of MLC phosphorylation and VEGF inhibition decreases EC permeability (Becker et al. 2001, Mirzapoiazova et al. 2006, Thickett et al. 2001). Additional angiogenic factors with barrier-regulatory properties include the angiopoietin family (ANG-1, ANG-2) which are critical for normal vascular development. VEGF induces EC differentiation and migration, while ANG-1 stabilizes vascular networks (Gallagher et al. 2008, Li et al. 2008, Mura et al. 2004). ANG-1 and ANG-4 modulate EC permeability by altering the state of adherens junctions and specifically inhibit vascular leakage in response to VEGF or other barrier-disruptive agents, as well as promoting vessel maturation. ANG-2 antagonizes ANG-1 and promotes barrier dysregulation by blocking the ability of ANG-1 to activate its receptor (Mura et al. 2004).

Cytoskeletal Cortical Actin Formation and Barrier Enhancement and Restoration

Understanding the molecular mechanisms underlying endothelial barrier enhancement and restoration is essential in order to design therapeutic strategies that retard vascular barrier dysfunction. Historically, permeability-reducing strategies primarily consisted of cAMP augmentation, producing only modest barrier enhancement (Birukov et al. 2004a, Liu et al. 2001, Patan et al. 1996, Stelzner et al. 1989). More recently, barrier-promoting agents have been identified which share common signal transduction mechanisms that are distinct from cAMP signals and target the endothelial actin cytoskeleton to facilitate barrier-restorative processes (Fig. 1). The dynamic process of actin polymerization allows for the rapid reorganization of actin structures with profound functional consequences for barrier regulation that are highly dependent on the exact spatial location of this actin rearrangement, occurring as either barrier-disrupting cytosolic stress fibers or as a barrier-enhancing thickened cortical actin ring. We have demonstrated that the quiescent EC phenotype is characterized by a cortical actin ring and few stress fibers, a structure which favors cell-cell adhesion and cell matrix tethering. We have conceptualized a paradigm whereby barrier recovery after edemagenic agonists involves development of a cortical actin ring to anchor cellular junctions and a carefully choreographed (but poorly understood) gap-closing process via formation of RacGTPase-dependent lamellipodial protrusions into the paracellular space between activated endothelial cells. Within these lamellipodia, signals are transduced to actin-binding proteins (nmMLCK and cortactin) and phosphorylated MLCs in spatial-specific cellular locations. Lamellipodia also require formation of focal adhesions (regulated by the cytoskeleton) critical to establishment of linkage of the actin cytoskeleton to target effectors that restore cell-cell adhesion and cell-matrix adhesion. This process is essential to the barrier restoration as we have previously described following S1P, HGF, simvastatin, activated protein C, ATP, oxidized phospholipids, and hyaluronan (Birukov et al. 2004b, Finigan et al. 2005, Garcia et al. 1986, Jacobson and Garcia 2007, Liu et al. 2001, Singleton et al. 2006). Central to these events is the activation of small GTPases, Rac and cdc42 (Brown et al. 2010) that follows ligation of barrier-protective receptors and drives cortical actin remodeling and lamellipodia formation. In addition to lamellipodia, there is increased actin polymerization at the cell periphery (i.e., the cortical actin ring) that occurs with increased force driven by the actin-binding proteins, cortactin and nmMLCK, which also translocate to this spatially defined region. Like lamellipodia formation, RacGTPase-dependent increases in cortical actin follow exposure to multiple barrier-enhancing levels of shear stress or to potent barrier-enhancing agonists (Finigan et al. 2005, Jacobson and Garcia 2007) including S1P (Garcia et al. 1986, Jacobson and Garcia 2007), HGF (Liu et al. 2002), ATP (Jacobson and Garcia 2007), simvastatin (Jacobson and Garcia 2007), activated protein C (APC) (Finigan et al. 2005), prostaglandin PGE_2 (Birukov 2001), and oxidized phospholipid OxPAPC (Birukova et al. 2007). These observations serve to highlight the importance of the cellular location of cytoskeletal proteins in maintaining or enhancing EC barrier function with cortactin directly interacting with nmMLCK, an association which is increased by p60[src] tyrosine phosphorylation of either cortactin or nmMLCK (Dudek and Garcia 2003). Rac activation is in conjunction with Akt-mediated phosphorylation events known to be involved in EC proliferation and migration (Morales-Ruiz et al. 2000) and EC barrier enhancement. Akt-induced phosphorylation of the $S1P_1$ receptor is important in barrier enhancement produced by high molecular weight hyaluronan (HMW-HA) (Singleton et al. 2006).

Mechanisms of vascular barrier enhancement

The mechanisms by which sphingosine 1-phosphate (S1P) enhances the integrity of the endothelial barrier remain an active area of research with intense focus on the forces that regulate the integrity of the paracellular junction. Mechanistic approaches to endothelial cell barrier regulation have revealed the complexity of these processes. However, one valuable paradigm has described paracellular gap formation as represented by the balance of competing contractile forces, which generate centripetal tension, and of adhesive cell-cell and cell-matrix tethering forces, which together regulate cell shape changes (Dudek and Garcia 2001, Dudek et al. 2004). Both competing forces in this model are intimately linked to the cytoskeleton, a complex network of actin microfilaments, microtubules, and intermediate filaments, which combine to regulate shape change and transduce signals within and between neighboring endothelial cells. We examined S1P-mediated enhancement of endothelial junctional integrity via prominent cytoskeletal rearrangement involving the translocation of key actin binding proteins, such as cortactin and nmMLCK to the cell periphery and the formation of a strong cortical actin ring (Dudek et al. 2004, Liu et al. 2001). An emerging concept from this work is that a marked increase in polymerized cortical actin represents a common and essential feature of barrier-protective agents such as S1P and diverse stimuli such as shear stress, hepatocyte growth factor, simvastatin, and ATP (Birukova et al. 2004c, Jacobson et al. 2004, 2005, Liu et al. 2002). Overexpression of cofilin, an actin-severing protein, or treatment with cytochalasin B, an inhibitor of actin polymerization attenuates the barrier-enhancing effect of S1P (Garcia et al. 2001).

RacGTPase regulation of cortical actin and lamellipodial protrusions

Consistent with a critical role for the endothelial cytoskeleton in vascular barrier regulation, the Rho family of small GTPases, critical regulators of the non-muscle cytoskeleton, is intimately involved in S1P-mediated cytoskeletal rearrangement and distribution/assembly of intercellular adherens complexes and focal adhesions (FAs) (Wojciak-Stothard and Ridley 2002). The S1PR1 receptor is the key S1P receptor in endothelial barrier protection (English et al. 2000, Garcia et al. 2001) and exerts its action via coupling to Gi, while S1PR3 is coupled to Gi, Gq, and G12/13 (Spiegel and Milstien 2003). Ligation of cell surface S1P receptors triggers a complex signaling cascade mediated by intracellular G-proteins which activate RacGTPase and modulate molecular trafficking to and enzymatic activity

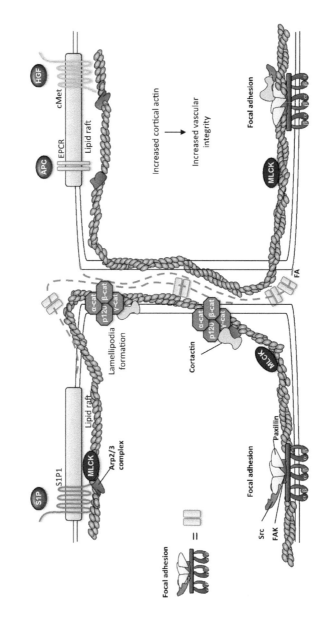

Figure 4. Schematic representation of lamellipodia formation by increased cortical actin at cell periphery and barrier restoration. Actin-binding proteins (i.e., MLCK and cortactin) translocate to this spatially defined region.

Color image of this figure appears in the color plate section at the end of the book.

at the cell periphery, resulting in peripheral cytoskeletal enhancement and the formation of functional adherens junction complexes.

Overexpression of constitutively active Rac enhances peripheral actin polymerization in the cortical ring (Garcia et al. 1986) and induces lamellipodia formation, membrane ruffling, the formation of cortical actin filaments, and the spreading of EC (Rodriguez et al. 2003). Conversely, inhibition of RacGTPase (with simultaneous Rho GTPase activation) leads to increased monolayer permeability and enhances the thrombin-mediated barrier dysfunction response through a variety of signaling proteins (Birukov et al. 2004, Dudek et al. 2004, Garcia et al. 1986, Vouret-Craviari et al. 1999). The downstream effector(s) that links Rac to cortical actin rearrangement remains unclear. Adenoviral overexpression of a Rac dominant/negative mutant attenuates barrier enhancement (TER elevation) by S1P. Rac activation of the p21-associated kinase (PAK) family is well described and is essential for S1P-mediated cortical actin rearrangement (Garcia et al. 1986). Activation of the Rac pathway also inactivates the actin-severing protein, cofilin, in the cell periphery through a signaling cascade involving PAK-1 and LIM kinase (Garcia et al. 1986). More recently, we have demonstrated that Rac activation is associated with the translocation of cortactin, essential for increasing actin polymerization in the cell periphery (Birukov et al. 2002, Dudek and Garcia 2001), with transfection with a dominant/negative Rac construct preventing the translocation of cortactin and subsequent cortical actin polymerization (Vouret-Craviari et al. 1999). Cortactin stimulates and stabilizes Arp2/3-mediated polymerization of branched actin filaments at peripheral sites of cytoskeletal rearrangement (Uruno et al. 2001, Weed and Parsons 2001); however, the spatial and temporal regulation of cortactin interaction with Arp2/3 is poorly understood. Our recent work has implicated a novel role for cortactin in S1P-mediated dynamic cortical actin rearrangement (Dudek and Garcia 2001) with antisense downregulation of cortactin expression significantly inhibiting S1P-induced barrier enhancement in EC (Dudek and Garcia 2001). A rapid increase in cortactin tyrosine phosphorylation is critical to subsequent barrier enhancement, as EC transfected with a tyrosine deficient mutant cortactin exhibit a blunted TER response (Dudek and Garcia 2001). Although cortactin has long been appreciated as a target for p60[src] (Wu et al. 1991), we were unable to link p60[src] to cortactin peripheral translocation after S1P, despite rapid increases in the phosphotyrosine content of cortactin in EC (Dudek and Garcia 2001). PP2, a specific p60[src] inhibitor, failed to significantly attenuate S1P-induced endothelial barrier enhancement (Garcia et al. 2001) and transient overexpression of mutant tyrosine deficient cortactin construct by site-directed mutation of three critical tyrosine residues (Y421,Y466, Y482) accounting for ~90% of p60[src] cortactin phosphorylation (Huang et al. 1998), definitively demonstrated that p60[src] signaling is not required for S1P-induced cortactin translocation. Tyrosine phosphorylation of cortactin appears to be necessary, however, for peak S1P-induced endothelial barrier enhancement as ECs overexpressing the tyrosine-deficient mutant cortactin exhibit a blunted TER response (Dudek et al. 2004). Significant inhibition of the S1P response was observed after incubation with the nonspecific tyrosine kinase inhibitors genistein, herbimycin A or erbstatin, suggesting that tyrosine kinases other than p60[src] phosphorylate cortactin at these three critical residues (Y421, Y466, Y482) during S1P-induced barrier enhancement (Dudek et al. 2004). Although multiple tyrosine kinases may be involved in cortactin phosphorylation, a recent report indicates that Rac-mediated translocation to the cell periphery is required for this phosphorylation to occur (Head et al. 2003). We have described increased MLC phosphorylation in a cortical distribution during S1P-induced endothelial barrier enhancement and more recently, we detailed the translocation of the endothelial MLCK isoform to the cell periphery as well (Brown et al. 2010, Dudek et al. 2004) indicating possible spatially defined MLCK activation and differentially regulated permeability responses. S1P triggers MLCK-associated cortical actin rearrangement with cellular spreading and flattening, which decreases gap formation, increases TER, and decreases fluid and solute flux across the monolayer. S1P-induced barrier enhancement likely involves cortactin binding to MLCK at the site of cortical actin polymerization, localizing the actomyosin interaction peripherally facilitating the stabilization of intercellular and cell–matrix adhesions (Dudek and Garcia 2001). Our recent immunofluorescent studies on nmMLCK and cortactin localization showed co-localization of nmMLCK and cortactin at specially defined lamellipodia at the basal resting state, which is significantly reduced upon thrombin-induced EC contraction. This dissociation is restored by introduction of the potent EC barrier-protective agonist, S1P (Brown et al. 2010) further highlighting the role of nmMLCK/cortactin interactions in the vascular barrier regulation.

Consistent with the role of MLCK and cortactin in S1P-mediated barrier regulation, cortactin blocking peptide targeting the SH-3 domain of cortactin interrupts the binding of MLCK, thus reducing peripheral MLC phosphorylation and the barrier protection induced by S1P (Dudek et al. 2004). In our recent co-localization study with cortactin mutant lacking SH3 domain, there was a significant reduction in cortactin co-localization with nmMLCK within lamellipodia, verifying that nmMLCK interact with cortactin through SH3 domain (Brown et al. 2010). Immunofluorescence studies reveal rapid, Rac-dependent translocation of cortactin to the expanded cortical actin band where co-localization with MLCK occurs within five minutes following S1P challenge. We also recently reported an increased direct association of Src kinase-mediated phosphorylated cortactin with MLCK, relative to non-phosphorylated cortactin (Dudek and Garcia 2003). MLCK co-immunoprecipitates with both cortactin and $p60^{src}$ in lung endothelial cells (Garcia et al. 1997), and the interaction of cortactin with MLCK attenuates cortactin-stimulated, Arp2/3-catalyzed actin polymerization *in vitro* (Dudek and Garcia 2003, Weaver et al. 2001). Tyrosine phosphorylation increases cortactin affinity for MLCK, which again downregulates actin polymerization, providing a potential mechanism through which cortactin tyrosine phosphorylation may modulate the cytoskeleton.

We recently showed that an actin-binding tyrosine kinase, c-Abl, directly interacts with nmMLCK specifically within lipid rafts and identified four c-Abl specific phosphorylation sites on nmMLCK via mass-spectrometry analysis (Dudek et al. 2010). Site-specific phosphorylation significantly alters nmMLCK function and cytoskeletal dynamics within barrier-promoting cortical actin structure. c-Abl phosphorylate nmMLCK at different sites than $p60^{src}$. Whereas $p60^{src}$ catalyzed nmMLCK, phosphorylation stimulates actin polymerization through Arp2/3 (thus increasing vascular permeability); c-Abl phosphorylated nmMLCK has an opposite effect and augments cortactin-augmented Arp2/3-catalyzed actin polymerization, enhancing the vascular barrier. Together, these data support novel roles for the cytoskeletal proteins cortactin and nmMLCK in mediating lung vascular barrier augmentation evoked by S1P. Given that cortactin fails to directly stimulate nmMLCK activity *in vitro* (Dudek and Garcia 2003), we speculate that cortactin increases MLC phosphorylation by either linking nmMLCK to enzymatic co-factors such as $p60^{src}$, which binds directly with cortactin and increases splice variant nmMLCK-1 activity *in vitro* (Birukov et al. 2003), or by facilitating substrate (MLC) targeting in this locale. The functional role of peripheral MLC phosphorylation in promoting S1P-induced barrier enhancement remains unclear, but phosphorylating MLC at the periphery may stimulate actomyosin interactions that produce cell spreading, cell flattening, or other changes that strengthen the cortical actin ring. Alternatively, nmMLCK may serve as a scaffolding protein that stabilizes the cortical actin ring through multiple actin binding sites or through other potential, yet undefined, cytoskeletal interactions.

Conclusion

Cytoskeletal components are involved in a wide range of biological processes critical for endothelial cell function. We have demonstrated the multifunctional role of endothelial actomyosin microfilaments with emphasis on vascular barrier regulation. The association of actin filaments with myosin, a key motor protein, followed by nmMLCK-mediated phosphorylation of the myosin subunit MLC, is essential for myosin conformational changes leading to functional contractile activity and stress fiber formation that in turn lead to paracellular gap formation. Increase in stress fiber assembly is signaled by permeability-increasing factors such as inflammatory mediators and cytokines which allow cells to make necessary response to outside stimuli. In contrast, vascular barrier enhancement is regulated by nmMLCK-mediated, cortactin-dependent distribution and polymerization of actin at the cell periphery in the cortical ring induced by barrier-enhancing factors such as S1P. A critical mediator in this balance is nmMLCK, a common gateway by multiple signaling, that distinguishes functional output by its spatial specific localization and phosphorylation. This tightly regulated balance between vascular permeability and barrier enhancement becomes impaired in a wide range of pulmonary diseases including ALI/ARDS, VILI, sepsis, asthma, and I/R-induced injury. The precise mechanisms of vascular barrier regulation remain elusive. Deeper understanding of the mechanisms underlying the role of actomyosin in vascular barrier regulation will provide novel therapeutic targets in inflammatory injuries.

References

Adyshev, D., N. Moldobaeva, V. Elangovan et al. 2011. Differential involvement of ezrin/radixin/moesin proteins in sphingosine 1-phosphate-induced human pulmonary endothelial cell barrier enhancement. Cell Signal 23: 2086–96.

Amano, M., M. Ito, K. Kimura et al. 1996. Phosphorylation and activation of myosin by Rho-associated kinase (Rho-kinase). J Biol Chem 271: 20246–9.

Angelini, D., S.-W. Hyun, D. Grigoryev et al. 2006. TNF-alpha increases tyrosine phosphorylation of vascular endothelial cadherin and opens the paracellular pathway through fyn activation in human lung endothelia. Am J Physiol Lung Cell Mol Physiol 291: L1232–45.

Aschner, J., J. Lennon, J. Fenton et al. 1990. Enzymatic activity is necessary for thrombin-mediated increase in endothelial permeability. Am J Physiol 259: L270–5.

Becker, P., A. Verin, M. Booth et al. 2001. Differential regulation of diverse physiological responses to VEGF in pulmonary endothelial cells. Am J Physiol Lung. Cell Mol Physiol 281: L1500–11.

Becker, P., A. Kazi, R. Wadgaonkar et al. 2003. Pulmonary vascular permeability and ischemic injury in gelsolin-deficient mice. Am J Physiol Lung Cell Mol Biol 28: 478–84.

Birukov, K. 2001. Differential regulation of alternatively spliced endothelial cell myosin light chain kinase isoforms by p60^{Src}. J Biol Chem 276: 8567–73.

Birukov, K., C. Csortos, L. Marzilli et al. 2001. Differential regulation of alternatively spliced endothelial cell myosin light chain kinase isoforms by p60(Src). J Biol Chem 276: 8567–73.

Birukov, K., A. Birukova, S. Dudek et al. 2002. Shear stress-mediated cytoskeletal remodeling and cortactin translocation in pulmonary endothelial cells. Am J Respir Cell Mol Biol 26: 453–64.

Birukov, K., J. Jacobson, A. Flores et al. 2003. Magnitude-dependent regulation of pulmonary endothelial cell barrier function by cyclic stretch. Am J Physiol Lung Cell Mol Physiol 285: L785–97.

Birukov, K., V. Bochkov, A. Birukova et al. 2004a. Epoxycyclopentenone-containing oxidized phospholipids restore endothelial barrier function via Cdc42 and Rac Circul Res 95: 892–901.

Birukov, K., N. Leitinger, V. Bochkov et al. 2004b. Signal transduction pathways activated in human pulmonary endothelial cells by OxPAPC, a bioactive component of oxidized lipoproteins. Microvasc Res 67: 18–28.

Birukova, A., K. Birukov, K. Smurova et al. 2004a. Novel role of microtubules in thrombin-induced endothelial barrier dysfunction. FASEB J 18: 1879–90.

Birukova, A., K. Smurova, K. Birukov et al. 2004b. Role of Rho GTPases in thrombin-induced lung vascular endothelial cells barrier dysfunction. Microvasc Res 67: 64–77.

Birukova, A., K. Smurova, K. Birukov et al. 2004c. Microtubule disassembly induces cytoskeletal remodeling and lung vascular barrier dysfunction: role of Rho-dependent mechanisms. J Cell Physiol 201: 55–70.

Bogatcheva, N. and A. Verin. 2008. The role of cytoskeleton in the regulation of vascular endothelial barrier function. Microvasc Res 76: 202–7.

Broderick, M. and S. Winder. 2005. Spectrin, alpha-actinin, and dystrophin. Adv Prot Chem 70: 203–46.

Brown, M., D. Adyshev, V. Bindokas et al. 2010. Quantitative distribution and colocalization of non-muscle myosin light chain kinase isoforms and cortactin in human lung endothelium. Microvasc Res 80: 75–88.

Burridge, K., L. Molony and T. Kelly. 1987. Adhesion plaques: sites of transmembrane interaction between the extracellular matrix and the actin cytoskeleton. J Cell Sci 8: 211–29.

Carbajal, J. and R. Schaeffer. 1998. H2O2 and genistein differentially modulate protein tyrosine phosphorylation, endothelial morphology, and monolayer barrier function. Biochem Biophys Res Commun 249: 461–6.

Corada, M., M. Mariotti, G. Thurston et al. 1999. Vascular endothelial-cadherin is an important determinant of microvascular integrity *in vivo*. Proc Natl Acad Sci USA 96: 9815–20.

Dejana, E., G. Bazzoni and M. Lampugnani. 1999. Vascular endothelial (VE)-cadherin: only an intercellular glue? Exp Cell Res 252: 13–9.

DeMali, K., C. Barlow and K. Burridge. 2002. Recruitment of the Arp2/3 complex to vinculin: coupling membrane protrusion to matrix adhesion. J Cell Biol 159: 881–91.

Dubreuil, R. 1991. Structure and evolution of the actin crosslinking proteins. Bioessays 13: 219–26.

Dudek, S. and J. Garcia. 2001. Cytoskeletal regulation of pulmonary vascular permeability. J Appl Phys (Bethesda, Md: 1985) 91: 1487–500.

Dudek, S. and J. Garcia. 2003. Rho family of guanine exchange factors (GEFs) in cellular activation: who's dancing? And with whom? Circul Res 93: 794–5.

Dudek, S., J. Jacobson, E. Chiang et al. 2004. Pulmonary endothelial cell barrier enhancement by sphingosine 1-phosphate: roles for cortactin and myosin light chain kinase. J Biol Chem 279: 24692–700.

Dudek, S., E. Chiang, S. Camp et al. 2010. Abl tyrosine kinase phosphorylates nonmuscle Myosin light chain kinase to regulate endothelial barrier function. Mol Biol Cell 21: 4042–56.

Dull, R. and J. Garcia. 2002. Leukocyte-induced microvascular permeability: how contractile tweaks lead to leaks. Circul Res 90: 1143–4.

Dvorak, H. 2006. Discovery of vascular permeability factor (VPF). Exp Cell Res 312: 522–6.

English, D., Z. Welch, A. Kovala et al. 2000. Sphingosine 1-phosphate released from platelets during clotting accounts for the potent endothelial cell chemotactic activity of blood serum and provides a novel link between hemostasis and angiogenesis. FASEB journal: official publication of the FASEB J 14: 2255–65.

Finigan, J., S. Dudek, P. Singleton et al. 2005. Activated protein C mediates novel lung endothelial barrier enhancement: role of sphingosine 1-phosphate receptor transactivation. J Biol Chem 280: 17286–93.

Flores, C., S.-F. Ma, K. Maresso et al. 2007. A variant of the myosin light chain kinase gene is associated with severe asthma in African Americans. Genet Epidemiol 31: 296–305.

Fujita, H., P. Allen, P. Janmey et al. 1997. Characterization of gelsolin truncates that inhibit actin depolymerization by severing activity of gelsolin and cofilin. Eur J Biochem 248: 834–9.

Furuta, Y., D. Ilić, S. Kanazawa et al. 1995. Mesodermal defect in late phase of gastrulation by a targeted mutation of focal adhesion kinase, FAK. Oncogene 11: 1989–95.

Gallagher, D., S. Parikh, K. Balonov et al. 2008. Circulating angiopoietin 2 correlates with mortality in a surgical population with acute lung injury/adult respiratory distress syndrome. Shock (Augusta, Ga) 29: 656–61.

Gao, L., A. Grant, I. Halder et al. 2006. Novel polymorphisms in the myosin light chain kinase gene confer risk for acute lung injury. Am J Resp Lung Cell Mol Biol 34: 487–95.

Garcia, J., A. Siflinger-Birnboim, R. Bizios et al. 1986. Thrombin-induced increase in albumin permeability across the endothelium. J Cell Physiol 128: 96–104.

Garcia, J., C. Patterson, C. Bahler et al. 1993. Thrombin receptor activating peptides induce Ca2+ mobilization, barrier dysfunction, prostaglandin synthesis, and platelet-derived growth factor mRNA expression in cultured endothelium. J Cell Physiol 156: 541–9.

Garcia, J., H. Davis and C. Patterson. 1995. Regulation of endothelial cell gap formation and barrier dysfunction: role of myosin light chain phosphorylation. J Cell Physiol 163: 510–22.

Garcia, J., K. Schaphorst, S. Shi et al. 1997. Mechanisms of ionomycin-induced endothelial cell barrier dysfunction. Am J Physiol 273: L172–84.

Garcia, J., A. Verin, K. Schaphorst et al. 1999. Regulation of endothelial cell myosin light chain kinase by Rho, cortactin, and p60(src). Am J Physiol 276: L989–98.

Garcia, J., K. Schaphorst, A. Verin et al. 2000. Diperoxovanadate alters endothelial cell focal contacts and barrier function: role of tyrosine phosphorylation. J Appl Physiol (Bethesda, Md: 1985) 89: 2333–43.

Garcia, J., F. Liu, A. Verin et al. 2001. Sphingosine 1-phosphate promotes endothelial cell barrier integrity by Edg-dependent cytoskeletal rearrangement. J Clin Invest 108: 689–701.

Goeckeler, Z. and R. Wysolmerski. 1995. Myosin light chain kinase-regulated endothelial cell contraction: the relationship between isometric tension, actin polymerization, and myosin phosphorylation. J Cell Biol 130: 613–27.

Goode, B., D. Drubin and G. Barnes. 2000. Functional cooperation between the microtubule and actin cytoskeletons. Curr Opin Cell Biol 12: 63–71.

Gottlieb, A., B. Langille, M. Wong et al. 1991. Structure and function of the endothelial cytoskeleton. Lab Invest 65: 123–37.

Gunst, S. 2004. Actions by actin: reciprocal regulation of cortactin activity by tyrosine kinases and F-actin. Biochem J 380: e7–8.

Head, J., D. Jiang, M. Li et al. 2003. Cortactin tyrosine phosphorylation requires Rac1 activity and association with the cortical actin cytoskeleton. Mol Biol Cell 14: 3216–29.

Helfand, B., L. Chang and R. Goldman. 2003. The dynamic and motile properties of intermediate filaments. Annu Rev Cell Dev Biol 19: 445–67.

Helfand, B., L. Chang and R. Goldman. 2004. Intermediate filaments are dynamic and motile elements of cellular architecture. J Cell Sci 117: 133–41.

Hirata, A., P. Baluk, T. Fujiwara et al. 1995. Location of focal silver staining at endothelial gaps in inflamed venules examined by scanning electron microscopy. Am J Physiol 269: L403–18.

Huang, C., J. Liu, C. Haudenschild et al. 1998. The role of tyrosine phosphorylation of cortactin in the locomotion of endothelial cells. J Biol Chem 273: 25770–6.

Ilić, D., Y. Furuta, S. Kanazawa et al. 1995. Reduced cell motility and enhanced focal adhesion contact formation in cells from FAK-deficient mice. Nature 377: 539–44.

Jacobson, J. and J. Garcia. 2007. Novel therapies for microvascular permeability in sepsis. Curr Drug Targets 8: 509–14.

Jacobson, J., S. Dudek, K. Birukov et al. 2004. Cytoskeletal activation and altered gene expression in endothelial barrier regulation by simvastatin. Am J Resp Cell Mol Biol 30: 662–70.

Jacobson, J., J. Barnard, D. Grigoryev et al. 2005. Simvastatin attenuates vascular leak and inflammation in murine inflammatory lung injury. Am J Lung Cell Mol Biol 288: L1026–32.

Kamisoyama, H., Y. Araki and M. Ikebe. 1994. Mutagenesis of the phosphorylation site (serine 19) of smooth muscle myosin regulatory light chain and its effects on the properties of myosin. Biochem 33: 840–7.

Kimura, K., M. Ito, M. Amano et al. 1996. Regulation of myosin phosphatase by Rho and Rho-associated kinase (Rho-kinase). Science (New York, NY) 273: 245–8.

Klymkowsky, M. 1999. Weaving a tangled web: the interconnected cytoskeleton. Nat Cell Biol 1: E121–3.

Kolodney, M. and R. Wysolmerski. 1992. Isometric contraction by fibroblasts and endothelial cells in tissue culture: a quantitative study. J Cell Biol 117: 73–82.

Kornberg, L., H. Earp, J. Parsons et al. 1992. Cell adhesion or integrin clustering increases phosphorylation of a focal adhesion-associated tyrosine kinase. J Biol Chem 267: 23439–42.

Lazar, V. and J. Garcia. 1999. A single human myosin light chain kinase gene (MLCK; MYLK). Genomics 57: 256–67.

Li, X., M. Stankovic, C. Bonder et al. 2008. Basal and angiopoietin-1-mediated endothelial permeability is regulated by sphingosine kinase-1. Blood 111: 3489–97.

Linz-McGillem, L., J. Moitra and J. Garcia. 2004. Cytoskeletal rearrangement and caspase activation in sphingosine 1-phosphate-induced lung capillary tube formation. Stem Cells Dev 13: 496–508.

Liu, F., A. Verin, T. Borbiev et al. 2001. Role of cAMP-dependent protein kinase A activity in endothelial cell cytoskeleton rearrangement. Am J Lung Cell Mol Biol 280: L1309–17.

Liu, F., K. Schaphorst, A. Verin et al. 2002. Hepatocyte growth factor enhances endothelial cell barrier function and cortical cytoskeletal rearrangement: potential role of glycogen synthase kinase-3beta. FASEB J 16: 950–62.

Lo, S., P. Janmey, J. Hartwig et al. 1994. Interactions of tensin with actin and identification of its three distinct actin-binding domains. J Cell Biol 125: 1067–75.

Lollar, P. and W. Owen. 1980. Clearance of thrombin from circulation in rabbits by high-affinity binding sites on endothelium. Possible role in the inactivation of thrombin by antithrombin III. J Clin Invest 66: 1222–30.

Lum, H. and A. Malik. 1994. Regulation of vascular endothelial barrier function. Am J Lung Physiol 267: L223–41.

Lum, H. and A. Malik. 1996. Mechanisms of increased endothelial permeability. Can J Physiol Pharmacol 74: 787–800.

Majno, G. and G.E. Palade. 1961. Studies on inflammation. 1. The effect of histamine and serotonin on vascular permeability: an electron microscopic study. J Biophys Biochem Cytol 11: 571–605.

Mantovani, A., F. Bussolino and M. Introna. 1997. Cytokine regulation of endothelial cell function: from molecular level to the bedside. Immunol Today 18: 231–40.

Mehta, D. and A. Malik. 2006. Signaling mechanisms regulating endothelial permeability. Physiol Rev 86: 279–367.

Mehta, D., A. Rahman and A. Malik. 2001. Protein kinase C-alpha signals rho-guanine nucleotide dissociation inhibitor phosphorylation and rho activation and regulates the endothelial cell barrier function. J Biol Chem 276: 22614–20.

Minnear, F., M. DeMichele, D. Moon et al. 1989. Isoproterenol reduces thrombin-induced pulmonary endothelial permeability *in vitro*. Am J Phys 257: H1613–23.

Mirzapoiazova, T., I. Kolosova, P. Usatyuk et al. 2006. Diverse effects of vascular endothelial growth factor on human pulmonary endothelial barrier and migration. Am J Lung Cell Mol Biol 291: L718–24.

Moore, T., P. Chetham, J. Kelly et al. 1998. Signal transduction and regulation of lung endothelial cell permeability. Interaction between calcium and cAMP. Am J Physiol 275: L203–22.

Morales-Ruiz, M., D. Fulton, G. Sowa et al. 2000. Vascular endothelial growth factor-stimulated actin reorganization and migration of endothelial cells is regulated via the serine/threonine kinase Akt. Circl Res 86: 892–6.

Mura, M., C. dos Santos, D. Stewart et al. 2004. Vascular endothelial growth factor and related molecules in acute lung injury. J Appl Physiol (Bethesda, Md: 1985) 97: 1605–17.

Nwariaku, F., J. Chang, X. Zhu et al. 2002. The role of p38 map kinase in tumor necrosis factor-induced redistribution of vascular endothelial cadherin and increased endothelial permeability. Shock (Augusta, Ga) 18: 82–5.

Orfanos, S., A. Armaganidis, C. Glynos et al. 2000. Pulmonary capillary endothelium-bound angiotensin-converting enzyme activity in acute lung injury. Circulation 102: 2011–8.

Parker, J. 2000. Inhibitors of myosin light chain kinase and phosphodiesterase reduce ventilator-induced lung injury. J Appl Physiol (Bethesda, Md: 1985) 89: 2241–8.

Patan, S., B. Haenni and P. Burri. 1996. Implementation of intussusceptive microvascular growth in the chicken chorioallantoic membrane (CAM): 1. pillar formation by folding of the capillary wall. Microvasc Res 51: 80–98.

Patterson, C., H. Lum, K. Schaphorst et al. 2000. Regulation of endothelial barrier function by the cAMP-dependent protein kinase. Endothelium 7: 287–308.

Petrache, I., A. Verin, M. Crow et al. 2001. Differential effect of MLC kinase in TNF-alpha-induced endothelial cell apoptosis and barrier dysfunction. Am J Lung Cell Mol Biol 280: L1168–78.

Phillips, P., H. Lum, A. Malik et al. 1989. Phallacidin prevents thrombin-induced increases in endothelial permeability to albumin. Am J Physiol 257: C562–7.

Potier, M., E. Chelot, Y. Pekarsky et al. 1995. The human myosin light chain kinase (MLCK) from hippocampus: cloning, sequencing, expression, and localization to 3qcen-q21. Genomics 29: 562–70.

Pries, A., T. Secomb and P. Gaehtgens. 2000. The endothelial surface layer. Pflügers Archiv: Eur J Physiol 440: 653–66.

Rodriguez, O., A. Schaefer, C. Mandato et al. 2003. Conserved microtubule-actin interactions in cell movement and morphogenesis. Nat Cell Biol 5: 599–609.

Romer, L., K. Burridge and C. Turner. 1992. Signaling between the extracellular matrix and the cytoskeleton: tyrosine phosphorylation and focal adhesion assembly. Cold Spring Harbor Symp. Quant Biol 57: 193–202.

Rüdiger, M. 1998. Vinculin and alpha-catenin: shared and unique functions in adherens junctions. Bioessays 20: 733–40.

Schaller, M., C. Otey, J. Hildebrand et al. 1995. Focal adhesion kinase and paxillin bind to peptides mimicking beta integrin cytoplasmic domains. J Cell Biol 130: 1181–7.

Shasby, D., S. Shasby, J. Sullivan et al. 1982. Role of endothelial cell cytoskeleton in control of endothelial permeability. Circul Res 51: 657–61.

Shikata, Y., K. Birukov and J. Garcia. 2003. S1P induces FA remodeling in human pulmonary endothelial cells: role of Rac, GIT1, FAK, and paxillin. J Appl Physiol (Bethesda, Md: 1985) 94: 1193–203.

Singleton, P., S. Dudek, S.-F. Ma et al. 2006. Transactivation of sphingosine 1-phosphate receptors is essential for vascular barrier regulation. Novel role for hyaluronan and CD44 receptor family. J Biol Chem 281: 34381–93.

Spiegel, S. and S. Milstien. 2003. Exogenous and intracellularly generated sphingosine 1-phosphate can regulate cellular processes by divergent pathways. Biochem Soc Trans 31: 1216–9.

Spinardi, L. and W. Witke. 2007. Gelsolin and diseases. Sub-cellular biochemistry 45: 55–69.

Stelzner, T., J. Weil and R. O'Brien. 1989. Role of cyclic adenosine monophosphate in the induction of endothelial barrier properties. J Cell Phsiol 139: 157–66.

Taveau, J.-C., M. Dubois, O. Le Bihan et al. 2008. Structure of artificial and natural VE-cadherin-based adherens junctions. Biochem Soc Trans 36: 189–93.

Thickett, D., L. Armstrong, S. Christie et al. 2001. Vascular endothelial growth factor may contribute to increased vascular permeability in acute respiratory distress syndrome. Am J Respir Crit Care Med 164: 1601–5.

Thomas, G. 2001. Spectrin: the ghost in the machine. Bioessays 23: 152–60.

Tsukita, S. and S. Yonemura. 1999. Cortical actin organization: lessons from ERM (ezrin/radixin/moesin) proteins. J Biol Chem 274: 34507–10.

Tsukita, S., S. Yonemura and S. Tsukita. 1997. ERM proteins: head-to-tail regulation of actin-plasma membrane interaction. Trends Biochem Sci 22: 53–8.

Uetrecht, A. and J. Bear. 2006. Coronins: the return of the crown. Trends in Cell Biology 16: 421–6.

Uruno, T., J. Liu, P. Zhang et al. 2001. Activation of Arp2/3 complex-mediated actin polymerization by cortactin. Nat Cell Biol 3: 259–66.

Verin, A., V. Lazar, R. Torry et al. 1998. Expression of a novel high molecular-weight myosin light chain kinase in endothelium. Am J Respir Cell Mol Biol 19: 758–66.

Vogel, S., X. Gao, D. Mehta et al. 2000. Abrogation of thrombin-induced increase in pulmonary microvascular permeability in PAR-1 knockout mice. Physiol Genomics 4: 137–145.

Vouret-Craviari, V., D. Grall, G. Flatau et al. 1999. Effects of cytotoxic necrotizing factor 1 and lethal toxin on actin cytoskeleton and VE-cadherin localization in human endothelial cell monolayers. Infect Immun 67: 3002–8.

Vu, T., D. Hung, V. Wheaton et al. 1991. Molecular cloning of a functional thrombin receptor reveals a novel proteolytic mechanism of receptor activation. Cell 64: 1057–68.

Ware, L. and M. Matthay. 2000. The acute respiratory distress syndrome. N Engl J Med 342: 1334–49.

Weaver, A., A. Karginov, A. Kinley et al. 2001. Cortactin promotes and stabilizes Arp2/3-induced actin filament network formation. Curr Biol 11: 370–4.

Weed, S. and J. Parsons. 2001. Cortactin: coupling membrane dynamics to cortical actin assembly. Oncogene 20: 6418–34.

Welch, P. and J. Wang. 1993. A C-terminal protein-binding domain in the retinoblastoma protein regulates nuclear c-Abl tyrosine kinase in the cell cycle. Cell 75: 779–90.

Wettschureck, N. and S. Offermanns. 2002. Rho/Rho-kinase mediated signaling in physiology and pathophysiology. J Mol Med 80: 629–38.

Wojciak-Stothard, B. and A. Ridley. 2002. Rho GTPases and the regulation of endothelial permeability. Vascul Pharmacol 39: 187–99.

Wu, H., A. Reynolds, S. Kanner et al. 1991. Identification and characterization of a novel cytoskeleton-associated pp60src substrate. Mol Cell Biol 11: 5113–24.

Ziegler, W., R. Liddington and D. Critchley. 2006. The structure and regulation of vinculin. Trends Cell Biol 16: 453–60.

Ziegler, W., A. Gingras, D. Critchley et al. 2008. Integrin connections to the cytoskeleton through talin and vinculin. Biochem Soc Trans 36: 235–9.

2

Emerging Themes on Pulmonary Microvascular Endothelial Cell

Microtubules and Microtubule Associated Proteins

Cristiaan D. Ochoa,[1,4] Ron Balczon[3,4] and Troy Stevens[1,2,4,a],*

Introduction

Endothelium forms the inner lining of all blood and lymphatic vessels. Due to this strategic localization, endothelium is a major regulator of numerous physiological functions including blood coagulation, vascular tone, vascular permeability, angiogenesis, lipid metabolism, inflammation, and leukocyte trafficking. In the lung, microvascular endothelium forms a highly restrictive barrier to allow proper gas exchange (Chetham et al. 1999, Jacobson and Garcia 2010, Kelly et al. 1998, Ofori-Acquah et al. 2008, Stevens et al. 2008).

[1]Departments of Pharmacology, University of South Alabama, Mobile, AL 36688, USA.
[2]Internal Medicine, University of South Alabama, Mobile, AL 36688, USA.
[3]Cell Biology and Neuroscience, University of South Alabama, Mobile, AL 36688, USA.
[4]The Center for Lung Biology, University of South Alabama, Mobile, AL 36688, USA.
[a]Email: tstevens@southalabama.edu
*Corresponding author

Inflammatory mediators and vascular permeability-increasing compounds cause retraction of cell borders and inter-endothelial gaps by reorganizing the endothelial cytoskeleton, cell-cell, and cell matrix interactions, an effect that causes alveolar edema and impairs gas exchange.

Dynamic changes in endothelial shape control endothelial barrier function (Mehta and Malik 2006). Endothelial cell shape is determined by dynamic adjustments between inward-directed, centripetal tension in contraposition to outward-directed, centrifugal traction according to the environmental requirements (Prasain and Stevens 2009). Actomyosin and actin-associated proteins are largely responsible for the centripetal forces, while microtubules, microtubule-associated proteins, cell-cell as well as cell-extracellular matrix adhesions regulate the centrifugal pull (Lee and Gotlieb 2003, Mehta and Malik 2006). Either increased endothelial cell contraction or microtubule disassembly contributes to inter-endothelial gap formation. Indeed, microtubules critically regulate endothelial barrier function. Agents that disrupt microtubule dynamics are sufficient to induce inter-endothelial gaps and precipitate increased vascular permeability (Birukova et al. 2004a, 2004b, 2005, Verin et al. 2001).

The importance of microtubules in control of endothelial cell barrier integrity has only recently been recognized. Microtubules were first identified using the newly developed electron microscope in the 1950s (Manton and Clarke 1950). Some investigators described them as long, rod-shape structures while others identified them as "[mitotic] spindle fibers" (Roth and Daniels 1962). The acceptance of these structures as true cellular elements was controversial largely because the cellular fixation preparations did not yield images with good resolution (Brinkley 1997, Roth and Daniels 1962). However, this changed once a prefixing step with glutaraldehyde was added to the common osmium textroxide electron microscopy preparations (Sabatini et al. 1963). This modification allowed Ledbetter and Porter to better visualize the fibrils and to observe that they were hollow structures; for the first time, they coined the term "microtubules" (Ledbetter and Porter 1963).

Today, we know that microtubules are ubiquitous in most eukaryotic cells, including endothelial cells; red blood cells represent an exception. They are hollow structures that originate in the microtubule-organizing center and extend throughout the cell cytoplasm. Each microtubule, a polymer of alternating α and β-tubulin heterodimers, is ~ 25 nm in diameter and consists of 13 parallel strands, termed protofilaments, that form a tubular structure (Kueh and Mitchison 2009) (Fig. 1). Microtubules have a "minus end" that is capped by the centrosome, and a "plus end" that extends toward the cell cortex. The plus end possesses a GTP cap that provides the necessary energy for growth or elongation. Loss of the GTP cap results in rapid microtubule

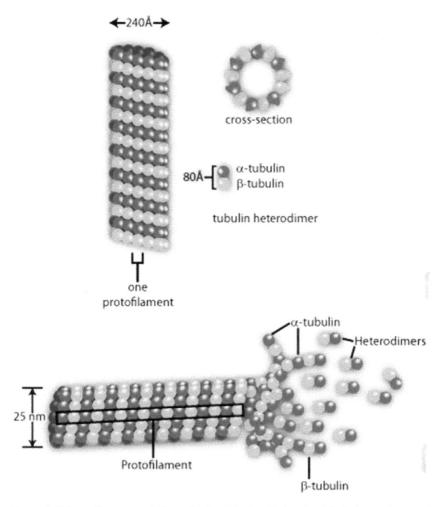

Figure 1. Schematic representation of microtubules. Each microtubule is a polymer of alternating α and β-tubulin heterodimers forming a diameter of about 25 nm.
Color image of this figure appears in the color plate section at the end of the book.

disassembly. Collectively, microtubules are dynamic structures that facilitate vesicular trafficking, chromosome movement, and organelle transport and provide structural benefit to cells.

Microtubules exhibit dynamic instability characterized by alternating phases of shortening and growth (Dimitrov et al. 2008). A balance between microtubule stabilizing and microtubule destabilizing proteins contributes to this dynamic behavior. Microtubule stabilizing proteins bind along microtubule fibers and raise the free energy state of the tubule leading to microtubule stabilization. Microtubule stabilizing proteins are known as Microtubule-associated Proteins, or MAPs. Although several MAPs bind to and stabilize microtubules, which MAPs do this in the pulmonary microvascular endothelium is largely unknown.

Tau and MAP6 are major MAPs. Tau is found primarily in the brain, although other tissues express it too, including the endothelium. In the brain, Tau is responsible for stabilizing and crosslinking microtubules. Some microtubule populations are not primarily stabilized by Tau (Baas et al. 1994), and instead are cold-stable (Bosc et al. 2003, Pirollet et al. 1983). Microtubule-associated protein 6 (MAP6) confers cold-stability (Bosc et al. 2003), and recent studies indicate that endothelium also expresses MAP6. How Tau and MAP6 interact with endothelial microtubules and coordinate their function remains incompletely understood.

Microtubule Regulation of Cell Shape is Essential for Lung Endothelial Barrier Integrity

The physiological importance of microtubules in control of cell shape is underscored in the pulmonary circulation. When pulmonary endothelial cell microtubules are disassembled—via nocodazole, thrombin, or bacterial exotoxins—endothelial cells retract forming inter-endothelial gaps, and endothelial cell permeability increases (Birukova et al. 2004a, 2004b, 2004c, 2005, Prasain et al. 2009, Verin et al. 2001). Studies addressing the pivotal role of microtubules in endothelial cell permeability were first made by Verin and colleagues in the 1990s (Verin et al. 2001). Nocodazole, an inhibitor of microtubule polymerization, induced inter-endothelial cell gaps and endothelial barrier dysfunction (Fig. 2). However, microtubule disassembly was also accompanied by actomyosin stress fiber formation. These findings were supported by Northhover, whose studies revealed that nocodazole increased small intestine microvessel filtration (Northover and Northover 1993). Together, these two manuscripts demonstrated that microtubules were instrumental for the barrier function of the endothelium, and also illustrated the necessary functional coupling between two key cytoskeletal elements that control cell shape: microtubules and actin. The nature of this interaction, and the signaling intermediates that regulate it, remain important areas of study today.

Separating microtubules from the actin cytoskeleton in control of cell shape has been a difficult task, as the molecular interactions between microtubules and actin are extensive. This problem is further complicated by evidence that increased actomyosin interaction is accompanied by microtubule breakdown, and microtubule breakdown is accompanied by

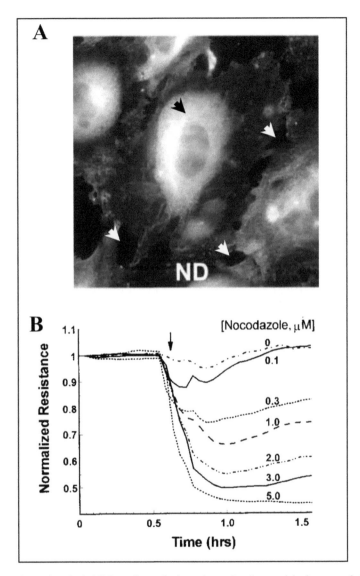

Figure 2. Nocodazole inhibits microtubule polymerization and induces endothelial barrier dysfunction. [A] Immunofluorescence against β-tubulin in human pulmonary artery endothelial cells treated with 200 nM nocodazole (ND) for 30 minutes. White arrowheads point to interendothelial cell gaps. Black arrowheads point to depolymerized tubulin. [B] Trans-endothelia electrical resistance measurements—a surrogate for endothelial barrier function —over 90 minutes in bovine pulmonary artery endothelial cells treated with nocodazole over a dose range. Modified from (Birukova et al. 2004b).

increased actomyosin interaction. Recently, however, a mechanism causing inter-endothelial cell gap formation due to microtubule breakdown—independent of actomyosin interaction and stress fiber formation—was identified (Prasain et al. 2009). Resolution of this mechanism arose from studies on the endothelial cell barrier *protective* actions of adenosine 3′,5′-cyclic monophosphate (cAMP). cAMP is a ubiquitous second messenger that in endothelium was well-recognized for its barrier enhancing actions; epinephrine activation of the endothelial cell β_2-adrenoreceptor activates Gs proteins, which stimulate transmembrane adenylyl cyclase to convert ATP to cAMP. This membrane-associated cAMP signal is intensely barrier protective. However, in 1982 Leppla and colleagues (Leppla 1982) recognized that edema factor, a virulence factor produced by *Bacillus anthracis*, is an adenylyl cyclase toxin that increases tissue permeability. While it was not initially recognized that the increase in tissue permeability was due to disruption of the endothelial cell barrier, work from the Leppla lab provided the impetus for endothelial cell biologists to address the apparent paradox: that cAMP elevations can both increase and decrease permeability.

Sayner et al. (Sayner et al. 2004) first addressed this issue directly, using *Pseudomonas aeruginosa* as a model system. *P. aeruginosa* introduces a cyclase toxin termed exotoxin Y, or ExoY, into mammalian cells through a type 3 secretion system (T3SS); this toxin shares some enzymatic properties with *Bacillus anthracis* edema factor (Ahuja et al. 2004, Shen et al. 2005). The cyclase activity of ExoY is sufficient to induce inter-endothelial cells gaps and increase endothelial cell permeability both *in vitro* and *in vivo* (Ochoa et al. 2012). Subsequent studies would reveal that the location of ExoY within the cytosol is responsible for endothelial cell barrier disruption. Whereas adenylyl cyclases located in the plasma membrane generate a cAMP signal that protects the barrier, soluble adenylyl cyclases located within the cytosol generate a cAMP signal that disrupts the barrier (Ochoa et al. 2012, Prasain et al. 2009, Sayner et al. 2006).

Although ExoY is a soluble enzyme, its precise intracellular locale is not well understood. It is introduced through the T3SS as a linear protein, and must re-fold into a tertiary structure once inside the mammalian cell. ExoY is not active until it binds to a cytosolic mammalian co-factor (Yahr et al. 1998); currently, the co-factor responsible for supporting ExoY activity is unknown. Immunocytochemical analysis suggested that ExoY associates with the centrosome or peripheral microtubules, leading us to examine whether ExoY activity impacts microtubule structure (Sayner et al. 2004). In support of this idea, both ExoY and other soluble mammalian chimeric adenylyl cyclases were shown to break down microtubules (Prasain et al. 2009), an effect that is necessary for inter-endothelial cell gap formation.

While these data demonstrated that soluble cyclases generate a cyclic nucleotide signal responsible for peripheral microtubule breakdown, they also provided insight into the relationship between actin and microtubules. In all previously reported cases, microtubule breakdown led to, or was associated with, actomyosin interaction and increased centripetal tension. However, activation of soluble adenylyl cyclases did not increase myosin light chain-20 phosphorylation, and it did not promote stress fiber formation (Prasain et al. 2009). Thus, ExoY and other soluble cyclases cause microtubule breakdown, but do not appear to increase actin-based centripetal tension. It is not clear how ExoY remodels sites of cell-cell and cell-matrix adhesion. Nonetheless, the enzymatic activity of soluble cyclases represents a mechanism of endothelial cell barrier disruption that is due to microtubule breakdown, and is independent from an increase in actin-based centripetal tension.

Lung Endothelial Cells have a Heterogeneous Population of Microtubules: Acetylation and Tyrosinylation

Studies with ExoY have also contributed to a growing appreciation of the heterogeneous populations of endothelial cell microtubules. The first evidence to suggest that there might be more than one population of endothelial cell microtubules was presented in 2004 when Birukova and collaborators showed that when thrombin caused endothelial barrier dysfunction, it decreased the levels of acetylated tubulin in human pulmonary endothelial cells (Birukova et al. 2004a). More recently, Alieva and colleagues (Alieva et al. 2010) built upon this observation and showed that human pulmonary endothelial cells have both tyrosinated and acetylated microtubules. Interestingly, acetylated microtubules were almost exclusively detected in the centrosome region, while tyrosinated microtubules were found throughout the cell (Alieva et al. 2010). The functional significance of distinct microtubule populations that are resolved by their acetylation and tyrosinylation patterns remains unclear.

Lung Endothelial Cells have a Heterogeneous Population of Microtubules: Microtubule-associated Proteins

Microtubules interact with an array of binding proteins that contribute to their organization and function, an effect that is largely absent when purified tubulin is studied in a test tube. These proteins can be enzymatically active (motor proteins), attach to the microtubule minus end (centrosome proteins), or bind along them (structural proteins). Collectively, proteins that bind microtubules in such a manner are called MAPs.

Microtubules exhibit dynamic instability that is characterized by alternating phases of shortening and growth—known as catastrophes and rescues—respectively (Fig. 3). A balance between microtubule stabilizing and microtubule destabilizing proteins contributes to this dynamic behavior (Dimitrov et al. 2008). Structural MAPs bind along microtubule fibers and either raise or lower the free energy state of the tubule, leading to stabilization or destabilization, respectively (Fig. 4). In mammals, structural MAPs are currently organized into multiple distinct groups: the MAP1 and the MAP2/Tau families, as well as additional MAPs that are not members of these two families. These proteins are incompletely studied in endothelium, but recent evidence suggests endothelial cells express proteins belonging to at least two different classes of MAPs.

The MAP1 family: Mammals contain three family members—MAP1A, MAP1B, and a shorter form called MAP1S (also known as MAP8) (Halpain and Dehmelt 2006). These proteins are the classical microtubule lattice

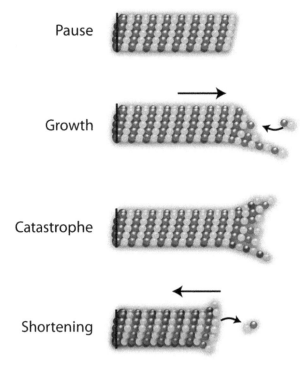

Figure 3. Schematic representation of the dynamic instability model. The alternating phases of shortening and growth—known as catastrophes and rescues—respectively are responsible for the dynamic instability behavior of microtubules. Importantly, MAPs contribute to this behavior.

Color image of this figure appears in the color plate section at the end of the book.

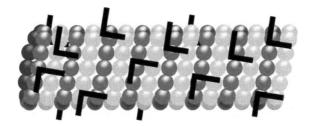

Structural MAPs bind to and stabilize MTs

β-tubulin
α-tubulin   MAP

Figure 4. Structural MAPs bind to and stabilize microtubules.
Color image of this figure appears in the color plate section at the end of the book.

binding proteins and are best known for their microtubule-stabilizing activity (Halpain and Dehmelt 2006). The mechanism of stabilization is believed to depend upon their ability to slow depolymerization rates. These proteins are predominantly found in neurons and are important for the formation of axons and dendrites. MAP1A is expressed in the adult brain and MAP1B is expressed in early brain development. MAP1S links microtubules and mitochondria (Xie et al. 2011). They are phylogenetically related and share one microtubule and one f-actin binding domain in their carboxy termini (Halpain and Dehmelt 2006). MAP1 family members have not been identified in the endothelium.

The MAP2/Tau family: In mammals, MAP2, MAP4, and Tau—and their splice variants—belong to this family (Dehmelt and Halpain 2005). While their amino-terminal domain varies in size, they contain a microtubule-binding sequence in their carboxy-terminal domain, and each has a KXGS conserved sequence that can be phosphorylated. Whereas MAP2 and Tau are preferentially expressed in the neuronal tissue, MAP4 (previously known as MAP3) is considered non-neuronal. Overall, these proteins stabilize the microtubule lattice by reducing the frequency and duration of catastrophes (MAP2) or by protecting microtubules from depolymerization (Tau). The binding of these proteins to microtubules is regulated by the phosphorylation of the KXGS motif. Of these, only Tau has been found in the endothelium (Birukova et al. 2004a, Creighton et al. 2008, Ochoa et al. 2012, Sayner et al. 2011, Tar et al. 2004, Tar et al. 2006, Zhu et al. 2010).

Other structural MAPs: There are other MAPs that do not meet the classification criteria of the previous two families. These include:

MAP6: This protein is also known as stable-tubule only protein or STOP. This protein confers microtubule cold-stability, and has just recently been identified in lung endothelium (Ochoa et al. 2011).

MAP7: This protein is expressed predominantly in cells of epithelial origin. MAP7 and TRPV4 co-localize in the lung and kidney; MAP7 has been shown to increase the membrane expression of the TRPV4 channel (Suzuki et al. 2003). MAP7 has also been reported to associate with the spermatid manchette playing a critical role in spermiogenesis (Penttila et al. 2003). Its presence in endothelium has not yet been investigated.

MAP9: MAP9 is also known as aster-associated protein or ASAP. It is required for spindle formation, mitosis, and cytokinesis (Eot-Houllier et al. 2010, Saffin et al. 2005). Like MAP7, this MAP has yet to be investigated in endothelial cells. Collectively, however, the large number of known MAPs suggests the endothelial microtubule cytoskeleton is complex.

Microtubule-associated Protein Tau

Microtubule-associated Protein Tau (MAPT), or simply Tau, was discovered in 1975 by Weingarten and collaborators (Weingarten et al. 1975) when they observed that phosphocellulose column chromatography removed a factor from porcine brain extracts that was required for microtubule assembly. This factor was trypsin sensitive, and to their astonishment, extremely heat stable. They concluded it was a protein and called it Tau (τ) for its ability to promote tubule polymerization.

Tau is found primarily in neurons and glial cells (Hanger et al. 2009, Querfurth and LaFerla 2010). More recently, Tau has been found in non-neuronal tissue like mammary and gastric epithelia (Rouzier et al. 2005) as well as endothelium (Birukova et al. 2004a, Creighton et al. 2008, Ochoa et al. 2012, Sayner et al. 2011, Tar et al. 2004, Tar et al. 2006, Zhu et al. 2010).

There is genomic evidence for nine natural variations of Tau in humans, with robust experimental evidence for only six. These natural variants arise from a single gene that is alternatively spliced. The *MAPT* gene was mapped to chromosome 17 in the position 17q21.3 (Andreadis et al. 1992). The base transcript contains 16 exons. Exons 2, 3, and 10 are differentially included or spliced out (Fig. 5) (Goedert et al. 1989, King et al. 2000). Tau isoforms are defined by the presence of two repeats in the N-terminus and the presence of the second microtubule-binding domain in the C-terminus. As a result, three of the isoforms contain three microtubule-binding domains (type I Tau) and three possess four microtubule-binding domains (type II Tau). Type I Tau is expressed in the adult and fetal brain whereas type II is only found in the adult brain.

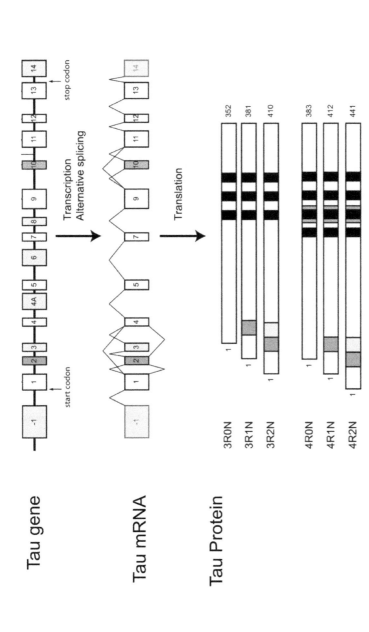

Figure 5. Organization of the Tau gene, Tau mRNA, and Tau proteins. The complex series of intronic regions are shown between exons (numbered). Splice sites are shown in the Tau mRNA that give rise to six Tau proteins. Tau proteins are defined by the number of their microtubule binding domains (R; black) and the number of N-terminal repeat domains (N; orange and yellow).

Color image of this figure appears in the color plate section at the end of the book.

When the long Tau isoforms are compared, there is 88% identity in humans and rodent Tau (Poorkaj et al. 2001). There is genomic evidence for eight natural variations of Tau in rats, with experimental evidence for six. These natural variants differ from each other by the presence or absence of up to three of the 16 exons. Two different C-termini are obtained either by the retention or the splicing of intron 13/14. Tau-like proteins have been reported in the central nervous system of goldfish (Liu et al. 1997), in the epidermis and mechanosensory neurons of *Caenorhabditis elegans* (Goedert et al. 1996), and in *Drosophila malanogaster* larvae (Cambiazo et al. 1995).

The C-terminus of Tau binds axonal microtubules and the N-terminus binds neural plasma membrane. When Tau binds to microtubule fibers, it raises the free energy state of the microtubule that decreases the rate of microtubule breakdown, reducing the frequency of catastrophic events, and increasing the probability for rescues (Binder et al. 2005). Overall, these factors render microtubule fibers more stable. Thus, Tau promotes assembly, stability, and bundling of axonal microtubules, making Tau essential for the proper development and maintenance of the nervous system (Dehmelt and Halpain 2005). In neurons, Tau ensures microtubule-dependent axonal transport of vesicles and organelles by motor proteins (Schneider and Mandelkow 2008).

Currently, two competing hypotheses have been advanced to explain how Tau binds to and stabilizes microtubules. On one hand, investigators argue that Tau binds microtubules outside the protofilaments, in longitudinal fashion (Fig. 6) (Al-Bassam et al. 2002). This was observed using cryo-electron microscopy, helical image analysis, and microtubules stabilized with taxol. Currently, this is the most accepted model of Tau binding to microtubules. However, using cryo-electron microscopy but in the absence of taxol, Kar et al. demonstrated that Tau binds microtubules in horizontal fashion—inside microtubules (Kar et al. 2003). In this last case, kinesin binding stabilized microtubules. This latter observation is still controversial. Since microtubules disassemble by the curving of the microtubule fiber inside out, longitudinal binding to microtubules is thought to function like "molecular glue", therefore preventing catastrophes (Fig. 7). Horizontal binding is proposed to induce a conformational change in microtubule fibers that help stabilize lateral bonds within tubulin dimers to provide stiffening to the protofilament.

Tau has a baseline phosphorylation state. Tau phosphorylation occurs at serine and/or threonine residues (Scott et al. 1993); in the normal brain, there are 2–3 moles of phosphate per mole of Tau protein (Gong and Iqbal 2008). However, when Tau is hyperphosphorylated, its ability to bind microtubules is compromised, causing destabilization of microtubule fibers and microtubule disassembly. Tau hyperphosphorylation means that a phosphate group is added to sites that are not usually phosphorylated (increasing the

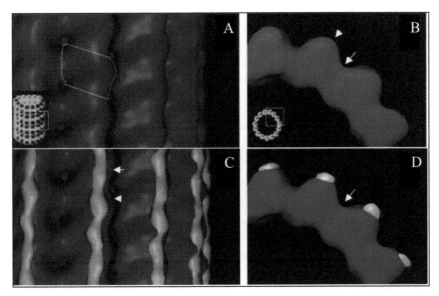

Figure 6. Tau binding to microtubules. [A] 3D reconstruction of cryo-electron microscopy of a microtubule stabilized by taxol. A tubulin monomer is outlined in white. Insert: microtubule. [B] Transversal view of four protofilaments seen from the minus side. Arrowheads: protofilaments. Arrows: ridge. [C] En face view of Tau (orange) decorating the microtubule. [D] View of C from the minus end showing Tau bound to microtubules. Modified from figures originally published in (Al-Bassam et al. 2002).

Color image of this figure appears in the color plate section at the end of the book.

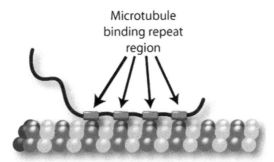

Microtubule
binding repeat
region

Figure 7. Tau functions as molecular glue. When Tau binds to microtubule fibers, it raises the free energy state of the microtubule that decreases the rate of microtubule breakdown, reducing the frequency of catastrophic events, and increasing the probability for rescues.

Color image of this figure appears in the color plate section at the end of the book.

moles of phosphate to 8–12 per mole of Tau protein) (Gong and Iqbal 2008). It has also been taken to mean that for a given phosphor residue, a higher than normal percentage of Tau molecules are phosphorylated. A majority of phosphorylation sites are grouped within the proline-rich region, in the microtubule-binding repeats or in the C-terminus (Fig. 8). Excessive kinase activity, reduced phosphatase activity, or both cause hyper-phosphorylated Tau (Querfurth and LaFerla 2010). Several protein kinases have been linked to the hyperphosphorylated state of Tau protein and several of these kinases can phosphorylate the same residue. The role of protein phosphatases in the control of the hyperphosphorylation state of Tau is less well understood, although decreased Protein Phosphatase 2A and Protein Phosphatase 5 activities has been implicated (Bennecib et al. 2000).

Among the many phosphorylation sites on Tau, phosphorylation at serine 214 critically affects Tau physiology. Tau ser-214 phosphorylation strongly prevents Tau binding to microtubules (Schneider et al. 1999), reduces microtubule assembly by 70% when compared to controls (Liu et al. 2007), and makes Tau a more favorable substrate for other kinases (Liu et al. 2004). The Tau ser-214 phosphorylation site is relevant to endothelial cell biology, since this site is phosphorylated by protein kinase A. ExoY and other soluble adenylyl cyclases that activate protein kinase A cause ser-214 phosphorylation, leading to Tau release from microtubules and microtubule breakdown (Creighton et al. 2008, Prasain et al. 2009, Zhu et al. 2010). These studies suggest that pharmacological approaches that target the ser-214 site could have therapeutic value.

Tau hyperphosphorylation is sufficient to cause neurodegenerative disease, so-called tauopathies (Brion 2006). Indeed, Tau hyperphosphorylation is a hallmark of neurodegenerative tauopathies like Alzheimer's disease (Grundke-Iqbal et al. 1986), argyrophilic grain disease, Pick's disease, and hereditary frontotemporal dementias (Frank et al. 2008, Spillantini et al. 1997). In these diseases, hyperphosphorylation is thought to generate a detergent insoluble form of Tau (Bouchard and Suchowersky 2011) that ultimately leads to its aggregation (Williams 2006). Aggregated Tau then forms paired helical filaments that are characteristic of neurofibrillary tangles (Fig. 9).

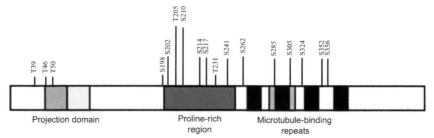

Figure 8. 4R2N Tau phosphorylation sites. The majority of phosphorylation sites are grouped within the proline-rich region (blue) and in the microtubule-binding repeats (black squares). T = Threonine. S = Serine.

Color image of this figure appears in the color plate section at the end of the book.

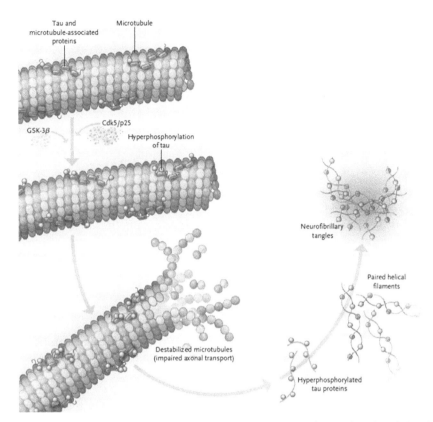

Figure 9. Schematic represents the current understanding of Tau phosphorylation in neurodegenerative diseases. When microtubule-bound Tau gets hyperphosphorylated, Tau detaches from microtubules. Subsequently, microtubules disassemble and Tau aggregates. GSK-3β refers to glycogen synthase kinase-3 beta and Cdk5/p25 refers to a cyclin dependent kinase 5 that is hyperactivated. Originally published in (Querfurth and LaFerla 2010).

Color image of this figure appears in the color plate section at the end of the book.

Spillantini and collaborators introduced the term tauopathy in 1997 (Spillantini et al. 1997). They reported a novel autosomal dominant disease characterized by presenile dementia and abundant fibrillary deposits of Tau protein in the absence of other pathological findings (Spillantini et al. 1997). Subsequent human and animal studies resolved that mutations in *MAPT* caused this type of dementia (Allen et al. 2002, Hutton et al. 1998, Lewis et al. 2000, Poorkaj et al. 1998). These manuscripts are considered hallmarks in the understanding of chronic neurodegenerative diseases.

Tauopathies have not to date been described outside the central nervous system. However, evidence that soluble toxins such as ExoY can increase Tau phosphorylation, especially at ser-214, raises the possibility that infectious agents can cause a form of tauopathy. Ochoa and colleagues (Ochoa et al. 2012) recently tested this idea in endothelium. ExoY increased intracellular levels of cAMP and cGMP, suggesting that ExoY does not just possess adenylyl cyclase activity, but also guanylyl cyclase activity. cAMP and cGMP independently activated protein kinases A and G, which caused Tau ser-214 phosphorylation. This phosphorylated Tau form was insoluble in sarkosyl detergent, consistent with a tauopathy. However, unlike other causes of tauopathy, ExoY did not induce intracellular Tau aggregates and neurofibrillary tangles. Although it has generally been thought that Tau hyperphosphorylation leads to aggregation and ultimately tangles, this concept has been challenged by de Calignon and colleagues (de Calignon et al. 2010), who suggest that non-aggregated forms of hyperphosphorylated, detergent insoluble Tau are deleterious. It remains to be seen whether toxins such as ExoY cause a characteristic tauopathy, or whether they induce a unique form of tauopathy that shares some features of neural disease. Nonetheless, these new data demonstrate a mechanism through which infectious agents cause dysfunction of Tau protein.

Microtubule-associated Protein 6

Although it is well established that Tau binds to and stabilizes microtubules, there is evidence that a subset of microtubules is not stabilized by Tau (Baas et al. 1994) and, moreover, that other MAPs are involved in the Tau-independent microtubule stabilization (Bosc et al. 2003, Pirollet et al. 1983); this appears to be the case in endothelium. The study that defined the role of microtubules in regulating cell shape relied on the ability of low temperature (< 15°C) to induce rapid polymer breakdown (Fig. 10) (Tilney and Porter 1967). However, Behnke and Forer demonstrated that in crane-fly spermatids, a subset of microtubules do not break down in response to low temperatures (Behnke and Forer 1967). In the 1970s, Brinkley and Cartwright extended these observations to mammalian cells when they reported cold-resistant microtubules in the mitotic spindle of rat kangaroo fibroblasts (Brinkley and Cartwright 1975).

Although these two populations of microtubules (cold-stable vs. cold-labile) had been clearly identified in mammals (Bosc et al. 2003, Wallin and Stromberg 1995), the mechanism by which this happened was a mystery. At the time, there were two competing hypotheses that explained this phenomenon. On one hand, some argued that cold-stability was a property of tubulin itself (see below). Alternatively, others posited that there was a microtubule-associated protein that conferred cold-stability.

In support of the latter idea, brain microtubules were shown to be cold-stable *in vivo*, while they were cold-labile *in vitro* (Webb and Wilson 1980). Indeed, the process of tubulin purification was determined to render cold-

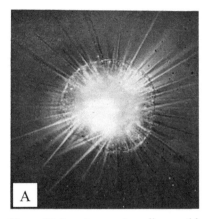 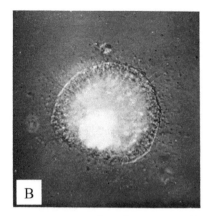

Figure 10. **Low temperature disassembles microtubules.** Adapted from figures originally published by Tilney et al. (Tilney and Porter 1967). [A] Polarizing micrograph of *Actinosphaerium nucleofilum* exposed to 4°C for three minutes showing axonems. [B] The same *A. nucleofilum* after exposure to 4°C for 135 minutes. Axonems were later discovered to consist of microtubule-based skeletons.

stable microtubules cold labile. A "low molecular weight substance" was responsible for cold-stability, and was coined as the term "cold-stabilizing factor (CSF)". Notably, Tau was the first MAP to be ruled out as having cold-stabilizing properties (Baas et al. 1994, Pirollet et al. 1983). MAP6 (previously known as Stable Tubulin-Only Polypeptides or (STOP) proteins) (Bosc et al. 2003, Wallin and Stromberg 1995) was eventually identified to be responsible for microtubule cold-stability in mammals. The first MAP6 was purified from rat brain extracts in 1986 (Margolis et al. 1986) and cloned in 1996 (Bosc et al. 1996). Two years later, non-neuronal cold-stable microtubules were found in NIH 3T3 fibroblasts and with them, a second MAP6 (Denarier et al. 1998b). In addition to neurons and fibroblasts, MAP6 family members have been detected in glial cells. MAP6 activity was also detected in lung tissue (Pirollet et al. 1989) (Fig. 11). Although the cell type(s) in lung tissue responsible for MAP6 expression are unclear, endothelium isolated from the pulmonary artery, capillaries and veins all express MAP6 (Ochoa et al. 2011).

Cold-induced microtubule breakdown is an intrinsic property of mammalian tubulin protofilaments, yet the mechanisms by which cold promotes microtubule disassembly are largely unexplored. Nevertheless, studies on arctic fish shed some light on how this could happen. Several reports have resolved that the microtubules of fish living in the polar seas are intrinsically resistant to cold-induced disassembly (Billger et al. 1994, Detrich et al. 2000, Himes and Detrich 1989). Comparative genomic analysis against bovine microtubules concluded that arctic fish α-tubulin and β-tubulin primary sequences present significant variation in the

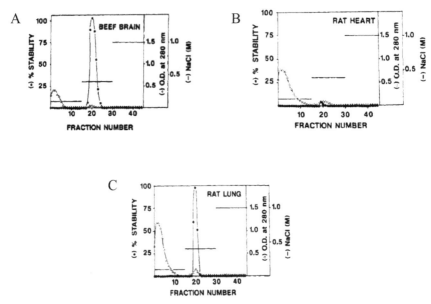

Figure 11. The lung has cold-stable microtubule activity. The cytosol fraction of these tissues was subjected to chromatography separation. Fractions were then assayed for microtubule stability using a protocols described in Pirollet et al. 1989. Heparin-Sepharose chromatography of MAP6 activity in the bovine brain [A], rat heart [B], and rat lung [C]. Microtubule stability was expressed as total possible stability. Modified from Pirollet et al. 1989. Of note, is it not clear what cell type(s) is responsible for MAP6 activity in the lung.

residues that map to the lateral, interprotofilament segments (Detrich et al. 2000). The authors speculated that these substitutions could: 1) increase the lateral interaction between adjacent protofilaments; 2) act to resist the conformational change induced by the nucleotide hydrolysis [from GTP to GDP]; or, 3) both. Indeed, a mutation in the cystine 354 residue of *Saccharomyces cerevisiae* β-tubulin to either alanine or serine (C354A or C354S) rendered the previously cold-labile microtubules, cold stable (Gupta et al. 2001). Therefore, one can envision a scenario in which low temperature induces a conformational change in the tubulin dimer that reduces the interprotofilament interactions equivalent to the loss of the GTP cap, and therefore promotes microtubule breakdown.

The MAP6 family consists of two proteins, MAP6 proper and MAP6d1. There is evidence for three natural variants of MAP6 in the mouse, two in the human, and two in the rat (Bosc et al. 2003, Denarier et al. 1998b). These variants have been termed MAP6-1, MAP6-2, and MAP6-3 when appropriate (Dacheux et al. 2012). *MAP6* is conserved in chimpanzee, dog, cow, mouse, rat, chicken, and zebrafish (Bosc et al. 2003). More recently, a MAP6-related protein was found in a protozoan, *Trypanosoma brucei* (Dacheux et al. 2012). In humans, *MAP6* is found within chromosome 11q14.

It consists of 4 exons that are alternatively spliced. MAP6d1 is the newest family member discovered (Gory-Faure et al. 2006). It is coded by *MAP6D1* and it has been found in human, mouse, and cow neurons. MAP6d1 is a 21 kDa protein that localizes to the Golgi apparatus, although MAP6d1 function is largely unexplored.

MAP6 natural variants bind to and stabilize microtubules rendering them cold-stable (Bosc et al. 1996, 2003, Denarier et al. 1998a, Guillaud et al. 1998). Since the core temperature in non-hibernating animals rarely falls below 35°C, cold is not a usual stress, and therefore the significance of having cold-stable microtubules is unclear. In this regard, Wallin and Stronberg have proposed that cold-stability in homeotherms has nothing to do with the need to have structurally sound microtubules at low temperatures. They posit that cold-stability and cold-instability simply reflect the existence of subsets of microtubules that require stability for very specific functions (organelle transport, cell shape maintenance, mitotic spindle architecture, etc.) (Wallin and Stromberg 1995). Cold could be seen as the first—and serendipitous—stimulus found to segregate and study these populations of microtubules.

In the brain, MAP6-bound microtubules are resistant to nocodazole-induced disassembly, sonication, and temperatures as low as –80°C (Job et al. 1987, Margolis et al. 1990). This evidence has led many to propose the term "super-stable microtubules" (Job et al. 1987). Of interest is the resistance to nocodazole, which was mapped to the Mn modules of MAP6 (Bosc et al. 2001). MAP6 variants with fewer Mn modules have a weakened ability to resist nocodazole, while preserving their cold-stability properties. Interestingly, cold-stable microtubules are not resistant to colchicine or vinblastine (Bershadsky et al. 1979). The mechanisms for this discrepancy are unclear.

Early evidence demonstrated that MAP6 proteins are regulated by Ca^{2+}-calmodulin (Job et al. 1981). In this case, Ca^{2+}-bound calmodulin blocks MAP6-microtubule interaction, yielding cold labile microtubules. Indeed, sequencing of the rat MAP6-1 revealed twelve calmodulin-binding sequences clustered within the microtubule-binding repeats or in their vicinity (Bosc et al. 2001). A second mechanism was discovered recently. This second pathway involves phosphorylation at ser-198 and ser-491 by the Ca^{2+}-calmodulin-dependent protein kinase II, which prevents MAP6 binding to microtubules (Baratier et al. 2006). These phosphorylation events also generated cold labile microtubules. Of note, there is a difference in how neuronal MAP6 and non-neuronal MAP6 bind microtubules (Pirollet et al. 1989). In neurons, MAP6 is constitutively bound to microtubules. In contrast, non-neuronal MAP6 is recruited to microtubules only after the cell is exposed to low temperatures (Denarier et al. 1998b); this is the case in endothelium. Non-neuronal MAP6 also binds to the microtubules of the mitotic spindle. The implications of this observation are unclear since cells that do not express MAP6 complete mitosis without apparent difficulty (Andrieux et al. 2002, Denarier et al. 1998b). Currently, it is not known what other stimuli promote MAP6-microtubule interaction.

The functional significance of MAP6-stabilized microtubules came to light when the MAP6 knock-out mouse was reported (Andrieux et al. 2002). First, these animals did not have developmental abnormalities. However, MAP6 null-mice demonstrated depletion of synaptic vesicles and severe behavioral disorders similar to those found in animal models of schizophrenia. These behaviors were in part alleviated by the chronic administration of neuroleptics (Begou et al. 2008). Supportive of this, MAP6 null mice have abnormalities in dopaminergic neurotransmission (Bouvrais-Veret et al. 2008).

Discovery that endothelial cells express MAP6 came from work on the effector targets of ExoY. Already knowing that ExoY causes Tau phosphorylation and microtubule breakdown, Balczon and colleagues designed experiments to determine whether ExoY increases the rate of microtubule disassembly, impairs centrosome nucleation, or decreases the rate of microtubule reassembly. To perform these experiments, a mechanism of abrupt microtubule disassembly was needed. Acute cold exposure was utilized to disassemble microtubules, which resulted in the rapid dissolution of microtubules into soluble tubulin, with an immunofluorescence image that appeared as a haze. Taken at first glance, it appeared that all microtubules had spontaneously disassembled. However, when the cell membrane was permeabilized and the soluble tubulin was washed away, intact microtubules could be seen within the "ghosts". Ochoa et al. demonstrated that the remaining microtubules were cold-stable due to the presence of MAP6, which upon cold exposure translocated from the cytosolic fraction to a microtubule-rich fraction, causing microtubule stabilization (Ochoa et al. 2011). At present, the physiological significance of MAP6, and cold-stable microtubules, in endothelium is unknown. However, it appears that these studies have resolved two, and perhaps three, different microtubule populations. One microtubule population is stabilized by Tau, and does not interact with MAP6. A second population binds both Tau and MAP6, while a third population binds only MAP6 following cold exposure. This latter microtubule population does not interact with Tau. Resolving the physiological significance of these discrete endothelial cell microtubule populations awaits more rigorous study.

Conclusion

The past 30 years have been characterized by an extensive expansion in our understanding of the sub-cellular biomechanical forces that control endothelial cell shape and barrier integrity, including the structural role played by microtubules. In just the past five years, a new level of complexity has been realized, where a diversity of endothelial cell microtubules has been defined from interactions with Tau and MAP6. Tau represents an effector target of bacterial virulence factors, like ExoY. These findings demonstrate that Tau-bound microtubules play an essential role in maintaining barrier integrity, and illustrate that loss of Tau-stabilized microtubules results in edema, perhaps with long-standing consequences. These Tau-stabilized microtubules can be compared and contrasted with MAP6-bound microtubules. Only a minor fraction of microtubules interact with MAP6 constitutively. In response to cold stress, MAP6 translocates from the cytosol to bind a population of microtubules and stabilize them; in contrast, Tau-stabilized microtubules spontaneously disassemble upon cold stress. Little is known about MAP6 in endothelium, and at present, a physiologically relevant stimulus for MAP6-microtubule binding is not known and is the focus of ongoing work. In the future, research focused on resolving the nature of the interaction between Tau- and MAP6-bound microtubules will provide important insight into how microtubules dynamically adjust endothelial cell barrier integrity.

Acknowledgements

This work was supported by National Institutes of Health grants R37HL60024 (T.S.), HL66299 (T.S. and R.B.), HL76125 (C.O.), and HL107122 (C.O.).

References

Ahuja, N., P. Kumar and R. Bhatnagar. 2004. The adenylate cyclase toxins. Crit Rev Microbiol 30: 187–196.

Al-Bassam, J., R.S. Ozer, D. Safer et al. 2002. MAP2 and tau bind longitudinally along the outer ridges of microtubule protofilaments. J Cell Biol 157: 1187–1196.

Alieva, I.B., E.A. Zemskov, I.I. Kireev et al. 2010. Microtubules growth rate alteration in human endothelial cells. J Biomed Biotechnol 2010: 671536.

Allen, B., E. Ingram, M. Takao et al. 2002. Abundant tau filaments and nonapoptotic neurodegeneration in transgenic mice expressing human P301S tau protein. J Neurosci 22: 9340–9351.

Andreadis, A., W.M. Brown and K.S. Kosik. 1992. Structure and novel exons of the human tau gene. Biochemistry 31: 10626–10633.

Andrieux, A., P.A. Salin, M. Vernet et al. 2002. The suppression of brain cold-stable microtubules in mice induces synaptic defects associated with neuroleptic-sensitive behavioral disorders. Genes Dev 16: 2350–2364.

Baas, P.W., T.P. Pienkowski, K.A. Cimbalnik et al. 1994. Tau confers drug stability but not cold stability to microtubules in living cells. J Cell Sci 107(Pt 1): 135–143.

Baratier, J., L. Peris, J. Brocard et al. 2006. Phosphorylation of microtubule-associated protein STOP by calmodulin kinase II. J Biol Chem 281: 19561–19569.

Begou, M., J. Volle, J.B. Bertrand et al. 2008. The stop null mice model for schizophrenia displays [corrected] cognitive and social deficits partly alleviated by neuroleptics. Neuroscience 157: 29–39.

Behnke, O. and A. Forer. 1967. Evidence for four classes of microtubules in individual cells. J Cell Sci 2: 169–192.

Bennecib, M., C.X. Gong, I. Grundke-Iqbal et al. 2000. Role of protein phosphatase-2A and -1 in the regulation of GSK-3, cdk5 and cdc2 and the phosphorylation of tau in rat forebrain. FEBS Lett 485: 87–93.

Bershadsky, A.D., V.I. Gelfand, T.M. Svitkina et al. 1979. Cold-stable microtubules in the cytoplasm of mouse embryo fibroblasts. Cell Biol Int Rep 3: 45–50.

Billger, M., M. Wallin, R.C. Williams, Jr. et al. 1994. Dynamic instability of microtubules from cold-living fishes. Cell Motil Cytoskeleton 28: 327–332.

Binder, L.I., A.L. Guillozet-Bongaarts, F. Garcia-Sierra et al. 2005. Tau, tangles, and Alzheimer's disease. Biochim. Biophys Acta 1739: 216–223.

Birukova, A.A., K.G. Birukov, K. Smurova et al. 2004a. Novel role of microtubules in thrombin-induced endothelial barrier dysfunction. FASEB J 18: 1879–1890.

Birukova, A.A., F. Liu, J.G. Garcia et al. 2004b. Protein kinase A attenuates endothelial cell barrier dysfunction induced by microtubule disassembly. Am J Physiol Lung Cell Mol Physiol 287: L86–93.

Birukova, A.A., K. Smurova, K.G. Birukov et al. 2004c. Microtubule disassembly induces cytoskeletal remodeling and lung vascular barrier dysfunction: role of Rho-dependent mechanisms. J Cell Physiol 201: 55–70.

Birukova, A.A., K.G. Birukov, D. Adyshev et al. 2005. Involvement of microtubules and Rho pathway in TGF-beta1-induced lung vascular barrier dysfunction. J Cell Physiol 204: 934–947.

Bosc, C., J.D. Cronk, F. Pirollet et al. 1996. Cloning, expression, and properties of the microtubule-stabilizing protein STOP. Proc Natl Acad Sci USA 93: 2125–2130.

Bosc, C., R. Frank, E. Denarier et al. 2001. Identification of novel bifunctional calmodulin-binding and microtubule-stabilizing motifs in STOP proteins. J Biol Chem 276: 30904–30913.

Bosc, C., A. Andrieux and D. Job. 2003. STOP proteins. Biochemistry 42: 12125–12132.

Bouchard, M. and O. Suchowersky. 2011. Tauopathies: one disease or many? Can J Neurol Sci 38: 547–556.

Bouvrais-Veret, C., S. Weiss, N. Hanoun et al. 2008. Microtubule-associated STOP protein deletion triggers restricted changes in dopaminergic neurotransmission. J Neurochem 104: 745–756.

Brinkley, B.R. and J. Cartwright, Jr. 1975. Cold-labile and cold-stable microtubules in the mitotic spindle of mammalian cells. Ann NY Acad Sci 253: 428–439.

Brinkley, W. 1997. Microtubules: a brief historical perspective. J Struct Biol 118: 84–86.

Brion, J.P. 2006. Immunological demonstration of tau protein in neurofibrillary tangles of Alzheimer's disease. J Alzheimers Dis 9: 177–185.

Cambiazo, V., M. Gonzalez and R.B. Maccioni. 1995. DMAP-85: a tau-like protein from Drosophila melanogaster larvae. J Neurochem 64: 1288–1297.

Chetham, P.M., P. Babal, J.P. Bridges et al. 1999. Segmental regulation of pulmonary vascular permeability by store-operated Ca^{2+} entry. Am J Physiol 276: L41–50.

Creighton, J., B. Zhu, M. Alexeyev et al. 2008. Spectrin-anchored phosphodiesterase 4D4 restricts cAMP from disrupting microtubules and inducing endothelial cell gap formation. J Cell Sci 121: 110–119.

Dacheux, D., N. Landrein, M. Thonnus et al. 2012. A MAP6-related protein is present in protozoa and is involved in flagellum motility. PLoS One 7: e31344.

de Calignon, A., L.M. Fox, R. Pitstick et al. 2010. Caspase activation precedes and leads to tangles. Nature 464: 1201–1204.

Dehmelt, L. and S. Halpain. 2005. The MAP2/Tau family of microtubule-associated proteins. Genome Biol 6: 204.

Denarier, E., M. Aguezzoul, C. Jolly et al. 1998a. Genomic structure and chromosomal mapping of the mouse STOP gene (Mtap6). Biochem Biophys Res Commun 243: 791–796.

Denarier, E., A. Fourest-Lieuvin, C. Bosc et al. 1998b. Nonneuronal isoforms of STOP protein are responsible for microtubule cold stability in mammalian fibroblasts. Proc Natl Acad Sci USA 95: 6055–6060.

Detrich, H.W., 3rd, S.K. Parker, R.C. Williams, Jr. et al. 2000. Cold adaptation of microtubule assembly and dynamics. Structural interpretation of primary sequence changes present in the alpha- and beta-tubulins of Antarctic fishes. J Biol Chem 275: 37038–37047.

Dimitrov, A., M. Quesnoit, S. Moutel et al. 2008. Detection of GTP-tubulin conformation *in vivo* reveals a role for GTP remnants in microtubule rescues. Science 322: 1353–1356.

Eot-Houllier, G., M. Venoux, S. Vidal-Eychenie et al. 2010. Plk1 regulates both ASAP localization and its role in spindle pole integrity. J Biol Chem 285: 29556–29568.

Frank, S., F. Clavaguera and M. Tolnay. 2008. Tauopathy models and human neuropathology: similarities and differences. Acta Neuropathol. 115: 39–53.

Goedert, M., M.G. Spillantini, R. Jakes et al. 1989. Multiple isoforms of human microtubule-associated protein tau: sequences and localization in neurofibrillary tangles of Alzheimer's disease. Neuron 3: 519–526.

Goedert, M., C.P. Baur, J. Ahringer et al. 1996. PTL-1, a microtubule-associated protein with tau-like repeats from the nematode Caenorhabditis elegans. J Cell Sci 109(Pt 11): 2661–2672.

Gong, C.X. and K. Iqbal. 2008. Hyperphosphorylation of microtubule-associated protein tau: a promising therapeutic target for Alzheimer disease. Curr Med Chem 15: 2321–2328.

Gory-Faure, S., V. Windscheid, C. Bosc et al. 2006. STOP-like protein 21 is a novel member of the STOP family, revealing a Golgi localization of STOP proteins. J Biol Chem 281: 28387–28396.

Grundke-Iqbal, I., K. Iqbal, Y.C. Tung et al. 1986. Abnormal phosphorylation of the microtubule-associated protein tau (tau) in Alzheimer cytoskeletal pathology. Proc Natl Acad Sci USA 83: 4913–4917.

Guillaud, L., C. Bosc, A. Fourest-Lieuvin et al. 1998. STOP proteins are responsible for the high degree of microtubule stabilization observed in neuronal cells. J Cell Biol 142: 167–179.

Gupta, M.L., Jr., C.J. Bode, C.A. Dougherty et al. 2001. Mutagenesis of beta-tubulin cysteine residues in Saccharomyces cerevisiae: mutation of cysteine 354 results in cold-stable microtubules. Cell Motil Cytoskeleton 49: 67–77.

Halpain, S. and L. Dehmelt. 2006. The MAP1 family of microtubule-associated proteins. Genome Biol 7: 224.

Hanger, D.P., B.H. Anderton and W. Noble. 2009. Tau phosphorylation: the therapeutic challenge for neurodegenerative disease. Trends Mol Med 15: 112–119.

Himes, R.H. and H.W. Detrich, 3rd. 1989. Dynamics of Antarctic fish microtubules at low temperatures. Biochemistry 28: 5089–5095.

Hutton, M., C.L. Lendon, P. Rizzu et al. 1998. Association of missense and 5'-splice-site mutations in tau with the inherited dementia FTDP-17. Nature 393: 702–705.

Jacobson, J.R. and J.G. Garcia. 2010. Endothelial cell function. *In:* R.J. Mason, B.C. Broaddus, T.R. Martin, T.E. King, D.E. Schraufnagel, J.F. Murray and J.A. Nadel (Eds.). Textbook of Respiratory Medicine. United Sates of America: Saunders, 2010. Chapter 6 Appendix, page 1.

Job, D., E.H. Fischer and R.L. Margolis. 1981. Rapid disassembly of cold-stable microtubules by calmodulin. Proc Natl Acad Sci USA 78: 4679–4682.

Job, D., C.T. Rauch and R.L. Margolis. 1987. High concentrations of STOP protein induce a microtubule super-stable state. Biochem Biophys Res Commun 148: 429–434.

Kar, S., J. Fan, M.J. Smith et al. 2003. Repeat motifs of tau bind to the insides of microtubules in the absence of taxol. EMBO J 22: 70–77.

Kelly, J.J., T.M. Moore, P. Babal et al. 1998. Pulmonary microvascular and macrovascular endothelial cells: differential regulation of Ca^{2+} and permeability. Am J Physiol 274: L810–819.

King, M.E., T.C. Gamblin, J. Kuret et al. 2000. Differential assembly of human tau isoforms in the presence of arachidonic acid. J Neurochem 74: 1749–1757.

Kueh, H.Y. and T.J. Mitchison. 2009. Structural plasticity in actin and tubulin polymer dynamics. Science 325: 960–963.

Ledbetter, M.C. and K.R. Porter. 1963. A "microtubule" in plant cell fine structure. J Cell Biol 19: 239–250.

Lee, T.Y. and A.I. Gotlieb. 2003. Microfilaments and microtubules maintain endothelial integrity. Microsc Res Tech 60: 115–127.

Leppla, S.H. 1982. Anthrax toxin edema factor: a bacterial adenylate cyclase that increases cyclic AMP concentrations of eukaryotic cells. Proc Natl Acad Sci USA 79: 3162–3166.

Lewis, J., E. McGowan, J. Rockwood et al. 2000. Neurofibrillary tangles, amyotrophy and progressive motor disturbance in mice expressing mutant (P301L) tau protein. Nat Genet 25: 402–405.

Liu, F., B. Li, E.J. Tung et al. 2007. Site-specific effects of tau phosphorylation on its microtubule assembly activity and self-aggregation. Eur J Neurosci 26: 3429–3436.

Liu, S. J., J.Y. Zhang, H.L. Li et al. 2004. Tau becomes a more favorable substrate for GSK-3 when it is prephosphorylated by PKA in rat brain. J Biol Chem 279: 50078–50088.

Liu, Y., J. Xia, D. Ma et al. 1997. Tau-like proteins in the nervous system of goldfish. Neurochem Res 22: 1511–1516.

Manton, I. and B. Clarke. 1950. Electron microscope observations on the spermatozoid of Fucus. Nature 166: 973–974.

Margolis, R.L., C.T. Rauch and D. Job. 1986. Purification and assay of a 145-kDa protein (STOP145) with microtubule-stabilizing and motility behavior. Proc Natl Acad Sci USA 83: 639–643.

Margolis, R.L., C.T. Rauch, F. Pirollet et al. 1990. Specific association of STOP protein with microtubules *in vitro* and with stable microtubules in mitotic spindles of cultured cells. EMBO J 9: 4095–4102.

Mehta, D. and A.B. Malik. 2006. Signaling mechanisms regulating endothelial permeability. Physiol Rev 86: 279–367.

Northover, A.M. and B.J. Northover. 1993. Possible involvement of microtubules in platelet-activating factor-induced increases in microvascular permeability *in vitro*. Inflammation 17: 633–639.

Ochoa, C.D., T. Stevens and R. Balczon. 2011. Cold exposure reveals two populations of microtubules in pulmonary endothelia. Am J Physiol Lung Cell Mol Physiol 300: L132–138.

Ochoa, C.D., M. Alexeyev, V. Pastukh et al. 2012. Pseudomonas aeruginosa exotoxin Y is a promiscuous cyclase that increases endothelial tau phosphorylation and permeability. J Biol Chem 287: 25407–25418.

Ofori-Acquah, S.F., J. King, N. Voelkel et al. 2008. Heterogeneity of barrier function in the lung reflects diversity in endothelial cell junctions. Microvasc Res 75: 391–402.

Penttila, T.L., M. Parvinen and J. Paranko. 2003. Microtubule-associated epithelial protein E-MAP-115 is localized in the spermatid manchette. Int J Androl 26: 166–174.

Pirollet, F., D. Job, E.H. Fischer et al. 1983. Purification and characterization of sheep brain cold-stable microtubules. Proc Natl Acad Sci USA 80: 1560–1564.

Pirollet, F., C.T. Rauch, D. Job et al. 1989. Monoclonal antibody to microtubule-associated STOP protein: affinity purification of neuronal STOP activity and comparison of antigen with activity in neuronal and nonneuronal cell extracts. Biochemistry 28: 835–842.

Poorkaj, P., T.D. Bird, E. Wijsman et al. 1998. Tau is a candidate gene for chromosome 17 frontotemporal dementia. Ann Neurol 43: 815–825.

Poorkaj, P., A. Kas, I. D'Souza et al. 2001. A genomic sequence analysis of the mouse and human microtubule-associated protein tau. Mamm Genome 12: 700–712.

Prasain, N. and T. Stevens. 2009. The actin cytoskeleton in endothelial cell phenotypes. Microvasc Res 77: 53–63.

Prasain, N., M. Alexeyev, R. Balczon et al. 2009. Soluble adenylyl cyclase-dependent microtubule disassembly reveals a novel mechanism of endothelial cell retraction. Am J Physiol Lung Cell Mol Physiol 297: L73–83.

Querfurth, H.W. and F.M. LaFerla. 2010. Alzheimer's disease. N Engl J Med 362: 329–344.

Roth, L.E. and E.W. Daniels. 1962. Electron microscopic studies of mitosis in amebae. II. The giant ameba Pelomyxa carolinensis. J Cell Biol 12: 57–78.

Rouzier, R., R. Rajan, P. Wagner et al. 2005. Microtubule-associated protein tau: a marker of paclitaxel sensitivity in breast cancer. Proc Natl Acad Sci USA 102: 8315–8320.

Sabatini, D.D., K. Bensch and R.J. Barrnett. 1963. Cytochemistry and electron microscopy. The preservation of cellular ultrastructure and enzymatic activity by aldehyde fixation. J Cell Biol 17: 19–58.

Saffin, J.M., M. Venoux, C. Prigent et al. 2005. ASAP, a human microtubule-associated protein required for bipolar spindle assembly and cytokinesis. Proc Natl Acad Sci USA 102: 11302–11307.

Sayner, S.L., D.W. Frank, J. King et al. 2004. Paradoxical cAMP-induced lung endothelial hyperpermeability revealed by Pseudomonas aeruginosa Exo Y Circ Res 95: 196–203.

Sayner, S.L., M. Alexeyev, C.W. Dessauer et al. 2006. Soluble adenylyl cyclase reveals the significance of cAMP compartmentation on pulmonary microvascular endothelial cell barrier. Circ Res 98: 675–681.

Sayner, S.L., R. Balczon, D.W. Frank et al. 2011. Filamin A is a phosphorylation target of membrane but not cytosolic adenylyl cyclase activity. Am J Physiol Lung Cell Mol Physiol 301: L117–124.

Schneider, A. and E. Mandelkow. 2008. Tau-based treatment strategies in neurodegenerative diseases. Neurotherapeutics 5: 443–457.

Schneider, A., J. Biernat, M. von Bergen et al. 1999. Phosphorylation that detaches tau protein from microtubules (Ser262, Ser214) also protects it against aggregation into Alzheimer paired helical filaments. Biochemistry 38: 3549–3558.

Scott, C.W., R.C. Spreen, J.L. Herman et al. 1993. Phosphorylation of recombinant tau by cAMP-dependent protein kinase. Identification of phosphorylation sites and effect on microtubule assembly. J Biol Chem 268: 1166–1173.

Shen, Y., N.L. Zhukovskaya, Q. Guo et al. 2005. Calcium-independent calmodulin binding and two-metal-ion catalytic mechanism of anthrax edema factor. EMBO J 24: 929–941.

Spillantini, M.G., M. Goedert, R.A. Crowther et al. 1997. Familial multiple system tauopathy with presenile dementia: a disease with abundant neuronal and glial tau filaments. Proc Natl Acad Sci USA 94: 4113–4118.

Stevens, T., S. Phan, M.G. Frid et al. 2008. Lung vascular cell heterogeneity: endothelium, smooth muscle, and fibroblasts. Proc Am Thorac Soc 5: 783–791.

Suzuki, M., A. Hirao and A. Mizuno. 2003. Microtubule-associated [corrected] protein 7 increases the membrane expression of transient receptor potential vanilloid 4 (TRPV4). J Biol Chem 278: 51448–51453.

Tar, K., A.A. Birukova, C. Csortos et al. 2004. Phosphatase 2A is involved in endothelial cell microtubule remodeling and barrier regulation. J Cell Biochem 92: 534–546.

Tar, K., C. Csortos, I. Czikora et al. 2006. Role of protein phosphatase 2A in the regulation of endothelial cell cytoskeleton structure. J Cell Biochem 98: 931–953.

Tilney, L.G. and K.R. Porter. 1967. Studies on the microtubules in heliozoa. II. The effect of low temperature on these structures in the formation and maintenance of the axopodia. J Cell Biol 34: 327–343.

Verin, A.D., A. Birukova, P. Wang et al. 2001. Microtubule disassembly increases endothelial cell barrier dysfunction: role of MLC phosphorylation. Am J Physiol Lung Cell Mol Physiol 281: L565–574.

Wallin, M. and E. Stromberg. 1995. Cold-stable and cold-adapted microtubules. Int Rev Cytol 157: 1–31.

Webb, B.C. and L. Wilson. 1980. Cold-stable microtubules from brain. Biochemistry 19: 1993–2001.

Weingarten, M.D., A.H. Lockwood, S.Y. Hwo et al. 1975. A protein factor essential for microtubule assembly. Proc Natl Acad Sci USA 72: 1858–1862.

Williams, D.R. 2006. Tauopathies: classification and clinical update on neurodegenerative diseases associated with microtubule-associated protein tau. Intern Med J 36: 652–660.

Xie, R., S. Nguyen, K. McKeehan et al. 2011. Microtubule-associated protein 1S (MAP1S) bridges autophagic components with microtubules and mitochondria to affect autophagosomal biogenesis and degradation. J Biol Chem 286: 10367–10377.

Yahr, T.L., A.J. Vallis, M.K. Hancock et al. 1998. ExoY, an adenylate cyclase secreted by the Pseudomonas aeruginosa type III system. Proc Natl Acad Sci USA 95: 13899–13904.

Zhu, B., L. Zhang, J. Creighton et al. 2010. Protein kinase A phosphorylation of tau-serine 214 reorganizes microtubules and disrupts the endothelial cell barrier. Am J Physiol Lung Cell Mol Physiol 299: L493–501.

3

Endothelial Actin Cytoskeleton and Angiogenesis

Sadiqa K. Quadri

Introduction

Vasculogenesis is the differentiation of endothelial progenitor cells and their assembly into a primary capillary plexus (Li et al. 2006). Vessel formation has been defined in several ways (Carmeliet and Jain 2011). Vasculogenesis, the de novo formation of vessels from mesoderm-derived endothelial precursor cells (angioblasts), is responsible for the formation of the first, primitive blood vessels in the embryo (Ferguson et al. 2005). The vascular tree develops early in embryogenesis to provide oxygen to the developing organism (Flamme et al. 1997, Hickey and Simon 2006, Robb and Elefanty 1998, Watson and Cross 2005). Physiological and pathological blood vessel growth in later life is predominantly achieved through angiogenesis. Unwanted angiogenesis has been associated with the expansion of atherosclerotic lesions, diabetic retinopathy, psoriasis, and cancer progression (Adams and Alitalo 2007, Carmeliet and Jain 2011). On the other hand, ischemic and cardiovascular diseases have been associated with insufficient angiogenesis (Adams and Alitalo 2007, Becker and D'Amato 2007, Fukumura and Jain 2007, Renault and Losordo 2007). Vessels fuel inflammatory and malignant diseases to promote tumor neo-vascularization, growth and progression to metastasis

Associate Research Scientist, Division of Pulmonary, Allergy and Critical Care Medicine, Columbia University, Medical center, New York, NY, 10032.
Email: Skq1@columbia.edu

in cancer (Folkman 2007). The blood vasculature surrounding and within the growing tumor not only delivers nutrients and oxygen required for sustained tumor growth but also provides access for macrophages to the tumor, which promotes further growth and metastasis (Condeelis and Pollard 2006). Blood vessels also provide the route for rapid tumor dissemination to distant organs (Wyckoff et al. 2007). The most important question in vascular developmental biology today is: what are the general mechanisms guiding angiogenesis?

Angiogenesis and arteriogenesis refer to the sprouting and distinct signals that specify arterial or venous differentiation (Swift and Weinstein 2009). In order to generate a mature functional vessel, no matter whether the blood vessels arise from vasculogenesis or endothelial cells (ECs), it is believed that Ecs first have to form a cord, which subsequently lumenizes, then forms sprouts and further develops into a network structure. During these processes, cellular junctions rearrange between adjacent ECs and are involved in sprouting and lumen formation. Angiogenesis is frequently accompanied by an increase in vessel junctional permeability (Carmeliet and Jain 2000, Dvorak et al. 1999). It is known that vascular endothelial growth factor (VEGF) induces tyrosine phosphorylation of VE-cadherin and β-catenin (Esser et al. 1998) that makes linkage with actin cytoskeleton and is accompanied by decreased junctional strength and increased permeability. Evidence shows that mice lacking the actin binding protein moesin, have delayed lumen formation in the dorsal aorta and a decreased amount of F-actin (Strilic et al. 2009). Similarly, knock down of focal adhesion kinase (FAK) is associated with embryonically lethal phenotypes. Moreover FAK forms a critical bidirectional linkage between the actin cytoskeleton and the cell-extracellular matrix interface (Critchley 2000, Quadri 2012). However, the role of cadherin, FAK, and actin in sprouting and lumen formation is still not completely understood.

Most of our knowledge is based on the morphological processes and molecular regulation of angiogenesis results from studies in developing zebrafish embryos (Siekmann and Lawson 2007a, 2007b), or the vascularization of the retina in the postnatal mouse (Fruttiger 2007, Monica et al. 2007, Uemura et al. 2006). The experimental accessibility and the possibility of live imaging are the main advantages of the zebrafish system. The important features of the mouse retina model are the progressive expansion of the growing retinal blood vessels from the center to the periphery, their initial organization as a flat two-dimensional network, the spatial separation of growing vs. maturing areas, and the relatively easy identification of sprouts (Monica et al. 2007, Uemura et al. 2006). Considering the substantial clinical benefits of therapeutically manipulating pathological angiogenesis, the mechanisms controlling vascular lumen and sprouting process have been areas of major focus for vascular research over the last two decades. Anti-angiogenic approaches are based on influencing the output of pro-angiogenic factors or directly targeting endothelial cells. This review discusses our understanding of blood vessel morphogenesis, with particular emphasis on actin cytoskeleton and mechanisms that control and coordinate ECs, cell–cell, and cell-matrix molecules in vascular lumen and sprout formation.

Vascular Development and Angiogenesis Overview

Patterning of the primary vascular plexus results from a collective action of primordial endothelial cells. Once an angioblast is recruited into forming a vascular "tube" or vessel, it differentiates into bona fide ECs. All vessels consist of interconnected, tube-forming ECs. ECs form the seamless lining of the entire circulatory system. However, ECs are highly heterogeneous in terms of functional properties and gene expression profiles, which depend on their differentiation and proliferation status (Gebb and Stevens 2004, Ofori-Acquah et al. 2008, Stevens et al. 2008). One known anatomical and physiological distinction between vessels is that of arteries and veins. Arteries and veins are surrounded by mural cells (pericytes in medium vessels and smooth muscle cells in large vessels) to the abluminal (basal) surface and help to stabilize the vessel wall. Small blood vessels consist only of ECs, and are hence in immediate contact with surrounding tissues. Pericytes ensheath endothelial cells, suppress endothelial cell proliferation, and release cell-survival signals such as VEGF and angiopoietin-1 (ANG-1) (Gerhardt and Semb 2008). When ECs sense angiogenic stimuli such as VEGF, VEGF-C, ANG-2, FGFs or chemokines released by a hypoxic, inflammatory or tumor cell, pericytes first detach from the vessel wall and from the basement membrane by matrix metalloproteinases (MMPs) proteolytic degradation (Deryugina and Quigley 2010, van Hinsbergh and Koolwijk 2008). VEGF loosen endothelial cell–cell junctions and the nascent vessel dilates, causing plasma proteins to extravasate and lay down a provisional extracellular matrix (ECM) scaffold. ECs loosen cell–cell junctional contacts, activate proteases that degrade the surrounding basement membrane and acquire extensively invasive and motile behavior to initiate new blood vessel sprouting (Adams and Alitalo 2007, Carmeliet and Jain 2011).

To build a perfusable tube, a quiescent endothelial cell (Fig. 1A) selects to lead the tip in the presence of factors such as vascular endothelial growth factor (VEGF) receptors, neuropilins (NRPs), the Notch ligands, Delta like ligand-4 (DLL4), and JAGGED1 (Carmeliet and Jain 2011). VEGF activated ECs induce expression of DLL4 by the endothelial "tip cells" (TCs, Fig. 1B). Cell surface associated DLL4 expressed by the TCs then ligate to the Notch receptor adjacent to the TCs termed as stalk cells (SCs) (Hellstrom et al. 2007) (Figs. 1B and 1C). TCs do not form a lumen. SCs release molecules such as

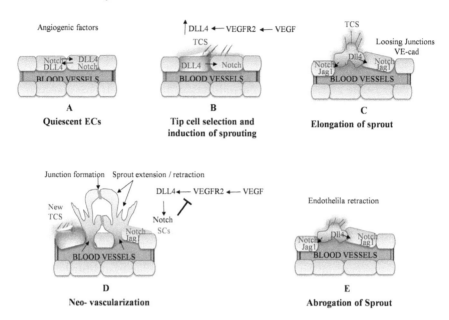

Figure 1. Sprouting and lumen formation. (A-C): Sprouting Quiescent ECs:DLL4 and Notch signaling are balanced in endothelial cells (pink). **(B)** *Tip cells (TCs) selection and induction of sprouting*: VEGF -mediated signals induce DLL4 expression in TCS (blue). TCS are selected by DLL4 expression. **(C)**. *Elongation of sprout*: Dll4-expressing tip cells react strongest to VEGF signaling and acquire a motile, invasive and sprouting phenotype by loosening the contact to adjacent endothelial stalk cells (SCs, pink) by endocytosis of VE cadherin. DLL4 activates Notch proteins in SCs. DLL4-NOTCH signaling is suppressed in Jag1-expressing, SCs, which form the base of the emerging sprout.

(D, E): Lumen formation (D) *Neo-vascularization*: VE-cadherin and other other junction molecules are expressed on filopodia of tip cells (green) and might promote the formation of new inter-endothelial connections between bridging sprouts and the establishment of the newly forming junction. VE-cadherin seal the contact between adjacent endothelial cells, antagonize pro-angiogenic signals by stabilizing cell–cell contacts. At the same time, new sprouts (blue) are induced, which involves stalk-to-tip conversions. **(E)** *Abrogation of sprout*: When conditions do not favor the formation of new endothelial tubules, like the presence of repulsive signals or inhibition of VEGF, vessels aborted.

Color image of this figure appears in the color plate section at the end of the book.

epidermal growth factor like domain and multiple 7 (EGFL7) into the ECM to convey spatial information about the position of their neighbors, so that the stalk elongates, stimulated by NOTCH-regulated signaling, ankyrin repeat protein (NRARP), wingless (WNTs), placental growth factor (PlGF) and fibroblast growth factor (FGF). Sprouts then convert into endothelial tubules and form connections with other vessels (Fig. 1D) which require the local suppression of motility. For a functional vessel, new junctions form, establish the lumen for optimal flow distribution (mediated by

VE-cadherin, CD34, VEGF, and hedgehog), while vessels regress if they are unable to perfuse (Carmeliet and Jain 2011) (Fig. 1E).

Cellular Mechanism of Angiogenesis

Role of Notch signaling

Notch signaling has been reviewed (Roca and Adams 2007). In brief, ECs express multiple Notch receptors (NOTCH1, NOTCH3, and NOTCH4) and transmembrane Notch ligands, (DLL1, Jagged 1 and Jagged 2). Notch signaling is a conserved pathway that plays a critical part in TC and SC fate decisions during angiogenesis (Phng and Gerhardt 2009, Roca and Adams 2007). TCs display a highly motile phenotype, extend many dynamic filopodial extensions (Fig. 1C) that sense and respond to environmental guidance cues such as ephrins and semaphorins and attractive or repulsive guidance signals within their immediate microenvironment to enable directionality and prevent unorganized, random vessel growth (De Smet et al. 2009, Gerhardt et al. 2003). Although ECs sense angiogenic stimuli, only a small proportion will be selective to TCs that lead newly sprouting vessels. SCs do not extend filopodia but rather re-establish junctions and prevent EC retraction (De Smet et al. 2009). SCs trail behind the tip, proliferate, elongate the stalk, form a lumen, and connect to the circulation (Fig. 1D). The Notch protein spans the cell membrane, with part of it inside and part outside. Ligand DLL4 binds to the extracellular domain of Notch receptor, induces proteolytic cleavage of Notch receptor and the release of an intracellular fragment (Fig. 2), known as the Notch intracellular domain (NICD), which enters the cell nucleus to modify gene expression, hence functioning as a key transcriptional regulator during cell-fate specification. A disintegrin and metalloprotease domain-containing protein 10 (ADAM10), ADAM17 and presenilins play a critical role in the cleavage of Notch receptor. In addition, ECspecific knockout of ADAM10 that inhibits Notch cleavage promotes enhanced EC sprouting (Glomski et al. 2011) and suppresses TC specification.

In response to VEGF, VEGFR-2 activation upregulates DLL4 expression in tip cells. In neighboring stalk cells, DLL4 then activates Notch, which downregulates VEGFR-2 but upregulates VEGFR-1 (Figs. 1D and 2); thus, the stalk cells become less responsive to the sprouting activity of VEGF. Overall, DLL4 and Notch signaling restricts branching but generates perfused vessels (Phng and Gerhardt 2009). Another Notch ligand, JAGGED1, expressed by stalk cells (Fig. 1C), promotes TC selection by interfering with the reciprocal DLL4 and Notch signaling from the SCs to the TCs (Benedito et al. 2009) (Fig. 2). Notch signaling in stalk cells is dynamic over time, because it upregulates its own inhibitor, ankyrin repeat-

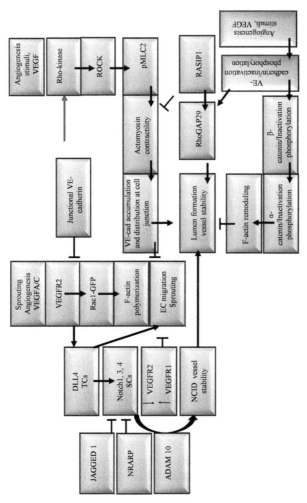

Figure 2. Key signalling pathways that control sprouting and vessel stabilization. General signalling pathways that control endothelial cell (EC) behaviour. vascular endothelial growth factors (VEGFs) bind homodimers and heterodimers of three VEGF receptors (VEGFRs). Homophilic VE-cadherin interactions maintain EC–EC junctions. Inhibiton of VE-cadherin, Rho-kinase, phospho-MLC2, and actomyosin contractility acting to suppress the VEGFR2-Rac1-dependent sprouting. Delta-like 4 (DLL4)-mediated activation of Notch receptors represses angiogenic cell behaviour and promotes vessel stability upon the proteolytic release of the Notch intracellular domain (NICD) by ADAM10. NOTCH signalling in stalk cells upregulates its own inhibitor, Ankyrine repeat protein (NRARP). SCs expressing, Jagged 1 competes with DLL4 for Notch to decrease DLL4–Notch-mediated signalling and promotes from SCs to the TCs. Ras interacting protein-1 (RASIP1) and RHOGAP29 influence lumen morphogenesis. RASIP1 may repress actomyosin contractility. Inactivation or VEGF induced phophorylation of VE-cadherin, β-catenin, markedly inhibits normal vascular development in the embryo. Possibly alteration in the F-actin cytoskeleton.

containing protein (NRARP) (Phng et al. 2009) (Fig. 2). An unanticipated complexity is that endothelial cells continuously compete for the TC position by fine-tuning their expression of VEGFR-2 versus VEGFR-1, indicating that this signaling circuit is constantly re-evaluated as cells meet new neighbors (Jakobsson et al. 2010).

DLL4–Notch signaling that controls TC specification shares similarities with the molecular mechanisms of epithelial tracheal branching in *D. melanogaster* (Affolter et al. 2009). Similar to the role of VEGF signaling in endothelial TC formation, tracheal TCs are specified by the fibroblast growth factor (FGF), ligand Branchless (BNL). Furthermore, in this system, Notch signaling via the ligand Delta limits excessive epithelial TC formation by repressing expression of the BNL receptor, breathless (btl), in SCs. Hence, the Notch-mediated lateral inhibition of receptor Tyr kinase expression controls the specification of TC during tissue branching morphogenesis in multiple systems.

Studies in mouse and zebrafish reveal that expression of DLL4 in TCs activates Notch in adjacent SCs to laterally inhibit TC fate and maintain the hierarchical organization of sprouting (Hellstrom et al. 2007, Leslie et al. 2007, Lobov et al. 2007, Suchting et al. 2007). Studies show that tumors are extensively vascularized on antibody-mediated reduction of DLL4–Notch interactions but vessels are hypoperfused and poorly functional, resulting in tumor hypoxia and decreased growth (Noguera-Troise et al. 2006, Ridgway et al. 2006). DLL4–Notch signaling in suppressing TC specification is not only critical for physiological blood vessel formation but also controls tumor angiogenesis. Hence, DLL4 is considered to be a promising target for anticancer therapeutics. However, this view has been challenged by studies showing that blockade of chronic DLL4 induces vascular tumors (Yan et al. 2010). Notch signaling controls whether specific endothelial cells become lead tip cells or trailing SCs. Sprouts then convert into endothelial lumen and form connections with other vessels requiring the local suppression of motility and the formation of new, stable cell–cell junctions. Whether these pro-angiogenic mechanisms are used selectively by TCs during sprouting is not clear.

WNT–β-catenin in TC and SC fate decision

Endothelial cells express various types of WNT ligand and their frizzled (FZD) receptors, of which several stimulate endothelial cell proliferation. Studies of angiogenesis using mice overexpressing β-catenin in ECs show that the WNT–β-catenin pathway positively regulates DLL4 expression in the vasculature (Corada et al. 2010). Consequently, β-catenin induces endothelial Notch signaling, characterized by branching defects (disrupts EC branching), although the role of this pathway in TC selection is unclear.

In parallel, DLL4–Notch activates WNT signaling in proliferating SCs, which maintains EC–EC contact during vessel branching (Phng et al. 2009). Notch signaling promotes vascular stability by inducing expression of the WNT regulator gene, Notch-regulated NRARP. NRARP limits Notch activity and stimulates lymphoid enhancer-binding factor 1 (LEF1)- and βcatenin-dependent WNT signaling to stabilize SCs. Hence, functional interplay between Notch and WNT signaling is critical for vascular development.

Role of junctional molecules and actin in angiogenesis

Quiescent endothelial cells form a monolayer of interconnected cells, whereas angiogenic endothelial cells dissociate their junctions to migrate. The signals received by endothelial cells (ECs) from their environment (cell-matrix) and their neighboring cells (cell–cell) control sprouting and lumen processes. ECs express multiple types of cell–cell interacting junctional proteins, such as adherens junctions (AJs) and tight junctions (TJs), which are required for cellular organization of complex networks in the process of angiogenesis, and FAK which interact to the cell-matrix through integrins. AJs and FAK-rich adhesions are complex multi-unit plasma membrane structures that assemble in a localized spatial and temporal organization to maintain structural tissue organization, and to provide cell signaling function. The TJ molecules claudins and occludins maintain barriers such as the blood brain barrier (BBB), whereas AJs molecules that establish cell–cell adhesion are linked to the cytoskeleton and are involved in remodeling of actin cytoskeleton and intracellular signaling (Dejana et al. 2009). Inactivation of the genes coding for certain AJs proteins, such as VE-cadherin (Carmeliet et al. 1999) or β-catenin (Cattelino et al. 2003), markedly inhibits normal vascular development in the embryo (Fig. 2). By contrast, the absence of certain TJ proteins, such as occludin (Saitou et al. 2000) or claudin-5 (Nitta et al. 2003), does not affect the vascular system. Loss of VE-cadherin does not prevent vessel development but induces defects in vascular remodeling and integrity (Carmeliet 2003).

Vessel sprouting results from the division of differentiated endothelial cells (ECs). Sprout elongation depends on a continuous supply of endothelial cells. Endothelial cells migrate along the sprout, towards its tip, in a vascular endothelial (VE) cadherin-dependent process. The observed abundance of multicellular sprout formation in *in vitro* and *in vivo* systems can be explained by a general mechanism based on preferential attraction to elongated structures. Evidence shows that VE-cadherin, which is responsible for endothelial adherens junction organization plays a crucial role in the cessation of sprouting (Carmeliet et al. 1999, Corada et al. 1999, Dejana et al. 2009, Stevens et al. 2000). Abrogating VE-cadherin function in an organotypic angiogenesis assay and in zebrafish embryos stimulates sprouting. Sprout contact neighboring extensions with stable junction is one of the mechanisms used to generate new blood vessels (Fig. 1D).

In response to VEGF, the adhesive function of VE-cadherin is reduced (Fig. 1C) by endocytosis (Dejana et al. 2009). At the same time, the localization of VE-cadherin at filopodia allows TCs to establish new contacts with cells on outreaching sprouts, while SCs behind the TCs proliferate, elongate and form a lumen (Fig. 1D). Lumen formation involves the coordination of many different cellular activities for the positioning of cells and for the establishment of tubular structures. During angiogenic sprouting, all functional blood vessels need to form a patent vascular lumen to establish blood flow. Although the molecular pathways controlling epithelial lumen morphogenesis are under intense investigation (Bryant and Mostov 2008), progress towards defining the mechanisms of vascular lumen formation is not clear.

Prior to lumen formation, blood vessels consist of multicellular rods of ECs that are interconnected by uniform EC–EC junctions and have yet to establish apicobasal polarity. It is known that in epithelial lumen formation, partitioning defective 3 (PAR3) is a critical determinant of cell polarity *in vitro* (Iruela-Arispe and Davis 2009). The first phase of blood vessel lumen morphogenesis appears to involve PAR3 mediated acquisition of EC apicobasal polarity and the lateral redistribution of junctional proteins (including zonula occludens 1 (ZO1), claudin 5, CD99 and vascular endothelial cadherin (VE-cadherin) from the apical EC surface to the vascular cord periphery (Xu et al. 2011, Zovein et al. 2010). Both β1 integrin and VE-cadherin regulate the accumulation of PODXL at the apical surface, which may be important for lumen formation *in vivo* (Strilic et al. 2009, Xu et al. 2011, Zovein et al. 2010). Following its redistribution, PODXL recruits moesin apically as the vascular lumen begins to form (Strilic et al. 2009). Similar to what is seen with PODXL and VE-cadherin, the apical accumulation of the actin binding protein, moesin may control lumen morphogenesis and appears to function by recruiting F-actin (Strilic et al. 2009, Wang et al. 2010).

To maintain a luminal structure, ECs must remain adherent to the matrix surface created during morphogenesis, and also maintain cell–cell interactions mediated via junctional contacts between ECs (Blum and Benvenisty 2008). VE-cadherin is also required for localizing CD34 to cell–cell contacts for lumen formation (Strilic et al. 2009). In quiescent phalanx endothelial cells, VE-cadherin promotes vessel stabilization by inhibiting vascular endothelial growth factor receptor-2 (VEGFR-2) signaling (Figs. 1D and 2). Mice deficient for VE-cadherin die in mid-gestation of vascular dysfunctions with defects in the extraembryonic placental vasculature (Gory-Faure et al. 1999). VE-cadherin blocking monoclonal antibody (mAb) effectively inhibits endothelial cell tube formation *in vitro* (Bach et al. 1998) and reduces tumor neovascularization in VE-cadherin null animal models (Liao et al. 2000). This effect is accompanied by signs of endothelial apoptosis in tumor vessels; however, the mechanism of VE-cadherin action in tubulogenesis is still obscure.

Cytoplasmic tails of adhesion proteins bind to cytoskeletal and signaling proteins that promote anchoring of junctions to actin microfilaments and transfer of intracellular signals to the inside of the cell. The interaction of junctional adhesion proteins with the actin cytoskeleton is also relevant in the maintenance of cell shape and polarity (Hartsock and Nelson 2008). VE-cadherin gets localized laterally and connects to the F-actin cytoskeleton or to p120 via catenins. VE-cadherin binds to β-catenin and plakoglobin (or γ-catenin) via its C-terminal cytoplasmic tail. β-catenin and plakoglobin in turn bind to α-catenin, which connects VE-cadherin to F-actin (Quadri 2012). Evidence shows that β-catenin-deficient mice die during embryonic development because of vascular defects. Early vasculogenesis and angiogenesis appear normal, but irregular and inconsistent vascular lumens develop, as do hemorrhages. This phenotype is probably due to alterations in the F-actin cytoskeleton as well as a reduced amount of α-catenin at the junctions (Cattelino et al. 2003). VEcadherin also binds to p120, which does not link to F-actin, but can modulate the F-actin cytoskeleton via interaction with RhoA (Cattelino et al. 2003). Thus, VE-cadherin anchors the cells to one another, while also connecting and influencing the contractile, cytoskeletal fibres necessary for lumen formation. Inhibition of cadherins (Abraham et al. 2009, Gory-Faure et al. 1999) blocks angiogenic responses, indicating that actin anchoring at these sites is critical to this process. To develop anti-angiogenic therapy, it is important to define the mechanism that is involved sprouting and vascular lumen development.

Role of GTPase and actin in angiogenesis

Small GTPases of the Rho family (Rho, Rac, and Cdc42) are regulatory proteins that coordinate remodeling of the actin cytoskeleton in response to various stimuli (Ridley 2001). In the active, GTP-bound form, they interact with and activate target proteins that regulate actin polymerization, cell motility, and gene expression. The Rho GTPases are involved in cell transformation, and some of their activators are oncogenes, including Vav, Net, and Dbl (van Aelst and D'Souza-Schorey 1997). Evidence also shows that VE-cadherin signals to Rho-kinase-dependent myosin light-chain-2 phosphorylation, leading to actomyosin contractility (van Nieuw Amerongen et al. 2007), regulates the distribution of VE-cadherin at cell–cell junctions, and hence stabilizes vessels junction (Fig. 2).

In vitro studies of blood vessel lumen formation have shown that potential molecules involved in this process include integrins, CDc42, Src, Rac1, p21activated kinase 2 (PAK2), PAK4, Raf, partitioning defective 3 (PAR3; also known as PARD3), PAR6 and a typical protein kinase C (Iruela-Arispe and Davis 2009). In addition, evidence indicates that Ras-interacting protein-1 (RASIP1) and its binding partner RhoA GTPase-activating protein-29 (RhoAGAP29) also influence lumen morphogenesis in mice (Xu et al. 2011) (Fig. 2). Vascular-specific expression of RASIP1 regulates RhoGTPase activity and also shifts junctional proteins laterally by activating integrins (including β1 integrin) and controlling PAR3 localization. It seems that acquisition of apicobasal polarity and spatial redistribution of EC-EC junctions initiate lumen formation. Small GTPases are also likely to be involved in regulating actin cytoskeletal organization and AJ assembly (Fukata and Kaibuchi 2001). However, the molecular mechanisms by which small GTPases affect angiogenesis are still poorly understood.

Lumenal expansion proceeds by a variety of mechanisms. For example, vascular endothelial growth factor receptor 2 (VEGFR2) signaling and activation of Rho-associated coiled-coil kinase (ROCK) may promote the association of non-muscle myosin II with apical F-actin to drive actomyosin-mediated cell shape changes. By contrast, RASIP1 may repress actomyosin contractility to fine tune this response. VE-cadherin antagonizes VEGFR2 signaling, inhibition of VE-cadherin and Rho-kinase, or actomyosin contractility leads to VEGF-driven, Rac1-dependent sprouting (Fig. 2). These findings suggest a sprouting mechanism by which cell–cell adhesion suppresses Rac-dependent migration and increasing actomyosin contractility at cell junctions. It has been difficult to confirm these observations *in vivo*. Studies in mice and zebrafish (Covassin et al. 2006, Ruhrberg et al. 2002, Siekmann and Lawson 2007b) are shedding new light on VEGF receptors and Notch signaling. Time-lapse analysis reveals that endothelial tip cells undergo a stereotypical pattern of proliferation and migration during sprouting. Notch signaling is necessary to restrict angiogenic cell behavior to tip cells in developing segmental arteries in the zebrafish embryo. flt4 (vegfr3) is expressed in segmental artery tip cells and becomes ectopically expressed throughout the sprout in the absence of Notch (Siekmann and Lawson 2007b). In zebrafish it has also been shown that angioblasts migrate as individual cells to form a vascular cord at the midline. This transient structure is stabilized by endothelial cell–cell junctions, and subsequently undergoes lumen formation to form a fully patent vessel (Jin 2005).

Although lumen formation in most multicellular vessels involves transition through a number of discrete phases, the molecular mechanism is not clear.

Role of FAK in angiogenesis

One of the issues in lumen formation is the capacity of ECs to interact with matrix (cell-matrix) that initiate signaling cascades leading to activation of non-receptor tyrosine kinase, FAK—the molecule involved in early cell-matrix signaling, and also a point of convergence for signals from the environment, soluble factors, and mechanical stimuli. The widely studied and well-understood FAK signaling pathway is mediated by integrins. Integrin-mediated cell adhesion events are considered a key element of angiogenesis. FAK signaling regulates fundamental aspects of neovascularization (Orr et al. 2004, Vadali et al. 2007, Zhao and Guan 2009). Global FAK knockout (FAK-/-) mice show embryonic lethality and impaired development of the vascular system (Ilic et al. 2003). Human umbilical vascular endothelial cells (HUVEC) deficient in FAK expression or function display a reduced ability to form tubule like structures in Matrigel, abrogated capillary formation and impaired development of vascular system (Ilic et al. 2003). Further, EC-specific knockout of FAK shows defective angiogenesis in the embryo, yolk sac and placenta, impaired vasculature, developmental delay, and late embryonic lethal phenotype (Shen et al. 2005).

FAK is also a critical component of the VEGF/VEGFR2 and Ang/Tie-2 receptor systems (Mitra and Schlaepfer 2006, Mitra et al. 2006). VEGF and the VEGF receptor-2 system (also known as fetal liver kinase (Flk-1) and kinase domain receptor (KDR)) play fundamental roles in the formation of blood vessels (Park et al. 2004, Shalaby et al. 1997). Vessels exposed to pro-angiogenic signals such as VEGFA induce the activation of tyrosine kinase receptor VEGF receptor-2 (Flk1) that promotes endothelial elongation, motility, and proliferation. Moreover, VEGF released from lung metastasizing cancer cells can activate the Src-FAK complex in lung endothelial cells and promote vascular hyperpermeability, up-regulation of endothelial adhesion molecules, and cancer cell homing (Eliceiri et al. 2002, Kim et al. 2009). In addition, VEGF induces FAK and activates multiple downstream pathways via signaling intermediates, such as mitogen-activated protein kinases (MAPKs), phosphoinositide 3-kinases (PI3Ks), Akt, phospholipase Cγ and small GTPases (Coultas et al. 2005, Lohela et al. 2009, Shibuya 2006). Thus FAK not only serves as a critical component of integrin signaling, but is also a downstream element of the tumor cells over expressing factors (VEGF, TGF-β, TNF-α, and angiopoietin-2), that increase permeability (Hiratsuka et al. 2002, 2006, Huang et al. 2009, Kim et al. 2009) and also VEGF/VEGF–receptor and Ang/Tie-2 receptor systems that regulate neovascularization.

VEGF stimulates TCs, ligates to the Notch receptor on the adjacent ECs, and induces intracellular signaling to activate the Hes and Hey transcriptional target genes that promotes SCs cell proliferation (Benedito et al. 2009, Hellstrom et al. 2007, Hofmann and Iruela-Arispe 2007, Jakobsson et al. 2010, Kume 2009, Nakatsu et al. 2003, Siekmann and Lawson 2007b). All of these processes may involve FAK signaling, either directly or indirectly, in the context of key molecules and pathways controlling cell migration, VEGF production, cell proliferation and the formation of new vessels. During angiogenesis, endothelial cells receive biochemical cues from soluble and extracellular matrix proteins to elongate, migrate, proliferate, and resist anoikis. It could also be worthwhile to re-examine the role of FAK in endothelial cell proliferation *in vivo* using cutting-edge technology such as real time confocal microscopy to determine if there is a connection between the formation of "SCs" (proliferative endothelial cell phenotype) and FAK activation in tumor angiogenesis. A clearer understanding of how FAK gene expression regulates in sprouting endothelial versus quiescent endothelial cells may aid in the development of molecules targeting FAK at the level of gene transcription. Continued research on FAK may therefore yield novel therapies to improve treatment modalities for the pathological neovascularization associated with diseases.

Actin and angiogenesis

Filopodial extensions—thin cytoplasmic protrusions containing bundled actin filaments that dynamically extend from the leading edge of migrating cells and explore the surrounding microenvironment—show the importance of actin and migration in angiogenesis. It is known that endostatin, thrombospondin-1, fumagillin, and its synthetic derivative, TNP-470, are potent inhibitors of endothelial cell proliferation and migration in culture and of angiogenesis *in vivo*. In response to thrombospondin-1, fumagillin, cofilin phosphorylated and localize to hsp27. Actin stress fiber and focal adhesion densities are also increased, indicating that angiogenesis inhibitors induce endothelial cells to assume an adhesive state that is not conducive to motility, suggesting a possible common mechanism used by a panel of angiogenesis inhibitors to inhibit endothelial cell migration (Keezer et al. 2003).

Several proteins get recruited to the junctions at lateral positions; the same holds true for the apical cell surface. More specifically, ezrin, radixin, moesin (ERM) proteins are recruited to the apical cell surface of ECs (Niggli and Rossy 2008). Mice lacking moesin have delayed lumen formation in the dorsal aorta and a decreased amount of F-actin beneath the apical surface of the endothelium, which is where lumen formation is initiated (Strilic et al. 2009). Deletion of moesin, however, is not lethal and does not cause major vascular defects but causes a slight delay in aortic lumen formation, probably due to insufficient recruitment of F-actin to the apical CD34-sialomucins (Doi et al. 1999). The mild phenotype could be due to the presence of the other ERM proteins in ECs and/or alternative mechanisms of linking the apical cell surface to the F-actin cytoskeleton (Doi et al. 1999). In contrast, model organisms that harbor only a single ERM protein (e.g., Drosophila melanogaster) display more severe cell polarity and lumen formation defects upon ERM protein deletion (Gobel et al. 2004, van Furden et al. 2004). This supports the notion that ERM proteins that directly link F-actin to apical transmembrane glycoproteins in their open, phosphorylated conformation are critical for lumen formation.

Cell–cell adhesion complex that contains cadherins-catenins are attached to cytoplasmic actin filaments. Through their cytoplasmic tail, junctional adhesion proteins bind to cytoskeletal and signaling proteins, which allows the anchoring of the adhesion proteins to actin microfilaments and the transfer of intracellular signals inside the cell (Braga 2002, Wheelock and Johnson 2003). Actin association with cadherin complex is required not only for stabilization of the junctions, but also for the dynamic regulation of junction opening and closure. In addition, the interaction of junctional adhesion proteins with the actin cytoskeleton might be relevant in the maintenance of lumen formation in angiogenesis.

Evidence shows the importance of the rigidity of the adhesive support and the participation of the cytoskeleton in tubulogenesis of endothelial cells. Agents such as cytochalasin D and latrunculin B alter the actin cytoskeleton-focal adhesion plaque complex; reduced tension between the endothelial cells and the extracellular matrix is sufficient to trigger an intracellular signaling cascade leading to tubulogenesis which consequently enhances angiogenesis (Deroanne et al. 2001). In HUVEC, actin stress fibers and focal adhesions are enhanced in response to angiogenesis inhibitors thrombospondin-1, or EMAP II which also decrease the number of lamellipodia and increase the number of focal adhesions in microvascular as well as HUVECs (Keezer et al. 2003), suggesting that alteration of the actin cytoskeleton to strengthen endothelial cell attachment to the extracellular matrix is a common mechanism for inhibition of angiogenesis by these agents.

The actin cytoskeleton forms a cortical band that stabilizes cell–cell interactions (Quadri 2012) and is therefore likely to be critical for the formation of the vascular wall of the neo-capillaries. MLCK isoform 1 and 2 highly expresses mainly in the endothelium (Ohlmann et al. 2005). MLCK phosphorylates the light chain of myosine II, allowing myosin to bind to actin filament. Evidence shows that inhibition of myosin light chain kinase is achieved by targeting MLCK 210. Given the substantial evidence linking the above-described molecules associated with actin, it follows that the actin network and its associated proteins would be viable targets for anti-angiogenic therapies. However, it is not clear whether such actin-related intercellular interactions are critical for the network specifically in the sprouting and lumen forming processes in growing capillaries. Since endothelial cell motility is driven by the dynamics of cytoskeleton and the disruption of the angiogenesis process by targeting the actin, cytoskeleton is a promising strategy in the treatment of various angiogenic diseases. However, our understanding of the cellular and molecular mechanisms of angiogenesis in cancer and other diseases is still limited.

Problems in Anti-angiogenic Drug Design (anti-angiogenic therapy)

Evidence shows that inhibition of actin related protein 2/3 (ARP2/3) activation by Wiskostatin disrupted actin polymerization and prevented angiogenesis related properties in EC (Peterson et al. 2004). However, Wiskostatin exhibited side effects by altering intracellular ATP levels (Guerriero and Weisz 2007). Two leading angiogenesis laboratories present intriguing, almost perplexing evidence that VEGF-targeted drugs inhibit primary tumor growth yet may shorten survival of mice by promoting tumor invasiveness and metastasis (Ebos et al. 2009a, 2009b). Evidence shows that pre-treatment of healthy (non-tumor bearing) mice with VEGF inhibitors prior to intravenous injection of tumor cells also promotes metastasis. The VEGF effect on tumor cells extravasation mechanism is not clear.

Newer anti-angiogenic agents, such as DLL4 inhibitors inhibit primary tumor growth but also cause hypoxia via formation of hypoperfused vessels (Noguera-Troise et al. 2006). VEGF inhibitors induce vessel normalization during a particular time window (Batchelor et al. 2007, Jain 2005). Since this process has been related to increased drug delivery, it has been questioned whether pretreatment of cancer patients with VEGF inhibitors would improve chemotherapy. Further studies will be required to develop strategies that will allow us to optimally exploit the potential of VEGF inhibitors to block primary tumor growth while at the same time suppressing prometastatic effects, without having to choose between increased primary tumor growth in untreated conditions and induction of metastasis in treated conditions. Inhibitors that target F-actin are potential anti-angiogenic agents. This research provides evidence that actin is a critical component in angiogenesis; however, the role of the cytoskeleton in angiogenesis is not clear.

Conclusion

The past few years have brought tremendous insight into the molecular regulation of angiogenesis and, in particular, TC formation. However, many fundamental questions remain unsolved. Although during vertebrate angiogenesis, Notch regulates the cell-fate decision between vascular TCs versus SCs, it is not clear how actin regulates Notch signaling pathways. While many of the recently discovered molecules are connected to VEGF and Notch signaling, we still lack a good understanding of the dynamic nature of many processes and the molecular mechanism controlling various aspects of angiogenic growth. The exception is some live imaging studies in zebrafish embryos, which provide highly detailed snapshots but leave us in the dark about dynamic alterations. Finally, it is also unclear whether the highly successful model systems of zebrafish embryo and mouse retina deliver information of general relevance. Some processes might be distinct from other vascular beds and generalization should be done only with great caution. This last aspect is particularly relevant for the translation of findings into the clinical context and the development of new therapies for the much better pathology involving blood vessels.

Acknowledgments

I am grateful to Dr. Bhattacharya for helpful advice and Dr. Naim Islam for critical reading of the manuscript. This study was supported by National Heart, Lung and Blood Institutes Grant HL-36024 and HL064896 (PI: JB).

References

Abraham, S., M. Yeo, M. Montero-Balaguer et al. 2009. VE-Cadherin-mediated cell–cell interaction suppresses sprouting via signaling to MLC2 phosphorylation. Curr Biol 19: 668–674.

Adams, R.H. and K. Alitalo. 2007. Molecular regulation of angiogenesis and lymphangiogenesis. Nat Rev Mol Cell Biol 8: 464–478.

Affolter, M., R. Zeller and E. Caussinus. 2009. Tissue remodelling through branching morphogenesis. Nat Rev Mol Cell Biol 10: 831–842.

Bach, T.L., C. Barsigian, D.G. Chalupowicz et al. 1998. VE-Cadherin mediates endothelial cell capillary tube formation in fibrin and collagen gels. Exp Cell Res 238: 324–334.

Batchelor, T.T., A.G. Sorensen, E. di Tomaso et al. 2007. AZD2171, a pan-VEGF receptor tyrosine kinase inhibitor, normalizes tumor vasculature and alleviates edema in glioblastoma patients. Cancer Cell 11: 83–95.

Becker, C.M. and R.J. D'Amato. 2007. Angiogenesis and antiangiogenic therapy in endometriosis. Microvasc Res 74: 121–130.

Benedito, R., C. Roca, I. Sorensen et al. 2009. The notch ligands Dll4 and Jagged1 have opposing effects on angiogenesis. Cell 137: 1124–1135.

Blum, B. and N. Benvenisty. 2008. The tumorigenicity of human embryonic stem cells. Adv Cancer Res 100: 133–158.

Braga, V.M. 2002. Cell–cell adhesion and signalling. Curr Opin Cell Biol 14: 546–556.

Bryant, D.M. and K.E. Mostov. 2008. From cells to organs: building polarized tissue. Nat Rev Mol Cell Biol 9: 887–901.

Carmeliet, P. 2003. Angiogenesis in health and disease. Nat Med 9: 653–660.

Carmeliet, P. and R.K. Jain. 2000. Angiogenesis in cancer and other diseases. Nature 407: 249–257.

Carmeliet, P. and R.K. Jain. 2011. Molecular mechanisms and clinical applications of angiogenesis. Nature 473: 298–307.

Carmeliet, P., M.G. Lampugnani, L. Moons et al. 1999. Targeted deficiency or cytosolic truncation of the VE-cadherin gene in mice impairs VEGF-mediated endothelial survival and angiogenesis. Cell 98: 147–157.

Cattelino, A., S. Liebner, R. Gallini et al. 2003. The conditional inactivation of the beta-catenin gene in endothelial cells causes a defective vascular pattern and increased vascular fragility. J Cell Biol 162: 1111–1122.

Condeelis, J. and J.W. Pollard. 2006. Macrophages: obligate partners for tumor cell migration, invasion, and metastasis. Cell 124: 263–266.

Corada, M., M. Mariotti, G. Thurston et al. 1999. Vascular endothelial-cadherin is an important determinant of microvascular integrity *in vivo*. Proc Natl Acad Sci USA 96: 9815–9820.

Corada, M., D. Nyqvist, F. Orsenigo et al. 2010. The Wnt/beta-catenin pathway modulates vascular remodeling and specification by upregulating Dll4/Notch signaling. Dev Cell 18: 938–949.

Coultas, L., K. Chawengsaksophak and J. Rossant. 2005. Endothelial cells and VEGF in vascular development. Nature 438: 937–945.

Covassin, L.D., J.A. Villefranc, M.C. Kacergis et al. 2006. Distinct genetic interactions between multiple Vegf receptors are required for development of different blood vessel types in zebrafish. Proc Natl Acad Sci USA 103: 6554–6559.

Critchley, D.R. 2000. Focal adhesions—the cytoskeletal connection. Curr Opin Cell Biol 12: 133–139.

De Smet, F., I. Segura, K. De Bock et al. 2009. Mechanisms of vessel branching: filopodia on endothelial tip cells lead the way. Arterioscler Thromb Vasc Biol 29: 639–649.

Dejana, E., E. Tournier-Lasserve and B.M. Weinstein. 2009. The control of vascular integrity by endothelial cell junctions: molecular basis and pathological implications. Dev Cell 16: 209–221.

Deroanne, C.F., C.M. Lapiere and B.V. Nusgens. 2001. *In vitro* tubulogenesis of endothelial cells by relaxation of the coupling extracellular matrix-cytoskeleton. Cardiovasc Res 49: 647–658.

Deryugina, E.I. and J.P. Quigley. 2010. Pleiotropic roles of matrix metalloproteinases in tumor angiogenesis: contrasting, overlapping and compensatory functions. Biochim Biophys Acta 1803: 103–120.

Doi, Y., M. Itoh, S. Yonemura et al. 1999. Normal development of mice and unimpaired cell adhesion/cell motility/actin-based cytoskeleton without compensatory up-regulation of ezrin or radixin in moesin gene knockout. J Biol Chem 274: 2315–2321.

Dvorak, H.F., J.A. Nagy, D. Feng et al. 1999. Vascular permeability factor/vascular endothelial growth factor and the significance of microvascular hyperpermeability in angiogenesis. Curr Top Microbiol Immunol 237: 97–132.

Ebos, J.M., C.R. Lee, W. Cruz-Munoz et al. 2009a. Accelerated metastasis after short-term treatment with a potent inhibitor of tumor angiogenesis. Cancer Cell 15: 232–239.

Ebos, J.M., C.R. Lee and R.S. Kerbel. 2009b. Tumor and host-mediated pathways of resistance and disease progression in response to antiangiogenic therapy. Clin Cancer Res 15: 5020–5025.

Eliceiri, B.P., X.S. Puente, J.D. Hood et al. 2002. Src-mediated coupling of focal adhesion kinase to integrin alpha(v)beta5 in vascular endothelial growth factor signaling. J Cell Biol 157: 149–160.

Esser, S., M.G. Lampugnani, M. Corada et al. 1998. Vascular endothelial growth factor induces VE-cadherin tyrosine phosphorylation in endothelial cells. J Cell Sci 111(Pt 13): 1853–1865.

Ferguson, J.E., 3rd, R.W. Kelley and C. Patterson. 2005. Mechanisms of endothelial differentiation in embryonic vasculogenesis. Arterioscler Thromb Vasc Biol 25: 2246–2254.

Flamme, I., T. Frolich and W. Risau. 1997. Molecular mechanisms of vasculogenesis and embryonic angiogenesis. J Cell Physiol 173: 206–210.

Folkman, J. 2007. Angiogenesis: an organizing principle for drug discovery? Nat Rev Drug Discov 6: 273–286.

Fruttiger, M. 2007. Development of the retinal vasculature. Angiogenesis 10: 77–88.

Fukata, M. and K. Kaibuchi. 2001. Rho-family GTPases in cadherin-mediated cell–cell adhesion. Nat Rev Mol Cell Biol 2: 887–897.

Fukumura, D. and R.K. Jain. 2007. Tumor microvasculature and microenvironment: targets for anti-angiogenesis and normalization. Microvasc Res 74: 72–84.

Gebb, S. and T. Stevens. 2004. On lung endothelial cell heterogeneity. Microvasc Res 68: 1–12.

Gerhardt, H. and H. Semb. 2008. Pericytes: gatekeepers in tumour cell metastasis? J Mol Med (Berl) 86: 135–144.

Gerhardt, H., M. Golding, M. Fruttiger et al. 2003. VEGF guides angiogenic sprouting utilizing endothelial tip cell filopodia. J Cell Biol 161: 1163–1177.

Glomski, K., S. Monette, K. Manova et al. 2011. Deletion of Adam10 in endothelial cells leads to defects in organ-specific vascular structures. Blood 118: 1163–1174.

Gobel, V., P.L. Barrett, D.H. Hall et al. 2004. Lumen morphogenesis in C. elegans requires the membrane-cytoskeleton linker erm-1. Dev Cell 6: 865–873.

Gory-Faure, S., M.H. Prandini, H. Pointu et al. 1999. Role of vascular endothelial-cadherin in vascular morphogenesis. Development 126: 2093–2102.

Guerriero, C.J. and O.A. Weisz. 2007. N-WASP inhibitor wiskostatin nonselectively perturbs membrane transport by decreasing cellular ATP levels. Am J Physiol Cell Physiol 292: C1562–1566.

Hartsock, A. and W.J. Nelson. 2008. Adherens and tight junctions: structure, function and connections to the actin cytoskeleton. Biochim Biophys Acta 1778: 660–669.

Hellstrom, M., L.K. Phng, J.J. Hofmann et al. 2007. Dll4 signalling through Notch1 regulates formation of tip cells during angiogenesis. Nature 445: 776–780.

Hickey, M.M. and M.C. Simon. 2006. Regulation of angiogenesis by hypoxia and hypoxia-inducible factors. Curr Top Dev Biol 76: 217–257.

Hiratsuka, S., K. Nakamura, S. Iwai et al. 2002. MMP9 induction by vascular endothelial growth factor receptor-1 is involved in lung-specific metastasis. Cancer Cell 2: 289–300.

Hiratsuka, S., A. Watanabe, H. Aburatani et al. 2006. Tumour-mediated upregulation of chemoattractants and recruitment of myeloid cells predetermines lung metastasis. Nat Cell Biol 8: 1369–1375.

Hofmann, J.J. and M.L. Iruela-Arispe. 2007. Notch signaling in blood vessels: who is talking to whom about what? Circ Res 100: 1556–1568.

Huang, Y., N. Song, Y. Ding et al. 2009. Pulmonary vascular destabilization in the premetastatic phase facilitates lung metastasis. Cancer Res 69: 7529–7537.

Ilic, D., B. Kovacic, S. McDonagh et al. 2003. Focal adhesion kinase is required for blood vessel morphogenesis. Circ Res 92: 300–307.

Iruela-Arispe, M.L. and G.E. Davis. 2009. Cellular and molecular mechanisms of vascular lumen formation. Dev Cell 16: 222–231.

Jain, R.K. 2005. Normalization of tumor vasculature: an emerging concept in antiangiogenic therapy. Science 307: 58–62.

Jakobsson, L., C.A. Franco, K. Bentley et al. 2010. Endothelial cells dynamically compete for the tip cell position during angiogenic sprouting. Nat Cell Biol 12: 943–953.

Jin, S.W., D. Beis et al. 2005. Cellular and molecular analyses of vascular tube and lumen formation in zebrafish. Development 132(23): 5199–209.

Keezer, S.M., S.E. Ivie, H.C. Krutzsch et al. 2003. Angiogenesis inhibitors target the endothelial cell cytoskeleton through altered regulation of heat shock protein 27 and cofilin. Cancer Res 63: 6405–6412.

Kim, M.P., S.I. Park, S. Kopetz et al. 2009. Src family kinases as mediators of endothelial permeability: effects on inflammation and metastasis. Cell Tissue Res 335: 249–259.

Kume, T. 2009. Novel insights into the differential functions of Notch ligands in vascular formation. J Angiogenes Res 1: 8.

Leslie, J.D., L. Ariza-McNaughton, A.L. Bermange et al. 2007. Endothelial signalling by the Notch ligand Delta-like 4 restricts angiogenesis. Development 134: 839–844.

Li, B., E.E. Sharpe, A.B. Maupin et al. 2006. VEGF and PlGF promote adult vasculogenesis by enhancing EPC recruitment and vessel formation at the site of tumor neovascularization. FASEB J 20: 1495–1497.

Liao, F., Y. Li, W. O'Connor et al. 2000. Monoclonal antibody to vascular endothelial-cadherin is a potent inhibitor of angiogenesis, tumor growth, and metastasis. Cancer Res 60: 6805–6810.

Lobov, I.B., R.A. Renard, N. Papadopoulos et al. 2007. Delta-like ligand 4 (Dll4) is induced by VEGF as a negative regulator of angiogenic sprouting. Proc Natl Acad Sci USA 104: 3219–3224.

Lohela, M., M. Bry, T. Tammela et al. 2009. VEGFs and receptors involved in angiogenesis versus lymphangiogenesis. Curr Opin Cell Biol 21: 154–165.

Mitra, S.K. and D.D. Schlaepfer. 2006. Integrin-regulated FAK-Src signaling in normal and cancer cells. Curr Opin Cell Biol 18: 516–523.

Mitra, S.K., D. Mikolon, J.E. Molina et al. 2006. Intrinsic FAK activity and Y925 phosphorylation facilitate an angiogenic switch in tumors. Oncogene 25: 5969–5984.

Monica, B., S. Ramani, I. Banerjee et al. 2007. Human caliciviruses in symptomatic and asymptomatic infections in children in Vellore, South India. J Med Virol 79: 544–551.

Nakatsu, M.N., R.C. Sainson, J.N. Aoto et al. 2003. Angiogenic sprouting and capillary lumen formation modeled by human umbilical vein endothelial cells (HUVEC) in fibrin gels: the role of fibroblasts and Angiopoietin-1. Microvasc Res 66: 102–112.

Niggli, V. and J. Rossy. 2008. Ezrin/radixin/moesin: versatile controllers of signaling molecules and of the cortical cytoskeleton. Int J Biochem Cell Biol 40: 344–349.

Nitta, T., M. Hata, S. Gotoh et al. 2003. Size-selective loosening of the blood-brain barrier in claudin-5-deficient mice. J Cell Biol 161: 653–660.

Noguera-Troise, I., C. Daly, N.J. Papadopoulos et al. 2006. Blockade of Dll4 inhibits tumour growth by promoting non-productive angiogenesis. Nature 444: 1032–1037.

Ofori-Acquah, S.F., J. King, N. Voelkel et al. 2008. Heterogeneity of barrier function in the lung reflects diversity in endothelial cell junctions. Microvasc Res 75: 391–402.

Ohlmann, P., A. Tesse, C. Loichot et al. 2005. Deletion of MLCK210 induces subtle changes in vascular reactivity but does not affect cardiac function. Am J Physiol Heart Circ Physiol 289: H2342–2349.

Orr, A.W., M.A. Pallero, W.C. Xiong et al. 2004. Thrombospondin induces RhoA inactivation through FAK-dependent signaling to stimulate focal adhesion disassembly. J Biol Chem 279: 48983–48992.

Park, C., I. Afrikanova, Y.S. Chung et al. 2004. A hierarchical order of factors in the generation of FLK1- and SCL-expressing hematopoietic and endothelial progenitors from embryonic stem cells. Development 131: 2749–2762.

Peterson, J.R., L.C. Bickford, D. Morgan et al. 2004. Chemical inhibition of N-WASP by stabilization of a native autoinhibited conformation. Nat Struct Mol Biol 11: 747–755.

Phng, L.K. and H. Gerhardt. 2009. Angiogenesis: a team effort coordinated by notch. Dev Cell 16: 196–208.

Phng, L.K., M. Potente, J.D. Leslie et al. 2009. Nrarp coordinates endothelial Notch and Wnt signaling to control vessel density in angiogenesis. Dev Cell 16: 70–82.

Quadri, S.K. 2012. Cross talk between focal adhesion kinase and cadherins: role in regulating endothelial barrier function. Microvasc Res 83: 3–11.

Renault, M.A. and D.W. Losordo. 2007. Therapeutic myocardial angiogenesis. Microvasc Res 74: 159–171.

Ridgway, J., G. Zhang, Y.Wu. et al. 2006. Inhibition of Dll4 signalling inhibits tumour growth by deregulating angiogenesis. Nature 444: 1083–1087.

Ridley, A.J. 2001. Rho family proteins: coordinating cell responses. Trends Cell Biol 11: 471–477.

Robb, L. and A.G. Elefanty. 1998. The hemangioblast—an elusive cell captured in culture. Bioessays 20: 611–614.

Roca, C. and R.H. Adams. 2007. Regulation of vascular morphogenesis by Notch signaling. Genes Dev 21: 2511–2524.

Ruhrberg, C., H. Gerhardt, M. Golding et al. 2002. Spatially restricted patterning cues provided by heparin-binding VEGF-A control blood vessel branching morphogenesis. Genes Dev 16: 2684–2698.

Saitou, M., M. Furuse, H. Sasaki et al. 2000. Complex phenotype of mice lacking occludin, a component of tight junction strands. Mol Biol Cell 11: 4131–4142.

Shalaby, F., J. Ho, W.L. Stanford et al. 1997. A requirement for Flk1 in primitive and definitive hematopoiesis and vasculogenesis. Cell 89: 981–990.

Shen, T.L., A.Y. Park, A. Alcaraz et al. 2005. Conditional knockout of focal adhesion kinase in endothelial cells reveals its role in angiogenesis and vascular development in late embryogenesis. J Cell Biol 169: 941–952.

Shibuya, M. 2006. Differential roles of vascular endothelial growth factor receptor-1 and receptor-2 in angiogenesis. J Biochem Mol Biol 39: 469–478.

Siekmann, A.F. and N.D. Lawson. 2007a. Notch signalling and the regulation of angiogenesis. Cell Adh Migr 1: 104–106.

Siekmann, A.F. and N.D. Lawson. 2007b. Notch signalling limits angiogenic cell behaviour in developing zebrafish arteries. Nature 445: 781–784.

Stevens, T., J.G. Garcia, D.M. Shasby et al. 2000. Mechanisms regulating endothelial cell barrier function. Am. J. Physiol. Lung Cell Mol Physiol 279: L419–422.

Stevens, T., S. Phan, M.G. Frid et al. 2008. Lung vascular cell heterogeneity: endothelium, smooth muscle, and fibroblasts. Proc Am Thorac Soc 5: 783–791.

Strilic, B., T. Kucera, J. Eglinger et al. 2009. The molecular basis of vascular lumen formation in the developing mouse aorta. Dev Cell 17: 505–515.

Suchting, S., C. Freitas, F. le Noble et al. 2007. The Notch ligand Delta-like 4 negatively regulates endothelial tip cell formation and vessel branching. Proc Natl Acad Sci USA 104: 3225–3230.

Swift, M.R. and B.M. Weinstein. 2009. Arterial-venous specification during development. Circ Res 104: 576–588.

Uemura, A., S. Kusuhara, H. Katsuta et al. 2006. Angiogenesis in the mouse retina: a model system for experimental manipulation. Exp Cell Res 312: 676–683.

Vadali, K., X. Cai and M.D. Schaller. 2007. Focal adhesion kinase: an essential kinase in the regulation of cardiovascular functions. IUBMB Life 59: 709–716.

van Aelst, L. and C. D'Souza-Schorey. 1997. Rho GTPases and signaling networks. Genes Dev 11: 2295–2322.

van Furden, D., K. Johnson, C. Segbert et al. 2004. The C. elegans ezrin-radixin-moesin protein ERM-1 is necessary for apical junction remodelling and tubulogenesis in the intestine. Dev Biol 272: 262–276.

van Hinsbergh, V.W. and P. Koolwijk. 2008. Endothelial sprouting and angiogenesis: matrix metalloproteinases in the lead. Cardiovasc. Res 78: 203–212.

van Nieuw Amerongen, G.P., C.M. Beckers, I.D. Achekar et al. 2007. Involvement of Rho kinase in endothelial barrier maintenance. Arterioscler Thromb Vasc Biol 27: 2332–2339.

Wang, Y., M.S. Kaiser, J.D. Larson et al. 2010. Moesin1 and Ve-cadherin are required in endothelial cells during *in vivo* tubulogenesis. Development 137: 3119–3128.

Watson, E.D. and J.C. Cross. 2005. Development of structures and transport functions in the mouse placenta. Physiology (Bethesda) 20: 180–193.

Wheelock, M.J. and K.R. Johnson. 2003. Cadherin-mediated cellular signaling. Curr Opin Cell Biol 15: 509–514.

Wyckoff, J.B., Y. Wang, E.Y. Lin et al. 2007. Direct visualization of macrophage-assisted tumor cell intravasation in mammary tumors. Cancer Res 67: 2649–2656.

Xu, K., A. Sacharidou, S. Fu et al. 2011. Blood vessel tubulogenesis requires Rasip1 regulation of GTPase signaling. Dev Cell 20: 526–539.

Yan, M., C.A. Callahan, J.C. Beyer et al. 2010. Chronic DLL4 blockade induces vascular neoplasms. Nature 463: E6–7.

Zhao, J. and J.L. Guan. 2009. Signal transduction by focal adhesion kinase in cancer. Cancer Metastasis Rev 28: 35–49.

Zovein, A.C., A. Luque, K.A. Turlo et al. 2010. Beta1 integrin establishes endothelial cell polarity and arteriolar lumen formation via a Par3-dependent mechanism. Dev Cell 18: 39–51.

4

Adherens Junctions and Endothelial Cytoskeleton

Younes Smani[1], * and *Tarik Smani-Hajami[2],* *

Introduction

The vascular endothelium is comprised of a single endothelial cell monolayer with a thickness of 15 μm. The vascular endothelium can maintain the integrity of blood/tissue barrier, allowing important cellular processes such as nutrients permeability and impermeability to toxins. The endothelium also relays adaptive messages from the arterial wall to sustained pressures (vasoreactivity and haemostasis). The endothelium is considered a tight barrier; nevertheless, it handles a continuous flow of small molecules by transcytosis, a process by which various macromolecules, for example glucose, small peptides, high-density lipoprotein (HDL) are transported across the interior of the cells (Tuma and Hubbard 2003). An important feature of the endothelium is the presence of cholesterol-rich lipid microdomains, caveolae and rafts in the membrane which are relay points to throw up or to transmit messages from or to the vascular wall (Touyz 2006).

[1]Grupo de Enfermedades Infecciosas, Instituto de Biomedicina de Sevilla, Hospital Universitario Virgen del Rocío/CSIC/Universidad de Sevilla. Sevilla. Spain.
Email: ysmani-ibis@us.es
[2]Departamento de Fisiología Médica y Biofísica, Grupo de Fisiopatología Cardiovascular, Instituto de Biomedicina de Sevilla, Hospital Universitario Virgen del Rocío/CSIC/ Universidad de Sevilla. Sevilla. Spain.
Email: tasmani@us.es
*Corresponding authors

The endothelium lines the inside of blood vessels and is separated from the smooth muscle cells by the internal elastic lamina with a thickness of 40 to 80 μm. Endothelial cells allow the endothelium to function as an active secretory system secreting various vasoactive agents (O'Brien et al. 1987) and vasodilators agents including nitric oxide (NO), prostacyclin (PGI$_2$), kallikrein-kinin adhesion molecules like ICAM-1 and V-CAM-1, endothelin and platelet-activating factor (PAF) that act in the local or general circulation. Indeed, the endothelium is a real organ that secretes numerous vasoactive substances regulating vascular tone such as endothelins or NO and vascular growth such as vascular endothelial growth factor (VEGF). The discovery of PGI$_2$ in 1976 and NO in 1987 as endothelium-derived relaxing factors and later, the endothelin and platelet-activating factor (PAF), highlighted the fundamental role of the endothelium in cardiovascular physiology and pathology by regulating blood pressure and blood flow.

Endothelial cells form cell layers, in which neighboring cells are closely connected through their cell–cell junctions such as adherens, tight and gap junctions (Lampugnani and Dejana 1997). Molecules with low molecular weight can cross these junctions to enter the sub-endothelial space. These junctions systems must be tightly regulated to maintain endothelial integrity and to protect the vessels from any uncontrolled increase in permeability, inflammation or thrombotic reactions (Mehta and Malik 2006). Moreover, there are sites of attachment between endothelial cells which function as signaling structures that communicate cell position, limit growth and apoptosis, and in general regulate vascular homeostasis. Therefore, any change in junctional organization might have complex consequences, which could compromise endothelial reactions with blood elements or modify the normal architecture of the vessel wall (Mehta and Malik 2006). This chapter seeks to give an overview of recent works related to adherens junctions and cytoskeleton in endothelial cells and their physiopathological role.

Endothelial Cytoskeletons

The cytoskeleton is composed of an intracellular network of biopolymers that combine to form three types of filaments: microtubules, intermediate filaments, and actin microfilaments (Gotlieb 1990). Microtubules are tubulin polymers, which are formed in the perinuclear region and extend to the periphery, conditioning cell shape and guiding intracellular transport. Microtubules are hollow cylinders with internal and external diameters of 15 nm and 25 nm respectively. The intermediate filaments are connected to the nuclear envelope, intercellular junctions (such as desmosomes and hemidesmosomes) and other elements of the cytoskeleton, providing the structure with resistance. Actin microfilaments participate in the generation of contractile forces and are connected to the intracellular proteins of the adherens junctions (in the case of cell–cell contact) or of the focal adhesion plaques (in the case of cell-extracellular matrix contact), stabilizing the cell in relation to the forces that act on the membrane and determining the shape of the plasma membrane (Rogers et al. 1992). Since cells are dynamic structures, they actively remodel in order to adapt to the different tensions and stimuli. There are a large number of studies that have highlighted the critical role of endothelial cytoskeleton, for example in inflammation, oxidative stress responses, or in cells remodeling due to hypoxia and reperfusion (Hirase and Node 2012, Kumar et al. 2009, Weigand et al. 2012). The cytoskeleton is an integrated network—as the cell structure that is responsible for enduring cell deformations and the resulting tensions, it degrades and reorganizes itself, providing the system with plasticity. The cytoskeleton is far from being a rigid and static structure, and is considered highly dynamic (Trepat et al. 2007). It is able to restructure itself rapidly, adapting to different tensions and situations; it also changes cell viscosity so that the cell can adapt better to the environment (Wang et al. 1993).

Endothelial Junctions

The junctional structures located at the endothelial intercellular cleft are related to those found in epithelia; however their organization is more variable and in most vascular beds their topology is less restricted than in epithelial cells. Similar to epithelial cells, endothelial cells use mainly two types of junctions to mediate cell–cell interactions, tight and adherens junctions (Wallez and Huber 2008).

Tight junctions

The tight junctions markers are frequently located with adherens junctions markers all the way along the intercellular cleft (Wallez and Huber 2008). They mediate adhesion between endothelial cells, and form tight seals between cells to create the major barrier in the paracellular pathway. Tight junctions show considerable variability among different segments of the vascular tree (Franke et al. 1988). Tight junctions are well developed in brain microvessels where they contribute to the blood brain barrier and also in arteries, whereas they are less organized in veins, capillaries and organs that are characterized by a high rate of exchange (Bazzoni and Dejana 2004). This disparity constitutes a major evidence of vascular bed differentiation of endothelial cells and has a strong impact on vascular permeability and leukocyte extravasation. Tight junctions are highly dynamic structures consisting of integral membrane proteins such as occludin, claudins, junctional adhesion molecules (JAMs) and membrane-associated proteins, e.g., zonula occludens (ZO-1) and cingulin, which collectively form a complex protein network that spans junctions and connects membranes on adjacent cells to form a seal and maintain cell polarity.

Claudin family members are the major tight junction transmembrane constituents. They are found at the most apical part of the lateral surface of endothelial cells and serve as a continuous paracellular seal between the apical and basolateral sections. Claudin proteins exhibit homophilic and heterophilic adhesive activities through their extracellular domains and compose the complex network of strands (Furuse and Tsukita 2006). Occludin is another transmembrane component of tight junction, although it is not necessary for tight junction strand formation. Occludin is associated with increased tight junction barrier function (Cummins 2012). Claudins and occludin are linked to numerous intracellular partners, which form a molecular complex at the sub-membrane side of the tight junction that plays a key role in vascular physiopathology (Harhaj and Antonetti 2004). On the other hand, JAMs constitute a family of transmembrane adhesive proteins that co-localize with tight junctions. They develop homophilic adhesive activity, suggesting that they mediate endothelial cell–cell interaction (Muller 2003).

Adherens junctions

As indicated above, adherens junctions are frequently located with tight junctions, and participate in multiple functions, including establishment and maintenance of cell–cell adhesion, intracellular signaling, transcriptional regulation, and actin cytoskeleton remodeling. The protein compositions of adherens junctions in endothelial cells are similar to those in epithelial cells (Hartsock and Nelson 2008) with a few exceptions such as vascular endothelial cadherin (VE-cadherin) and N-cadherin.

VE-cadherin

Three decades ago cadherins were identified as calcium-dependent transmembrane glycoproteins that mediate adhesion through homophilic interactions of their extracellular domain with cadherins on adjacent cells (Franke 2009, Hulpiau and van Roy 2008, Takeichi 2007, Volk et al. 1987). The initial step in the formation of intercellular adhesive complexes appears to be the binding of two identical monomeric cadherin ectodomains on opposing cells (Hulpiau and Van Roy 2008, Pokutta and Weis 2007, Troyanovsky 2005, Troyanovsky et al. 2007, Zhang et al. 2009). More cadherin monomers are subsequently recruited to the site of first contact leading to stable adhesive complexes. Furthermore, cadherins are linked to the microfilaments of the F-actin (or actomyosin) cytoskeleton through the interaction of their cytoplasmic domain with a supramolecular complex of intracellular proteins (Dejana et al. 2009, Nelson 2008, Pokutta and Weis 2007, Van Roy and Berx 2008). Type I and II cadherins are capable of forming intercellular adhesive complexes (e.g., adherens junctions) that are essential for the maintenance of tissue integrity (Dejana et al. 2009, Hartsock and Nelson 2008, Hulpiau and van Roy 2008, May et al. 2005, Niessen 2007, Tinkle et al. 2008, Vestweber 2008). In adult tissues, these complexes usually contain identical cadherins. For example, only the type I cadherin, known as E-cadherin is found in epithelial adherens junctions (Van Roy and Berx 2008), whereas only the type II cadherin, known as VE-cadherin is present in endothelial adherens junctions (Lampugnani et al. 1992, Lampugnani et al. 2006, Liebner et al. 2006, Vestweber et al. 2008).

VE-cadherin binds through its cytoplasmic tail to members of the armadillo repeat family of proteins, including p120-catenin, β-catenin, and plakoglobin. p120-Catenin and β-catenin can also shuttle to the nucleus to regulate gene expression. α-Catenin associates indirectly with VE-cadherin by binding to β-catenin or plakoglobin. Both β-catenin and plakoglobin are highly homologous and bind to the same site in VE-cadherin, which may further interact with α-actinin, vinculin (Huveneers et al. 2012), ZO-1, formin-1 and Rho (Bazzoni and Dejana 2004, Maiden and Hardin 2011). In addition to these core components of the adherens junction, several other proteins bind VE-cadherin and regulate its activity such as β-arrestin, golgi-associated phospholipase A2α and myosin-X (Harris and Nelson 2010). Next to its interaction with the F-actin cytoskeleton, the VE–cadherin complex may associate with the vimentin cytoskeleton in some vascular locations (Calkins et al. 2003, Hammerling et al. 2006, Kowalczyk et al. 1998, Schmelz and Franke 1993, Schmelz et al. 1994, Venkiteswaran et al. 2002). Consistently, VE–cadherin clusters forming artificial adherens junctions were observed by assembly of VE–cadherin extracellular domains at the surface of two liposomes (Lambert et al. 2005).

The precise mechanism of VE–cadherin-based junction formation remains unclear. However, a model has been proposed for VE–cadherin lateral clustering that may be achieved by membrane tension and free VE–cadherin molecule diffusion rather than a cytoskeleton driven mechanism (Delanoë-Ayari et al. 2004). In contrast, a key role of the actin cytoskeleton in the formation and maintenance of adherens junctions is well recognized and molecular linkages between cadherins and actin filaments are largely deciphered (Harris and Nelson 2010, Harris and Tepass 2010, Yonemura 2011). Recently, Hoelzle and Svitkina have showed that myosin II activity is important for bridge formation and accumulation of VE-cadherin in adherens junctions of endothelial cells (Hoelzle and Svitkina 2012). There are many mechanisms proposed to regulate VE-cadherin including modulating VE-cadherin activity through phosphorylation and controlling the amount of VE-cadherin available for engagement at adherens junctions at both the protein and mRNA level.

N-cadherin

Besides VE-cadherin, endothelial cells also express high levels of neuronal (N)-cadherin, also known as cadherin II. However N-cadherin is not clustered at cell–cell contacts between endothelial cells (Gerhardt et al. 2000, Salomon et al. 1992). During development, N-cadherin mediates contacts between endothelial cells and pericytes and smooth muscle cells (Gerhardt et al. 2000, Sabatini et al. 2008), and VE-cadherin was reported to compete with N-cadherin in co-transfected CHO cells for localization at junctions (Navarro et al. 1998). In one report, N-cadherin was also found at junctions of endocardial cells (Luo and Radice 2005). Loss of N-cadherin in endothelial cells results in embryonic lethality at mid-gestation due to severe vascular defects. Intriguingly, loss of N-cadherin caused a significant decrease in VE-cadherin. The down-regulation of VE-cadherin was confirmed in cultured endothelial cells using small interfering RNA to knockdown N-cadherin. The same authors have also showed that N-cadherin is important for endothelial cell proliferation and motility. These findings provide a novel paradigm by which N-cadherin regulates angiogenesis, in part, by controlling VE-cadherin expression at the cell membrane (Luo and Radice 2005).

Catenins

Not long after the discovery of cadherins, three distinct proteins termed catenins were found to be associated with their cytoplasmic tails (Ozawa et al. 1989, Reynolds et al. 1994). β-Catenin, which was soon realized to be a member of a highly conserved family of proteins that includes Drosophila Armadillo (Peifer et al. 1992), contains 12 α-helical armadillo (arm) repeats and binds directly to the tail of cadherin (Huber et al. 1997b, Stappert and Kemler 1994). β-Catenin also binds directly to α-catenin (Aberle et al. 1994, Huber et al. 1997a) and in addition to its role in cell–cell adhesion, has a function in Wnt signaling (Nelson and Nusse 2004). A fourth conserved member of the cytoplasmic cell-adhesion complex (CCC) is p120-catenin, an arm repeat protein that binds a juxtamembrane region of the cadherin cytoplasmic tail and is known to be a regulator of Rho GTPases (small guanosine triphosphatases (GTPases) from the ras superfamily) and a predominant Src substrate, and to be involved in regulating cadherin stability (Anastasiadis 2007). α-Catenin is an actin-binding and bundling protein (Rimm et al. 1995) that contains three domains (VH for Vinculin Homology domains) related to another actin-binding and -bundling protein, vinculin (Herrenknecht et al. 1991, Nagafuchi et al. 1991). The N terminus of α-catenin contains a β-catenin–binding site (Aberle et al. 1994, Huber et al. 1997a, Koslov et al. 1997, Nieset et al. 1997, Pokutta and Weis 2000), whereas the C terminus contains the actin-binding domain (Imamura et al. 1999, Nagafuchi et al. 1994, Pokutta et al. 2002). Other binding partners include an assortment of actin-associated proteins, including vinculin itself (Imamura et al. 1999, Watabe-Uchida et al. 1998, Weiss et al. 1998), EPLIN "epithelial protein lost in neoplasm" (Abe and Takeichi 2008), α-actinin (Knudsen et al. 1995, Nieset et al. 1997) and the tight junction protein, ZO-1 (Imamura et al. 1999, Itoh et al. 1997); and the nectin junction protein afadin. These features made α-catenin a logical choice for linking the CCC directly to the actin cytoskeleton. A simple model predicted that cadherin, β-catenin, and α-catenin bind one another as a ternary complex with 1:1:1 stoichiometry, which in turns binds F-actin (Pappas and Rimm 2006, Pokutta et al. 2002, Rimm et al. 1995).

Plakoglobin

Plakoglobin, also known as γ-catenin (Ozawa et al. 1989), another vertebrate catenin, is highly homologous to β-catenin (Butz et al. 1992, McCrea et al. 1991). Like β-catenin, plakoglobin is closely related to the Drosophila protein armadillo, and these junctional components were among the founding members of a family of proteins defined by a 42-amino acid repeated motif termed an armadillo domain (Peifer 1993, Peifer et al. 1992, 1994). Interestingly, plakoglobin not only assembles into desmosomes but also binds to classical cadherins and assembles into adherens junctions (Sacco et al. 1995). Because of this dual targeting of plakoglobin to both actin and intermediate filament-based junctions in endothelial cells, plakoglobin is thought to play a key role in the cross talk between adherens junctions and desmosomes and likely participates in the coordinated assembly of these discrete adhesive complexes in a variety of epithelial tissues (Collins and Garrod 1994, Cowin et al. 1986, Lewis et al. 1997). The role of both β-catenin and plakoglobin in the assembly of intercellular junctions hinges on their ability to bind to both the cadherin tail and to downstream linking proteins such as β-catenin, which couples the cadherin-catenin complex to the actin cytoskeleton. Different studies indicate that plakoglobin may play a unique role in the Wnt signaling pathway, one that is different from that of β-catenin (Charpentier et al. 2000, Simcha et al. 1996, Zhurinsky et al. 2000).

Adherens Junctions and Endothelial Cytoskeletons

Cell to cell junction formation has profound effects on the state of the microtubule (Stehbens et al. 2009) and actin based cytoskeleton (Cavey and Lecuit 2009), which in turns allows cells to integrate into coherent tissue sheets. Classical cadherins bind directly and indirectly to many cytoplasmic proteins, particularly members of the catenin family. In endothelial cells, VE-cadherin associates with the intracellular catenins, an interaction that greatly enhances the cell-adhesive activity (Kemler 1993). VE-Cadherins bind with their C terminus to either β-catenin or plakoglobin, which bind in turn to α-catenin. It is a widely accepted concept that the dissociation of the cadherin-catenin complex could destabilize cell adhesion. Instead of a simple bridge between junctional cadherins and the actin microfilament cytoskeleton, α-catenin was involved as a linker in the actin polymerization which anchors the cadherins to the actin cytoskeleton to stabilize cadherin adhesion (Perez-Moreno and Fuchs 2006). However, this linker concept was challenged when it was found that α-catenin could not simultaneously bind to β-catenin and actin (Yamada et al. 2005). Dimeric α-catenin binds with high affinity to actin filaments, and by inhibiting Arp2/3 activity suppresses actin branching. In addition, dimeric α-catenin promotes the formation of linear actin cables by activating formin (Pokutta et al. 2008) and modulates actin filament assembly, indicating that cadherin-associated α-catenin might indirectly influence the actin cytoskeleton by serving as a pool of α-catenin molecules that, once dissociated from cadherin clusters, would be available to influence actin branching (Drees et al. 2005). These results suggested an important new mechanism by which cadherin clusters and cytoskeletal dynamics could influence each other. More studies seem needed to support this hypothesis.

The lack of simultaneous binding of α-catenin to β-catenin and actin triggered a search for the missing link that would enable the cadherin-catenin complex to bind to actin. Afadin, a cytoplasmic molecule localized at adherens junctions bound to the nectin adhesive receptors, can anchor actin to the plasma membrane and could contribute to the stabilization of the peripheral actin border (Niessen and Gottardi 2008, Pokutta et al. 2008). Other junctional components that might tether actin filaments to adherens junctions remain to be determined. Besides this, it was identified that the actin-binding protein EPLIN can link actin to α-catenin, whereas α-catenin is simultaneously linked to β-catenin and VE-cadherin, providing important evidence that catenins are able to support cadherin function through anchoring cadherin clusters to the actin cytoskeleton (Abe and Takeichi 2008). It will be important to test whether EPLIN functions in endothelial cells similar to the way it does in epithelial cells or whether other proteins anchor the VE-cadherin-catenin complex in endothelium.

β-catenin also has an important role in regulating the behaviour of actin. Small GTPases of the Rho family that have crucial effects on the control of actin polymerization (Ridley 2006) and on the contractility of actomyosin, are recruited and regulated at adherens junctions through β-catenin (Samarin and Nusrat 2009). Remodeling or de novo polymerization of actin filaments can be controlled by the GTPases Rac and Rho, while Rho also affects the contractility of actomyosin, the complex of actin and myosin that makes up the contractile stress fibers. Rac signaling at adherens junctions is likely to be downstream of the small GTPase Rap1 (Pannekoek et al. 2009), which in endothelial cells can be localized at adherens junctions in a complex with VE-cadherin through cerebral cavernous malformation protein 1 (CCM1) and β-catenin (Glading et al. 2007). In addition, Rap1 can be activated at junctions by PDZ-GEF, recruited locally by the protein MAGI, which is linked to VE-cadherin through β-catenin (Sakurai et al. 2006). Rap1 can also convey signals from cAMP through its guanine nucleotide exchange factor (GEF) Epac, to stabilize the barrier properties of the endothelium both *in vitro* and *in vivo* (Pannekoek et al. 2009). Exaggerated actomyosin contractility has been proposed to be detrimental to the correct function of the endothelium as a barrier to the passage of cells and molecules. This idea is supported by the fact that to maintain junctional stability in the endothelium, the activity of Rho has to be restricted. The specific Rho inhibitor p190Rho-GAP is required in endothelial cells to maintain endothelial integrity and endothelial barrier function. In this case p120-catenin acts as a scaffold (Wildenberg et al. 2006).

In a physiological context, an appreciation of the correspondence between *in vitro* models of endothelial tissue and the endothelium *in vivo* is important. Adherens junctions and stress fibers are readily observed in endothelial cells *in vitro*. Adherens junctions are present in the endothelium of all types of blood vessels *in vivo* as observed by *in situ* immunofluorescence and immuno-electron microscopy for junction-specific molecules (Baluk et al. 2007). Reports on the organization of actin in the endothelium *in vivo* are, however, very limited and somewhat controversial (Nehls and Drenckhahn 1991, Thurston et al. 1995). While suggesting that actin stress fibers can be observed *in vivo* in endothelial cells, the available data indicate that great care is needed in interpreting the arrangement of actin and its molecular regulation by adherens junctions *in vivo*, both in resting and activated endothelium. Nevertheless, endothelial adherens junctions seem to be equipped with the molecular components to regulate actin polymerization and actomyosin contraction. The results of Millan and colleagues (Millan et al. 2010) suggest that local tuning of the actin cytoskeleton at adherens junction sub-domains could allow endothelial cells in the monolayer to behave as a functional unit, withstanding the stress of continuous changes in blood flow that take place in the organism and also locally restricting the stress of junctional and actin rearrangements that occur during inflammation, leukocyte transmigration and angiogenic responses.

Adherens Junction and Cytoskeleton Role in Vascular Integrity: Endogenous Pathways

Vascular damage can happen when the endothelial layer is seriously compromised and endothelial cells are retracted. The resultant increase in vascular leakage defined as endothelial permeability can be accompanied by obvious disruption of the vessels, along with haemorrhages, adhesion of leukocytes, and the formation of small thrombi. One of the critical roles of adherens junctions is their regulation of increased endothelial permeability to plasma proteins and other solutes, for example upon inflammatory stimulus. Increased endothelial permeability has been considered the hallmark of inflammatory vascular oedema (Lum and Malik 1994).

Inflammatory mediators such as thrombin, histamine, bradykinin and oxidants evoke endothelial permeability by the activation of dynamic opening and closure of cell–cell adherens junctions, largely composed by VE-cadherin (Dejana et al. 2008). Few years ago, Rho GTPases (RhoA, Rac1, Cdc42) and Rap1, have been recognized as important regulators of endothelial barrier functions (Wojciak-Stothard and Ridley 2002). They regulate cell adhesion, in part by reorganization of the junction-associated cortical actin cytoskeleton (Popoff and Geny 2009). The main mechanism of RhoA-mediated barrier destabilization involves actin-myosin contractility. Rho kinase is known to inhibit myosin light chain (MLC) phosphatase and additionally directly targets MLC, both mechanisms leading to increased MLC phosphorylation and thus increased actin–myosin contractility (Kimura et al. 1996, Van Nieuw Amerongen et al. 2000). Earlier studies have shown that RhoA activation results in the formation of stress fibers composed of filamentous actin and myosin II (Ridley and Hall 1992, Wójciak-Stothard et al. 1998). In addition, mediators such as thrombin cause VE-cadherin disorganization, and the resulting loss of functional adherens junctions have been proposed as the basis of increased endothelial permeability (Corada et al. 1999, Rabiet et al. 1996). Several works have shown that the disassembly of VE-cadherin at the adherens junctions depend on the rise in the intracellular Ca^{2+} concentration ($[Ca^{2+}]_i$) induced by inflammatory mediators binding to heptahelical G protein-coupled receptors (GPCR) (Rabiet et al. 1996, Sandoval et al. 2001, Winter et al. 1999). Inflammatory mediators can elicit an increase in $[Ca^{2+}]_i$ by activating sequential events, which involve generation of inositol 1,4,5-trisphosphate (IP_3), IP_3-induced ER store Ca^{2+} release and Ca^{2+} release activated Ca^{2+} entry in endothelial cells. The resulted Ca^{2+} entry was related to the store operated Ca^{2+} channels as suggested previously (Mehta et al. 2005, Tiruppathi et al. 2002). Other components, such as PKCα and Ca^{2+}-dependent protein kinase C (PKC) isoform, play a critical role in initiating endothelial cell contraction and disassembly of VE-cadherin junctions. Rabiet et al. have shown that Calphostin C, a PKC inhibitor, prevents thrombin-induced disorganization of the VE-cadherin complex (Rabiet et al. 1996).

Additional studies have demonstrated the implication of other endogenous pathways which modulate vascular permeability such as tyrosine phosphorylation of VE-cadherin, which was associated with weak junctions and impaired barrier function. Inflammatory agents as histamine (Andriopoulou et al. 1999, Shasby et al. 2002), tumor necrosis factor-α (TNFα) (Angelini et al. 2006), PAF and VEGF (Esser et al. 1998) induce the tyrosine phosphorylation of VE-cadherin and its binding partners β-catenin, plakoglobin and p120 catenin (Komarova and Malik 2010).

Conclusion

Dysfunction of endothelial cell-to-cell junctions, which disrupts the endothelial barrier and increases vascular permeability, is involved in the pathogenesis of vascular failure, including atherosclerosis. Adherens junction and particularly VE-cadherin play a key role in maintaining a careful balance between intercellular junction plasticity and integrity, a requirement for endothelial cells to maintain proper barrier function of blood vessels while still being capable of dynamically responding to inflammatory and growth factor signals (Harris and Nelson 2010). Cell to cell junctions represent valuable targets of therapeutic intervention since modulation of junction adhesion and signaling may strongly influence endothelial growth, apoptosis and permeability. The study of molecular mechanisms of different regulatory and signaling pathways initiated by adherens junction might contribute to the discovery of new therapeutic targets for many pathological diseases related to endothelial dysfunction.

Acknowledgments and Sources of Funding

The Cardiovascular Physiopathology group is financed by grants from Spanish Ministry MICINN [BFU2010-21043-C02-02; PI12/00941], Red RECAVA [RD12/0042/0041], and the Government of Andalucía (CICE): [P10-CVI-6095]. The Infectious Diseases group is financed by grants from the Government of Andalucía (CICE): [CTS 6317/11]. Younes Smani is funded by Red REIPI [RD06/0008].

References

Abe, K. and M. Takeichi. 2008. EPLIN mediates linkage of the cadherin catenin complex to F-actin and stabilizes the circumferential actin belt. Proc Natl Acad Sci USA 105: 13–19.

Aberle, H., S. Butz, J. Stappert et al. 1994. Assembly of the cadherin-catenin complex *in vitro* with recombinant proteins. J Cell Sci 107: 3655–3663.

Anastasiadis, P.Z. 2007. p120-ctn: A nexus for contextual signaling via Rho GTPases. Biochim Biophys Acta 1773: 34–46.

Andriopoulou, P., P. Navarro, A. Zanetti et al. 1999. Histamine induces tyrosine phosphorylation of endothelial cell-to-cell adherens junctions. Arterioscler Thromb Vasc Biol 19: 2286–2297.

Angelini, D.J., S.W. Hyun, D.N. Grigoryev et al. 2006. TNF-alpha increases tyrosine phosphorylation of vascular endothelial cadherin and opens the paracellular pathway through fyn activation in human lung endothelia. Am J Physiol Lung Cell Mol Physiol 291: L1232-1245.

Baluk, P., J. Fuxe, H. Hashizume et al. 2007. Functionally specialized junctions between endothelial cells of lymphatic vessels. J Exp Med 204: 2349–2362.

Bazzoni, G. and E. Dejana. 2004. Endothelial cell-to-cell junctions: molecular organization and role in vascular homeostasis. Physiol Rev 84: 869–901.

Butz, S., J. Stappert, H. Weissig et al. 1992. Plakoglobin and β-catenin: distinct but closely related. Science 257: 1142–1144.

Calkins, C.C., B.L. Hoepner, C.M. Law et al. 2003. The Armadillo family protein p0071 is a VE-cadherin- and desmoplakin- binding protein. J Biol Chem 278: 1774–1783.

Cavey, M. and T. Lecuit. 2009. Molecular bases of cell–cell junctions stability and dynamics. Cold Spring Harb Perspect Biol 1: a002998.

Charpentier, E., R. Lavker, E. Acquista et al. 2000. Plakoglobin suppresses epithelial proliferation and hair growth *in vivo*. J Cell Biol 149: 503–520.

Collins, J.E. and D.R. Garrod. 1994. Hemidesmosomes. In Molecular Biology of Desmosomes and Hemidesmosomes. R.G. Landes Co. Austin, Texas, USA 53–68.

Corada, M., M. Mariotti, G. Thurston et al. 1999. Vascular endothelial-cadherin is an important determinant of microvascular integrity *in vivo*. Proc Natl Acad Sci USA 96: 9815–9820.

Cowin, P., H. Kapprell, W.W. Franke et al. 1986. Plakoglobin: a protein common to different kinds of intercellular adhering junctions. Cell 46: 1063–1073.

Cummins, P.M. 2012. Occludin: one protein, many forms. Mol Cell Biol 32: 242–250.

Dejana, E., F. Orsenigo and M.G. Lampugnani. 2008. The role of adherens junctions and VE-cadherin in the control of vascular permeability. J Cell Sci 121: 2115–2122.

Dejana, E., E. Tournier-Lasserve and B.M. Weinstein. 2009. The control of vascular integrity by endothelial cell junctions: molecular basis and pathological implications. Dev Cell 16: 209–221.

Delanoë-Ayari, H., R. Al Kurdi, M. Vallade et al. 2004. Membrane and acto-myosin tension promote clustering of adhesion proteins, Proc Natl Acad Sci USA 101: 2229–2234.

Drees, F., S. Pokutta, S. Yamada et al. 2005. apha-catenin is a molecular switch that binds E- cadherin-beta-catenin and regulates actin-filament assembly. Cell 123: 903–915.

Esser, S., M.G. Lampugnani, M. Corada et al. 1998. Vascular endothelial growth factor induces VE-cadherin tyrosine phosphorylation in endothelial cells. J Cell Sci 111: 1853–1865.

Franke, W.W. 2009. Discovering the molecular components of intercellular junctions-a historical view. Cold Spring Harb Perspect Biol 1: a003061.

Franke, W.W., P. Cowin, C. Grund et al. 1988. The endothelial junction: the plaque and its component. *In*: N. Simionescu and M. Simionescu [eds.]. Endothelial Cell Biology in Health and Diseases. Plenum Publishing Corporation, New York, USA, pp. 147–166.

Furuse, M. and S. Tsukita. 2006. Claudins in occluding junctions of humans and flies. Trends Cell Biol 16: 181–188.

Gerhardt, H., H. Wolburg and C. Redies. 2000. N-cadherin mediates pericytic-endothelial interaction during brain angiogenesis in the chicken. Dev Dyn 218: 472–479.

Glading, A., J. Han, R.A. Stockton et al. 2007. KRIT-1/CCM1 is a Rap1 effector that regulates endothelial cell cell junctions. J Cell Biol 179: 247–254.

Gotlieb, A.I. 1990. The endothelial cytoskeleton: organization in normal and regenerating endothelium. Toxicol Pathol 18: 603–617.

Hammerling, B., C. Grund, J. Boda-Heggemann et al. 2006. The complexus adhaerens of mammalian lymphatic endothelia revisited: a junction even more complex than hitherto thought. Cell Tissue Res 324: 55–67.

Harhaj, N.S. and D.A. Antonetti. 2004. Regulation of tight junctions and loss of barrier function in pathophysiology. Int J Biochem Cell Biol 36: 1206–1237.

Harris, E.S. and W.J. Nelson. 2010. VE-cadherin: at the front, center, and sides of endothelial cell organization and function. Curr Opin Cell Biol 22: 651–658.

Harris, T.J. and U. Tepass. 2010. Adherens junctions: from molecules to morphogenesis. Nat Rev Mol Cell Biol 11: 502–514.

Hartsock, A. and W.J. Nelson. 2008. Adherens and tight junctions: structure, function and connections to the actin cytoskeleton. Biochim Biophys Acta 1778: 660–669.

Herrenknecht, K., M. Ozawa, C. Eckerskorn et al. 1991. The uvomorulin-anchorage protein alpha catenin is a vinculin homologue. Proc Natl Acad Sci USA 88: 9156–9160.

Hirase, T. and K. Node. 2012. Endothelial dysfunction as a cellular mechanism for vascular failure. Am J Physiol Heart Circ Physiol 302: H499–H505.

Hoelzle, M.W. and T. Svitkina. 2012. The cytoskeletal mechanisms of cell–cell junction formation in endothelial cells. Mol Biol Cell 23: 310–323.

Huber, A.H., W.J. Nelson and W.I. Weis. 1997a. Three-dimensional structure of the armadillo repeat region of beta-catenin. Cell 90: 871–882.

Huber, O., M. Krohn and R. Kemler. 1997b. A specific domain in alphacatenin mediates binding to beta-catenin or plakoglobin. J Cell Sci 110: 1759–1765.

Hulpiau, P. and F. van Roy. 2008. Molecular evolution of the cadherin superfamily. Int J Biochem Cell Biol 41: 349–369.

Huveneers, S., J. Oldenburg, E. Spanjaard et al. 2012. Vinculin associates with endothelial VE-cadherin junctions to control force-dependent remodeling. J Cell Biol 196: 641–652.

Imamura, Y., M. Itoh, Y. Maeno et al. 1999. Functional domains of alpha-catenin required for the strong state of cadherin-based cell adhesion. J Cell Biol 144: 1311–1322.

Itoh, M., A. Nagafuchi, S. Moroi et al. 1997. Involvement of ZO-1 in cadherin-based cell adhesion through its direct binding to alpha catenin and actin filaments. J Cell Biol 138: 181–192.

Kemler, R. 1993. From cadherins to catenins: cytoplasmic protein interactions and regulation of cell adhesion. Trends Genet 9: 317–321.

Kimura, K., M. Ito, M. Amano et al. 1996. Regulation of myosin phosphatase by Rho and Rho-associated kinase (Rho-kinase). Science 273: 245–248.

Knudsen, K.A., A.P. Soler, K.R. Johnson et al. 1995. Interaction of alpha-actinin with the cadherin/catenin cell–cell adhesion complex via alpha-catenin. J Cell Biol 130: 67–77.

Komarova, Y. and A.B. Malik. 2010. Regulation of endothelial permeability via paracellular and transcellular transport pathways. Annu. Rev Physiol 72: 463–493.

Koslov, E.R., P. Maupin, D. Pradhan et al. 1997. Alpha-catenin can form asymmetric homodimeric complexes and/or heterodimeric complexes with beta-catenin. J Biol Chem 272: 27301–27306.

Kowalczyk, A.P., P. Navarro, E. Dejana et al. 1998. VE-cadherin and desmoplakin are assembled into dermal microvascular endothelial intercellular junctions: a pivotal role for plakoglobin in the recruitment of desmoplakinto intercellular junctions. J Cell Sci 111: 3045–3057.

Kumar, P., Q. Shen, C.D. Pivetti et al. 2009. Molecular mechanisms of endothelial hyperpermeability: implications in inflammation. Expert Rev Mol Med. 11: e19.

Lambert, O., J.C. Taveau, J.L. Him et al. 2005. The basic framework of VE–cadherin junctions revealed by cryo-EM. J Mol Biol 346: 1193–1196.

Lampugnani, M.G. and E. Dejana. 1997. Interendothelial junctions: structure, signalling and functional roles. Curr Opin Cell Biol 9: 674–682.

Lampugnani, M.G., M. Resnati, M. Raiteri et al. 1992. A novel endothelial-specific membrane protein is a marker of cell–cell contacts. J Cell Biol 118: 1511–1522.

Lampugnani, M.G., F. Orsenigo, M.C. Gagliani et al. 2006. Vascular endothelial cadherin controls VEGFR-2 internalization and signaling from intracellular compartments. J Cell Biol 174: 593–604.

Lewis, J.E., J.K. Wahl, J.M. Sass et al. 1997. Cross-talk between adherens junctions and desmosomes depends on plakoglobin. J Cell Biol 136: 919–934.

Liebner, S., U. Cavallaro and E. Dejana. 2006. The multiple languages of endothelial cell- to cell communication. Arterioscler Thromb Vasc Biol 26: 1431–1438.

Lum, H. and A.B. Malik. 1994. Regulation of vascular endothelial barrier function. Am J Physiol Lung Cell Mol Physiol 267: L223–L241.

Luo, Y. and G.L. Radice. 2005. N-cadherin acts upstream of VE-cadherin in controlling vascular morphogenesis. J Cell Biol 169: 29–34.

Maiden, S.L. and J. Hardin. 2011. The secret life of alpha-catenin: moonlighting in mophogeneis. J Cell Biol 195: 543–552.

May, C., J.F. Doody, R. Abdullah et al. 2005. Identification of a transiently exposed VE-cadherin epitope that allows for specific targeting of an antibody to the tumor neovasculature. Blood 105: 4337–4344.

McCrea, P., C. Turck and B. Gumbiner. 1991. A homolog of the armadillo protein in Drosophila (plakoglobin) associated with E-cadherin. Science 254: 1359–1361.

Mehta, D. and A.B. Malik. 2006. Signaling mechanisms regulating endothelial permeability. Physiol Rev 86: 279–367.

Mehta, D., M. Konstantoulaki, G.U. Ahmmed et al. 2005. Sphingosine 1-phosphate-induced mobilization of intracellular Ca2+ mediates rac activation and adherens junction assembly in endothelial cells. J Biol Chem 280: 17320–17328.

Millan, J., R.J. Cain, N. Reglero-Real et al. 2010. Adherens junctions connect stress fibers between adjacent endothelial cells. BMC Biol 8: 11.

Muller, W.A. 2003. Leukocyte–endothelial-cell interactions in leukocyte transmigration and the inflammatory response. Trends Immunol 24: 327–334.

Nagafuchi, A., M. Takeichi and S. Tsukita. 1991. The 102 kd cadherin-associated protein: similarity to vinculin and posttranscriptional regulation of expression. Cell 65: 849–857.

Nagafuchi, A., S. Ishihara and S. Tsukita. 1994. The roles of catenins in the cadherin-mediated cell adhesion: functional analysis of E-cadherin-alpha catenin fusion molecules. J Cell Biol 127: 235–245.

Navarro, P., L. Ruco and E. Dejana. 1998. Differential localization of VE- and N- cadherins in human endothelial cells: VE-cadherin competes with N-cadherin for junctional localization. J Cell Biol 140: 1475–1484.

Nehls, V. and D. Drenckhahn. 1991. Demonstration of actin filament stress fibers in microvascular endothelial cells *in situ*. Microvasc Res 42: 103–112.

Nelson, W.J. 2008. Regulation of cell–cell adhesion by the cadherin–catenin complex. Biochem Soc Trans 36: 149–155.

Nelson, W.J. and R. Nusse. 2004. Convergence of Wnt, beta-catenin, and cadherin pathways. Science 303: 1483–1487.

Nieset, J.E., A.R. Redfield, F. Jin et al. 1997. Characterization of the interactions of alpha-catenin with alpha-actinin and beta-catenin/plakoglobin. J Cell Sci 110: 1013–1022.

Niessen, C.M. 2007. Tight junctions/adherens junctions: basic structure and function. J Invest Dermatol 127: 2525–2532.

Niessen, C.M. and C.J. Gottardi. 2008. Molecular components of the adherens junction. Biochim Biophys Acta 1778: 562–571.

O'Brien, R.F., R.J. Robbins and I.F. McMurtry. 1987. Endothelial cells in culture produce a vasoconstrictor substance. J Cell Physiol 132: 263–270.

Ozawa, M., H. Baribault and R. Kemler. 1989. The cytoplasmic domain of the cell adhesion molecule uvomorulin associates with three independent proteins structurally related in different species. EMBO J 8: 1711–1717.

Pannekoek, W.J., M.R. Kooistra, F.J. Zwartkruis et al. 2009. Cell–cell junction formation: the role of Rap1 and Rap1 guanine nucleotide exchange factors. Biochim Biophys Acta 1788: 790–796.

Pappas, D.J. and D.L. Rimm. 2006. Direct interaction of the C-terminal domain of alpha-catenin and F-actin is necessary for stabilized cell–cell adhesion. Cell Commun Adhes 13: 151–170.

Peifer, M. 1993. The product of the Drosophila segment polarity gene armadillo is part of a multi-protein complex resembling the vertebrate adherens junction. J Cell Sci 105: 993–1000.

Peifer, M., P.D. McCrea, K.J. Green et al. 1992. The vertebrate adhesive junction proteins beta-catenin and plakoglobin and the Drosophila segment polarity gene armadillo form a multigene family with similar properties. J Cell Biol 118: 681–691.

Peifer, M., S. Berg and A.B. Reynolds. 1994. A repeating amino acid motif shared by proteins with diverse cellular roles. Cell 76: 789–791.

Perez-Moreno, M. and E. Fuchs. 2006. Catenins: keeping cells from getting their signals crossed. Dev Cell 11: 601–612.

Pokutta, S. and W.I. Weis. 2000. Structure of the dimerization and beta-catenin-binding region of alpha-catenin. Mol Cell 5: 533–543.

Pokutta, S. and W.I. Weis. 2007. Structure and mechanism of cadherins and catenins in cell–cell contacts. Annu Rev Cell Dev Biol 23: 237–261.

Pokutta, S., F. Drees, Y. Takai et al. 2002. Biochemical and structural definition of the l-afadin- and actin-binding sites of alphacatenin. J Biol Chem 277: 18868–18874.

Pokutta, S., F. Drees, S. Yamada et al. 2008. Biochemical and structural analysis of alpha-catenin in cell–cell contacts. Biochem Soc Trans 36: 141–147.

Popoff, M.R. and B. Geny. 2009. Multifaceted role of Rho, Rac, Cdc42 and Ras in intercellular junctions, lessons from toxins. Biochim Biophys Acta 1788: 797–812.

Rabiet, M.J., J.L. Plantier, Y. Rival et al. 1996. Thrombin-induced increase in endothelial permeability is associated with changes in cell-to-cell junction organization. Arterioscler Thromb Vasc Biol 16: 488–496.

Reynolds, A.B., J. Daniel, P.D. McCrea et al. 1994. Identification of a new catenin: the tyrosine kinase substrate p120cas associates with E-cadherin complexes. Mol Cell Biol 14: 8333–8342.

Ridley, A.J. 2006. Rho GTPases and actin dynamics in membrane protrusions and vesicle trafficking. Trends Cell Biol 16: 522–529.

Ridley, A.J. and A. Hall. 1992. The small GTP-binding protein rho regulates the assembly of focal adhesions and actin stress fibers in response to growth factors. Cell 70: 389–399.

Rimm, D.L., E.R. Koslov, P. Kebriaei et al. 1995. Alpha 1(E)-catenin is an actin-binding and -bundling protein mediating the attachment of F-actin to the membrane adhesion complex. Proc Natl Acad Sci USA 92: 8813–8817.

Rogers, K.R., C.J. Morris and D.R. Blake. 1992. The cytoskeleton and its importance as a mediator of inflammation. Ann Rheum Dis 51: 565–571.

Sabatini, P.J., M. Zhang, R. Silverman-Gavrila et al. 2008. Homotypic and endothelial cell adhesions via N-cadherin determine polarity and regulate migration of vascular smooth muscle cells. Circ Res 103: 405–412.

Sacco, P.A., T.M. McGranahan, M.J. Wheelock et al. 1995. Identification of plakoglobin domains required for association with Ncadherin and -catenin. J Biol Chem 270: 20201–20206.

Sakurai A, S. Fukuhara, A. Yamagishi et al. 2006. MAGI-1 is required for Rap1 activation upon cell–cell contact and for enhancement of vascular endothelial cadherin-mediated cell adhesion. Mol Biol Cell 17: 966–976.

Salomon, D., O. Ayalon, R. Patel-King et al. 1992. Extra junctional distribution of N-cadherin in cultured human endothelial cells. J Cell Sci 102: 7–17.

Samarin, S. and A. Nusrat. 2009. Regulation of epithelial apical junctional complex by Rho family GTPases. Front Biosci 14: 1129–1142.

Sandoval, R., A.B. Malik, R.D. Minshall et al. 2001. Ca^{2+} signalling and PKCalpha activate increased endothelial permeability by disassembly of VE-cadherin junctions. J Physiol 533: 433–445.

Schmelz, M. and W.W. Franke. 1993. Complexus adhaerentes, a new group of desmoplakin-containing junctions in endothelial cells: the syndesmos connecting retothelial cells of lymph nodes. Eur J Cell Biol 61: 274–289.

Schmelz, M., R. Moll, C. Kuhn et al. 1994. Complexus adhaerentes, a new group of desmoplakin-containing junctions in endothelial cells: II. Different types of lymphatic vessels. Diff Res Biol Div 57: 97–117.

Shasby, D.M., D.R. Ries, S.S. Shasby et al. 2002. Histamine stimulates phosphorylation of adherens junction proteins and alters their link to vimentin. Am J Physiol Lung Cell Mol Physiol 282: L1330–L1338.

Simcha, I., B. Geiger, S. Yehuda-Levenberg et al. 1996. Suppression of tumorigenicity by plakoglobin: an augmenting effect of N-cadherin. J Cell Biol 133: 199–209.

Stappert, J. and R. Kemler. 1994. A short core region of E-cadherin is essential for catenin binding and is highly phosphorylated. Cell Adhes Commun 2: 319–327.

Stehbens, S.J., A. Akhmanova and A.S. Yap. 2009. Microtubules and cadherins: a neglected partnership. Front Biosci 14: 3159–3167.

Takeichi, M. 2007. The cadherin superfamily in neuronal connections and interactions. Nat Rev Neurosci 8: 11–20.

Thurston, G., A.L. Baldwin and L.M. Wilson. 1995. Changes in endothelial actin cytoskeleton at leakage sites in the rat mesenteric microvasculature. Am J Physiol 268: H316–H329.

Tinkle, C.L., H.A. Pasolli, N. Stokes et al. 2008. New insights into cadherin function in epidermal sheet formation and maintenance of tissue integrity. Proc Natl Acad Sci USA 105: 15405–15410.

Tiruppathi, C., R.D. Minshall, B.C. Paria et al. 2002. Role of Ca^{2+} signaling in the regulation of endothelial permeability. Vascul Pharmacol 39: 173–185.

Touyz, R.M. 2006. Lipid rafts take center stage in endothelial cell redox signaling by death receptors. Hypertension 47: 16–18.

Trepat, X., L. Deng, S.S. An et al. 2007. Universal physical responses to stretch in the living cell. Nature 447: 592–595.

Troyanovsky, R.B., O. Laur and S.M. Troyanovsky. 2007. Stable and unstable cadherin dimers: mechanisms of formation and roles in cell adhesion. Mol Biol Cell 18: 4343–4352.

Troyanovsky, S.M. 2005. Cadherin dimers in cell–cell adhesion. Eur J Cell Biol 84: 225–233.

Tuma, P.L. and A.L. Hubbard. 2003. Transcytosis: crossing cellular barriers. Physiol Rev 83: 871–932.

van Nieuw Amerongen, G.P., S. van Delft, M.A. Vermeer et al. 2000. Activation of RhoA by thrombin in endothelial hyperpermeability: role of Rho kinase and protein tyrosine kinases. Circ Res 87: 335–340.

van Roy, F. and G. Berx. 2008. The cell–cell adhesion molecule E-cadherin. Cell Mol Life Sci 65: 3756–3788.

Venkiteswaran, K., K. Xiao, S. Summers et al. 2002. Regulation of endothelial barrier function and growth by VE- cadherin, plakoglobin, and beta-catenin. Am J Physiol Cell Physiol 283: C811–C821.

Vestweber, D. 2008. VE-cadherin: the major endothelial adhesion molecule controlling cellular junctions and blood vessel formation. Arterioscler Thromb Vasc Biol 28: 223–232.

Vestweber, D., M. Winderlich, G. Cagna et al. 2008. Cell adhesion dynamics at endothelial junctions: VE-cadherin as a major player. Trends Cell Biol 19: 8–15.

Volk, T., O Cohen and B. Geiger. 1987. Formation of heterotypic adherens-type junctions between L-CAM containing liver cells and A-CAM containing lens cells. Cell 50: 987–994.

Wallez, Y. and P. Huber. 2008. Endothelial adherens and tight junctions in vascular homeostasis, inflammation and angiogenesis. Biochim Biophys Acta 1778: 794–809.

Wang, N., J.P. Butler and D.E. Ingber. 1993. Mechanotransduction across the cell surface and through the cytoskeleton. Science 260: 1124–1127.

Watabe-Uchida, M., N. Uchida, Y. Imamura et al. 1998. alpha-Catenin-vinculin interaction functions to organize the apical junctional complex in epithelial cells. J Cell Biol 142: 847–857.

Weigand, J.E., J.N. Boeckel, P. Gellert et al. 2012. Hypoxia-induced alternative splicing in endothelial cells. PLoS One 7: e42697.

Weiss, E.E., M. Kroemker, A.H. Rüdiger et al. 1998. Vinculin is part of the cadherin-catenin junctional complex: complex formation between alpha-catenin and vinculin. J Cell Biol 141: 755–764.

Wildenberg, G.A., M.R. Dohn, R.H. Carnahan et al. 2006. p120-catenin and p190RhoGAP regulate cell–cell adhesion by coordinating antagonism between Rac and Rho. Cell 127: 1027–1039.

Winter, M.C., A.M. Kamath, D.R. Ries et al. 1999. Histamine alters cadherin-mediated sites of endothelial adhesion. Am J Physiol 277: L988–L995.

Wojciak-Stothard, B. and A.J. Ridley. 2002. Rho GTPases and the regulation of endothelial permeability. Vascul Pharmacol 39: 187–199.

Wójciak-Stothard, B., A. Entwistle, R. Garg et al. 1998. Regulation of TNF-a-induced reorganization of the actin cytoskeleton and cell–cell junctions by Rho, Rac, and Cdc42 in human endothelial cells. J Cell Physiol 176: 150–165.

Yamada, S., F., Pokutta, Drees et al. 2005. Deconstructing the cadherin-catenin-actin complex. Cell 123: 889–901.

Yonemura, S. 2011. Cadherin-actin interactions at adherens junctions. Curr Opin Cell Biol 23: 515–522.

Zhang, Y., S. Sivasankar, W.J. Nelson et al. 2009. Resolving cadherin interactions and binding cooperativity at the single-molecule level. Proc Natl Acad Sci USA 106: 109–114.

Zhurinsky, J., M. Shtutman and A. Ben-Ze'ev. 2000. Differential mechanisms of LEF/TCF-dependent transcriptional activation by b-catenin and plakoglobin. Mol Cell Biol 20: 4238–4252.

5

Mechanical Force Transmission via the Cytoskeleton in Vascular Endothelial Cells

Bori Mazzag,[1] Cecile L.M. Gouget,[2,a] Yongyun Hwang[3] and Abdul I. Barakat[2,b,]*

Mechanical Forces and the Vascular Endothelium

By virtue of its anatomic location, the vascular endothelium is constantly exposed to a highly dynamic mechanical stress environment due to the pulsatile nature of blood flow. The mechanical stress field on the vascular endothelial cell (EC) surface consists of three components: tangential shear forces due to the flow of viscous blood, normal pressure forces due to endovascular pressure, and circumferential stretch forces due to transmural pressure difference. The mechanical stresses to which a

[1]Department of Mathematics, Humboldt State University, 1 Harpst St., Arcata, CA 95521, USA.
Email: borim@humboldt.edu
[2]Hydrodynamics Laboratory (LadHyX), CNRS UMR 7646, Ecole Polytechnique, France.
[a]Email: gouget@ladhyx.polytechnique.fr
[b]Email: barakat@ladhyx.polytechnique.fr
[3]Department of Applied Mathematics and Theoretical Physics (DAMTP), University of Cambridge, United Kingdom.
Email: Y.Hwang@damtp.cam.ac.uk
*Corresponding author

particular population of ECs is subjected depends strongly on the vascular bed from which the cells are derived and the location of the cells within that bed. In the microvasculature, blood pulsatility is highly damped and the flow is quasi-steady. Consequently, ECs in microvessels experience a nearly constant level of shear stress in time, even though the magnitude of that shear stress depends strongly on microvessel size. In medium and large blood vessels, on the other hand, blood flow is highly unsteady, leading to large and periodic variations in the mechanical stresses experienced by the endothelium during the course of the cardiac cycle. Changes in activity level, for instance rest vs. exercise, further add to these temporal variations in the mechanical stress field. In addition, the complex vascular geometry, which includes branches, bifurcations and extensive curvature, induces disturbances in the vascular flow field which often take the form of flow separation and recirculation zones within which the mechanical stress field is severely altered. Therefore, the endothelium in medium and large blood vessels is additionally subjected to large spatial gradients in the mechanical stress field. In these vessels, there is also a difference in the mechanical stress environment between arteries and veins due to the large difference in transmural blood pressure between the two types of blood vessels.

Role of Mechanical Forces in Regulating Vascular Function and Dysfunction

Mechanical stimulation drives vascular development and regulates vascular structure and function. During morphogenesis, shear stress directs the formation of the vascular tree (Hahn and Schwartz 2009). In the arterial system, acute changes in blood flow regulate vessel diameter, and this flow-mediated vasoregulatory response requires the presence of an intact and normally functioning endothelium (Pohl et al. 1986). Chronic alterations in arterial blood flow elicit extensive remodeling of the arterial wall, and this response also requires the presence of the endothelium (Langille and O'Donnell 1986). These findings demonstrate the critical role that mechanical forces play in regulating normal vascular function and the central importance of the endothelium in modulating vascular responsiveness to mechanical stimulation.

Mechanical forces also play an important role in vascular dysfunction and, here again, the endothelium plays a central role. A prominent example is atherosclerosis, the arterial disease that leads to heart attacks and strokes. The initial stages of atherosclerosis involve sustained endothelial inflammation (Libby 2002, Tedgui and Mallat 2001). Early atherosclerotic lesions develop preferentially near arterial branches and bifurcations where, as described above, arterial flow patterns are highly disturbed. Therefore, it is believed that disturbances in arterial blood flow induce EC dysfunction that ultimately contributes to the development of the disease. The exact type of flow disturbance associated with the development of atherosclerotic lesions remains a topic of active research, although a number of studies have implicated low and/or oscillatory flow typically encountered within flow separation and recirculation zones (Asakura and Karino 1990, Chappell et al. 1998, Ku et al. 1985).

Mechanisms of Vascular Endothelial Mechanotransduction

Although numerous endothelial responses to mechanical stimulation have been documented (see reviews in Chien 2007, Davies 1995, Davies et al. 2005, Hahn and Schwartz 2009), how ECs sense mechanical forces and how they respond and adapt to these forces remain incompletely understood. Mechanical force sensing and rapid conversion into chemical signals often appears to involve specific subcellular structures at or near the cell membrane that act as mechanosensors. Candidate mechanosensors include mechanosensitive ion channels (Barakat et al. 1999, Gautam et al. 2006, Lieu et al. 2004, Olesen et al. 1988), integrins (Shyy and Chien 2002), the cellular glycocalyx (Tarbell and Pahakis 2006, Weinbaum et al. 2007), cell-cell adhesion complexes (Tzima et al. 2005), and G protein-coupled receptors (Meyer et al. 2000). How mechano-chemical conversion occurs in ECs has not yet been completely elucidated, but mechanisms that involve conformational changes of mechanoreceptive proteins and/or association of two molecules that are otherwise apart have been proposed. Mechano-chemical conversion triggers downstream signaling pathways with associated mobilization of transcription factors via reaction-diffusion processes, ultimately leading to changes in gene expression and protein synthesis.

Reaction-diffusion processes are relatively slow: typically, the transmission of a signal from the cell surface to the nucleus by these processes requires several seconds at a minimum. Some mechanically-induced cellular responses, however, occur over a considerably shorter time constant. For instance, mechanical activation of the tyrosine kinase Src in vascular smooth muscle cells has been reported to occur within ~300 milliseconds (Na et al. 2008). Therefore, in order to be effective at sensing mechanical signals and transmitting them to desired intracellular target locations, cells need to possess alternative mechanisms that enable very rapid and directional transmission of mechanical forces. There is mounting evidence that the cytoskeleton provides such a mechanism. The following describes the role of the cytoskeleton in cellular mechanotransduction and presents two complementary mathematical models that we have formulated to describe the dynamics of mechanical force transmission via the cytoskeleton.

Role of the Cytoskeleton in Mechanotransduction

It has long been known that mechanical forces regulate the organization of the three major components of cytoskeleton in ECs. For instance, exposure of ECs to steady flow leads to progressive remodeling of actin filaments and microtubules, which ultimately induces cellular elongation and alignment in the direction of flow (Malek and Izumo 1996, Nerem et al. 1981, Ookawa et al. 1992, Wechezak et al. 1985). This response becomes visible a few hours following the onset of flow and appears to require ~12–24 hours to be complete. Flow also leads to rapid displacement of EC intermediate filaments (Helmke et al. 2000). In addition to the effects of flow, both cyclic uniaxial stretch and hydrostatic pressure also induces cytoskeletal remodeling in ECs (Hsu et al. 2010, Salwen et al. 1998, Yamada and Ando 2007). These various findings clearly demonstrate that mechanical forces applied to the EC surface are transmitted to the cytoskeleton.

An important question is how mechanical force transmission from the EC surface to the cytoskeleton occurs. A number of studies have demonstrated that mechanical stimulation activates small GTPases such as Rho, Rac, and Cdc42 which participate in signaling pathways that regulate cytoskeletal remodeling (Li et al. 1999, Tzima et al. 2002). This can be considered as a form of indirect "biochemical" pathway for mechanical force transmission to the cytoskeleton.

Of particular relevance to the present focus is evidence that mechanical force transmission to the cytoskeleton may also occur via a "biophysical" pathway that relies on direct physical links between specific mechanosensors at the EC surface and cytoskeletal elements. Interestingly, this hard wiring appears to extend to a host of intracellular structures including mitochondria (Silberberg et al. 2008) and the nucleus (Maniotis et al. 1997, Wang et al. 1993), thereby potentially allowing specific spatiotemporal patterns of force distribution. In support of this notion, binding of ECs to microbeads coated with extracellular matrix ligands for integrin receptors induces rapid formation of focal adhesion complexes that mediate the transfer of the mechanical stress to the internal cytoskeleton (Wang et al. 1993). Mechanical tugging on these beads leads to nuclear deformation and elongation in the direction of the applied mechanical force (Maniotis et al. 1997). In contrast, mechanical tugging on beads coated with acetylated low density lipoprotein, a ligand for metabolic receptors that only physically connect to the submembranous cytoskeleton (but not to the internal cytoskeleton), fails to induce nuclear deformation and elongation (Maniotis et al. 1997). Mechanical tugging-induced deformation and elongation of the nucleus also occurs in the case of integrin-bound beads on ECs whose membranes and cytosolic components had been extracted (with Triton X-100), suggesting that force transmission from integrins to the nucleus occurs directly through the cytoskeletal lattice and not through more indirect pathways involving diffusion-based chemical signaling or protein polymerization (Maniotis et al. 1997). Actin filament disruption with cytochalasin-D abolishes mechanical force-induced nuclear deformation and elongation (Maniotis et al. 1997), suggesting that an intact actin network is required for force transmission to the nucleus.

Experimental advances over the past decade have greatly enhanced our understanding of force transmission within the endothelial cytoskeleton and its effect on intracellular force distribution (Helmke 2005, Helmke et al. 2001, 2003, Mott and Helmke 2007). A good example is the high-resolution and quantitative imaging of green fluorescent protein- (GFP-) expressing cytoskeletal constructs, which provides the spatiotemporal pattern of cytoskeletal displacements in response to mechanical stimulation in live cells. Such studies have shown that in response to flow, the displacement of vimentin, the primary intermediate filament in ECs, is spatially heterogeneous, with larger displacements in the downstream regions of the cell and strain focusing at the basal cell surface near focal adhesion sites or intercellular junctions (Helmke 2005, Helmke et al. 2003). Similar studies with GFP-actin (Li et al. 2002) in single ECs showed the formation of "actin ruffles" downstream and microfilament node displacement near the basal cell surface. A more comprehensive study, using GFP-labeled actin, vimentin, paxillin and vinculin allowed not only the visualization of the displacement of these cytoskeletal components but also an analysis of the coordination of the structural changes and the detection of the initially heterogeneous pattern of strain focusing at the onset of flow (Mott and Helmke 2007).

Force transmission via the cytoskeleton provides cells with unique capabilities (Wang et al. 2009). For instance, it allows cells to very rapidly transfer mechanical stimuli over long distances with limited loss of information as evidenced by the transmission of mechanical forces via actin stress fibers over a distance of ~50 μm in less than 300 ms (Na et al. 2008, Wang et al. 2009). Response times on the order of milliseconds are observed neither in signaling cascades mediated by diffusion-reaction processes which typically require more than 10 sec to travel the same distance, nor in force propagation through the viscous cytoplasm where the applied force gets rapidly damped. Another unique feature of mechanical force transmission via the cytoskeleton is that the filamentous nature of cytoskeletal architecture enables the spatially heterogeneous distribution of an applied mechanical force within cells (Hu et al. 2003, Na et al. 2008, Wang and Suo 2005), thereby allowing force focusing on particular intracellular sites where a particular effect is desired. In light of the myriad targets that cytoskeletal force distribution provides, the wide range of possible response time constants associated with these targets, and interactions among these various responses, force transmission via the cytoskeleton allows for a potentially very rich array of responses to mechanical stimuli to which the cell surface is subjected.

Models of Cytoskeletal Contributions to Cellular Rheology and Mechanotransduction

A number of mathematical models have been developed in an attempt to define the contribution of the cytoskeleton to cellular mechanotransduction and to quantify the dynamics of force transmission via cytoskeletal filaments. A primary attractiveness of these models is that they allow the study of the contributions of individual physical phenomena in isolation of all others, thus providing a more fundamental and mechanistic understanding of the processes at play. Naturally, these models need to be interpreted with caution because the biological phenomena they intend to describe are very complicated and often not completely understood. Therefore, it is essential for models to be validated to the extent possible against experimental data. It is only after such validation is performed that mathematical models can potentially fulfill their promise of providing powerful and predictive tools for studying the dynamics and spatiotemporal organization of mechanically-induced responses in cells.

Different modeling approaches have been pursued over the past several years. For instance, models that describe cells as elastic, viscoelastic, or poroelastic continua have been proposed (Kollmansberger and Fabry 2011, Lim et al. 2006, Mofrad 2009). In these models, the contribution of the cytoskeleton to cell mechanics is accounted for by deriving "averaged" constitutive laws. These models have been shown to accurately predict stress-strain relationships in cells; however, because they do not take into account the discrete nature of cytoskeletal architecture, they fail to accurately predict spatial heterogeneities in force transmission and distribution. Models that take into consideration the discrete nature of cytoskeletal fibers provide a better representation of force transmission via the cytoskeleton. For example, a model that accounts for both the interconnectedness of the different cytoskeletal elements and the major role that prestress generated by motor proteins plays in force transmission, the tensegrity model was proposed (Ingber 1997, 2008). In this model, a cell is viewed as a structure consisting of discrete elements that are either in tension (representing actin stress fibers and intermediate filaments) or in compression (representing microtubules and focal adhesions). These various elements are coupled to one another in such a way that the entire structure is mechanically stable. It has been argued that the tensegrity model effectively describes the coordinated response of the cellular cytoskeleton to external mechanical stimuli (Wang et al. 2009), explains several important features of cellular behavior including structural self-stabilization and reproduces important aspects of cellular mechanotransduction including heterogeneous force distribution and long-distance force transmission. The model remains, however, a conceptual framework that does not explicitly account for either the extremely complex cytoskeletal architecture or the active nature of cytoskeletal assembly and disassembly.

In addition to the above models, a number of new approaches have been proposed in recent years for elucidating the rheological properties of cells, which are driven by the cytoskeleton. Examples of these models include the soft glassy material model (Fabry et al. 2001), the semi-flexible network model (Gittes and MacKintosh 1998), the glassy semi-flexible network model (Semmrich et al. 2007), and the Brownian dynamics simulation model (Kim et al. 2009) (for further details the reader may refer to Mofrad (2009) and Kollmansberger and Fabry (2011)). These models are particularly useful for understanding fundamental features of cell rheology such as the power-law behavior of dynamic modulus. They also appear to be effective at reproducing the behavior of artificial polymer networks, albeit in a qualitative manner (Kollmansberger and Fabry 2011). However, modeling the cytoskeleton and its highly complex dynamics within this framework remains well beyond the current capabilities of these models.

While each of the models outlined thus far provides a framework to describe particular aspects of cytoskeletal behavior, most notably cellular rheology, they all have significant limitations. A particular limitation is the inability to provide a quantitative description of the dynamics of force transmission through the cytoskeleton. Determining these dynamics is a critical step in understanding rapid long-distance force transmission in cells. To address this gap, we have proposed two types of models that aim specifically to identify the physical parameters that determine how rapidly mechanical signals carried by the cytoskeleton are transmitted from the cell surface to the intracellular space and to shed light onto how the cytoskeleton modulates the spatiotemporal distribution of forces within cells. The present chapter describes these models and discusses the implications of their predictions to EC function and dysfunction.

Network Models of Cytoskeleton-mediated Mechanical Signal Transmission in Cells

We provide an overview of the mathematical formulation of our two models of force transmission via the cytoskeleton. The details of model development and analysis can be found in the original articles (Hwang and Barakat 2012, Hwang et al. 2012, Mazzag and Barakat 2011, Mazzag et al. 2003). After presenting the predictions of each model, we discuss the limitations of the present approaches and conclude with perspectives for future work.

Model 1: Temporal network model

Our first model for cytoskeleton-mediated force transmission in ECs assumes that a blood flow-derived mechanical force (shear or pressure force) is sensed by a mechanosensor at the cell surface and is subsequently transmitted via cytoskeletal elements that may be either actin stress fibers (A) or microtubules (M) to various intracellular transduction sites including the nucleus (N), cell-cell adhesion proteins (CCAP), and focal adhesion sites (FAS) as schematically depicted in Fig. 1A. All the structures involved in force transmission including the mechanosensor, the cytoskeletal transmission elements, and the intracellular transduction sites are assumed to comprise a network of interconnected linear viscoelastic solids with the connections corresponding directly to physical coupling among these structures. The model only considers mechanical signal transmission and does not include mechanically-induced activation of any biochemical signaling pathways. The model input is a force applied to the mechanosensor, and the model output is the time-evolution of the resulting deformations of the cytoskeletal elements and the intracellular transduction sites. The model also describes how the applied force gets distributed within the intracellular space by the

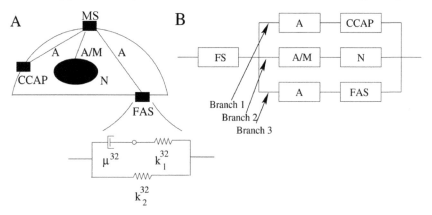

Figure 1. A. Schematic representation of an EC consisting of a mechanosensor (MS), cytoskeletal elements (either actin stress fibers (A) or microtubules (M)), a nucleus (N), cell-cell adhesion proteins (CCAP), and focal adhesion site (FAS). The inset shows a TPMM (or Kelvin body) representation and the viscoelastic parameters for FAS. The superscripts '32' on the parameters indicate that this element is the second element on the third branch (see text). **B.** Branching network representation of the EC components in panel A. Each cell component corresponds to a TPMM, coupled to other components according to the diagram shown. Actin stress fiber and CCAP connected in series are referred to as Branch 1, actin stress fiber/microtubule in series with the nucleus is Branch 2, and actin stress fiber in series with the FAS is Branch 3.

different cytoskeletal branches to which the mechanosensor is coupled. The mathematical framework in this model only describes the temporal dynamics of the network and does not provide any information about spatial aspects of force transmission, a limitation rectified in our second model described in a subsequent section. The viscoelastic solids within the network can be arranged in different ways, reflecting assumptions about the topology of intracellular cytoskeletal networks.

In the model, we represent the assumed linear viscoelastic behavior of each mechanical signaling transduction site by using a three-parameter Maxwell model (TPMM; Tschoegl 1989), also known as a Kelvin body (Fung 1981), which consists of a Maxwell body (spring with spring constant k_1 connected in series to a dashpot with a coefficient of viscosity μ) connected in parallel to a second spring with spring constant k_2 (see Fig. 1A inset for FAS example). The springs in this model produce the elastic portion of the viscoelastic behavior, whereas the dashpot produces the viscous portion of the response. Because the mechanical properties of the different structures in the network are different, each TPMM in the network is characterized by its own set of viscoelastic parameters k_1, k_2, and μ.

We take the example of the very simple network depicted in Fig. 1A and assume that a force F is applied to the mechanosensor. As this force propagates through the network, it splits among the three branches connected to the mechanosensor in accordance with the relative values of the mechanical properties in each branch. It is convenient to visualize the organization of this network in terms of rows and columns as depicted in Fig. 1B and to index the force on the ith branch as F_i and the deformation of the jth body on the ith row as u_{ij}. As detailed elsewhere (Fung 1981, Mazzag and Barakat 2011, Mazzag et al. 2003), the deformation u_{ij} is governed by the following differential equation:

$$F_i + \frac{\mu^{ij}}{k_2^{ij}} \frac{dF_i}{dt} = k_1^{ij} u_{ij} + \mu^{ij} \left(1 + \frac{k_1^{ij}}{k_2^{ij}}\right) \frac{du_{ij}}{dt}, \tag{1}$$

where k_1^{ij}, k_2^{ij}, and μ^{ij} are the viscoelastic parameters associated with the jth TPMM on the ith row (or branch). For example, FAS (second body on the third branch) has viscoelastic parameters k_1^{32}, k_2^{32}, and μ^{32}, as shown in Fig. 1. Note that the magnitude of the forces in the different branches is time-dependent and is determined by the viscoelastic properties of the bodies in the network; therefore, one must solve for the forces in each branch as well as the deformation of each of the TPMMs. This is possible by imposing the constraints that the sum of the forces within all rows must be the total applied force F and that the sum of deformations within each row must be equal to that in each other row and to the total deformation u of the entire network.

As detailed in our papers describing the model (Mazzag and Barakat 2011, Mazzag et al. 2003), combining the governing equations with the constraints described in the previous paragraph leads to a system of ordinary differential equations that can be solved subject to appropriate initial conditions. The viscoelastic parameter values for a number of the intracellular structures considered including the cytoskeletal elements, the nucleus, and some cell-surface receptors are available in the literature. It should be noted that the network shown in Fig. 1 is simply an example network and that the framework outlined above allows modeling a wide variety of possible network topologies.

Two types of cytoskeletal building blocks are considered in the model: actin stress fibers and microtubules. Both of these types of filaments have lengths that are considerably shorter than their persistence lengths, suggesting that their stochastic motion due to thermal effects can be neglected and that they can therefore be considered in a purely deterministic manner. Structurally, this is a consequence of the fact that stress fibers are rigid bundles of connected actin filaments and that microtubules have strong lateral reinforcements that render them quite rigid and capable of supporting enhanced mechanical loading (Brangwynne et al. 2006). Thus, in our framework, each actin stress fiber or microtubule connecting various intracellular structures should be viewed as a sufficiently large bundle of filaments for a deterministic (rather than statistical) description of cytoskeletal behavior to be valid. As such, each stress fiber or microtubule is represented by a single TPMM whose viscoelastic parameters are chosen to reflect the properties of a bundle (rather than those of a single filament).

The results of the TPMM network modeling provide insight into the dynamics of force transmission within ECs (Mazzag et al. 2003). For instance, we initially studied a simple scenario that focused on the dynamics of transmission of a constant force from a mechanosensor that is coupled to the nucleus via different types of cytoskeletal connections. More specifically, we considered the cytoskeletal connection to consist of either only actin stress fibers or of actin stress fibers in series with microtubules. Comparison of the results from these two configurations revealed that for the case where microtubules are present in the cytoskeletal connection, the microtubules attain an asymptotic deformation that is ~10 times larger than that of the actin stress fibers. Importantly, this asymptotic value is attained over a period of ~ 20 hr, which is considerably longer than the time constant that characterizes the steady state deformation of actin stress fibers alone (~ 10 min). These findings suggest that the nature of the cytoskeletal connection between the mechanosensor and the nucleus plays a critical role in determining the dynamics of force transmission. Furthermore, the results are consistent with experiments that have demonstrated that flow-induced cytoskeletal remodeling in ECs requires 12–24 hr to be complete and is critically dependent on microtubule reorganization (Malek and Izumo 1996).

As already described, oscillatory flow has been reported to induce EC dysfunction and to correlate with the development of early atherosclerotic lesions. Therefore, we also investigated the response of the simple networks described in the previous paragraph to oscillatory flow. The results revealed that the peak deformation (defined as the largest deformation within a period after the asymptotic time-periodic behavior is attained) of each of the structures in the network (mechanosensor, actin stress fibers, microtubules, and nucleus) is strongly frequency-dependent. At sufficiently low oscillatory frequencies, the peak deformations match those for constant forcing; however, above a threshold frequency, the peak deformations drop significantly. The analysis demonstrated that this threshold frequency is in the range of 10^{-5}–10^{-4} Hz for microtubules and 10^{-3}–10^{-2} Hz for actin stress fibers, suggesting that stress fibers can effectively transmit force over a wider frequency range.

Another application of the model was to investigate the effect of noise in the imposed mechanical signal on force transmission dynamics (Mazzag and Barakat 2011). An example of ECs subject to noisy mechanical signals is during the occurrence of turbulent blood flow. Turbulent flow is characterized by velocity and pressure fluctuations that occur in a random manner and that are characterized by a wide range of frequencies. Turbulence occurs in the arterial system under both physiological and pathological conditions. For instance, a transient burst of turbulence normally occurs at the end of systole in the thoracic aorta and at the entrance of the aorta downstream of the aortic valve. Turbulent jets are also often present downstream of advanced atherosclerotic stenoses. A key question we asked in the modeling is whether or not certain intracellular structures in the cytoskeleton-mediated force transmission pathway act as noise filters or amplifiers. For this analysis, we considered a slightly more elaborate cellular network where the connection between the mechanosensor and the nucleus was provided by either actin stress fibers or microtubules, whereas the other connections consisted of only actin stress fibers (see Fig. 1).

The simulation results demonstrated that in comparison with the all-actin network, replacing the mechanosensor-nucleus link (Branch 2 in Fig. 1) with a microtubule connection redistributes the force from the nucleus to the other downstream transduction sites, i.e., focal adhesions and cell-cell adhesion proteins (Figs. 2A-B). This force distribution leads to significantly smaller nuclear steady-state deformations (Figs. 2C-D). These

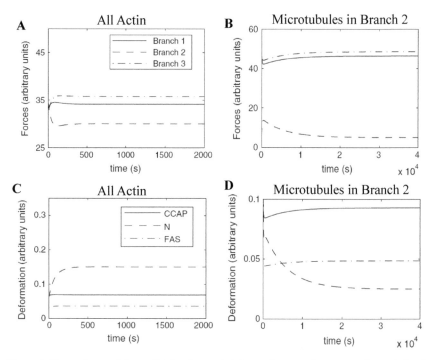

Figure 2. Dependence of force distribution and deformations on the nature of cytoskeletal connection between the mechanosensor and the nucleus for the cellular network in Fig. 1. **A** and **C.** Force distribution and deformations for the case where the cytoskeletal connections in all three branches are actin stress fibers. **B** and **D.** Force distribution and deformations for the case where the cytoskeletal connection in branch 2 consists of microtubules while branch 1 and 3 connections are maintained as actin stress fibers. N: nucleus; CCAP: cell-cell adhesion proteins; FAS: focal adhesion sites.

findings broadly imply that signaling pathways that depend on direct force transmission to the nucleus would be expected to behave differently depending on the detailed intracellular cytoskeletal architecture. Two other observations can be made when the flow sensor is coupled to the nucleus via microtubules: (1) The time evolution of the deformation of the nucleus is non-monotonic: nuclear deformation initially increases before declining to a small steady-state value (Fig. 2D), suggesting a biphasic behavior of signaling events that depend directly on nuclear deformation. (2) The deformations of the intracellular transduction sites are much slower (time scale of ~12 hr) when the nucleus is connected to the flow sensor via microtubules (vs. ~0.5 hr for an all-actin network), suggesting that the time required for steady-state signaling depends critically on cytoskeletal organization.

We also assessed the sensitivity to noise of each of the signaling sites within the network by determining the signal-to-noise ratio (SNR), defined as the mean amplitude of the mechanical signal at each of the signaling sites divided by its standard deviation. This SNR analysis was performed for a wide range of noise durations and noise amplitudes in the input mechanical signal (i.e., the signal imposed at the mechanosensor). The results demonstrated that the SNR was largest for the nucleus (regardless of whether or not the cytoskeleton included microtubules) and smallest for focal adhesion sites. Furthermore, the SNR of focal adhesion sites was shown to exhibit a higher sensitivity to noise amplitude and duration than the other signaling sites. These findings suggest that focal adhesion sites can act as effective "noise detectors" in ECs. In contrast, the nucleus acts as a much more effective force filter. The sensitivity of focal adhesion site SNR to noise was shown to be higher for an actin-only network than for a network that contains microtubules, suggesting that microtubules also act as effective noise filters.

Model 2: Spatiotemporal network model

The model described above provides insight into the temporal dynamics of force transmission via the cytoskeleton; however, it says nothing about the spatial distribution of stresses within cells. We have more recently expanded the analysis to consider the spatiotemporal dynamics of force transmission via the cytoskeleton (Hwang and Barakat 2012, Hwang et al. 2012). In order to gain a fundamental insight into the behavior, we began by considering the simplest case of an individual cytoskeletal filament. Because experiments have shown that actin stress fibers play a central role in force transmission through the cytoskeleton (Na et al. 2008, Wang and Suo 2005, Wang et al. 2009), we have placed particular emphasis on understanding force transmission through actin stress fibers.

We began by considering the highly simplified case where a single actin stress fiber links an integrin at the cell membrane to the nucleus as depicted in Fig. 3A. Stress fibers are in a state of prestress (i.e., pre-existing tension) due to tension-generating molecular motors; therefore, we assumed that a constant prestress σ_p is uniformly distributed along the stress fiber. The length of the stress fiber (L) was chosen as 10 μm, a representative value in many eukaryotic cells. This length is considerably shorter than the persistence length of a stress fiber (> 50 μm) (Boal 2002), which allowed us to neglect stochastic motion of the stress fiber due to thermal effects. The constitutive relation of the actin stress fiber was chosen as the Kelvin-Voigt model following recent experimental observations where the viscoelastic nature of stress fibers was measured using laser severing of individual fibers. An external damping force was considered due to cytosolic drag

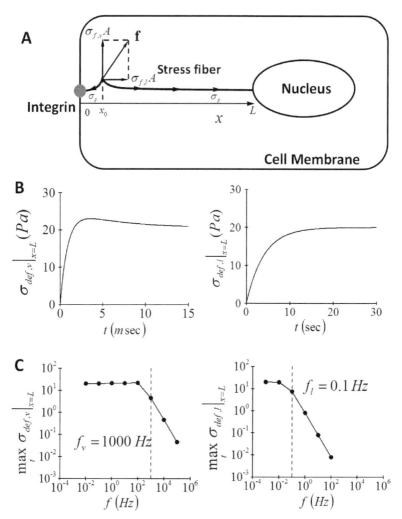

Figure 3. A. Schematic diagram of the actin stress fiber model. **B.** Temporal evolution of transverse (left) and axial (right) deformation-related stresses at the nucleus ($x=L$) with steady forcing. **C.** Effect of the forcing frequency on the amplitude of transverse (left) and axial (right) deformation-related stresses at the nucleus ($x=L$).

(assuming Stokes flow). Physiological mechanical stimuli often lead to only small deformations of the cytoskeleton (Na et al. 2008); thus, we assumed that the stress fiber deformation due to mechanical forcing is sufficiently small.

A force f applied at a point $x=x_0$ along the stress fiber length will lead to stress fiber motion both in the transverse direction (vertical direction in Fig. 3A) and in the longitudinal (or axial) direction. Therefore, one can write the equation of motion for this fiber in each of these two directions (Hwang and Barakat 2012). For transverse motion, the governing equation is as follows:

$$\rho A \frac{\partial w_v}{\partial t} = \frac{\partial}{\partial x}\left(V_P + V_E + V_{vis}\right) - C_v \mu \frac{\partial w_v}{\partial t} + \sigma_{f,v} A \delta\left(x - x_0\right), \tag{2a}$$

where

$$V_P = \sigma_P A \frac{\partial w_v}{\partial x}, \quad V_E = -EI \frac{\partial^3 w_v}{\partial x^3}, \quad V_{vis} = -\gamma I \frac{\partial^4 w_v}{\partial x^3 \partial t}. \tag{2b}$$

Here, w_v is the transverse displacement of the stress fiber, ρ the density of the stress fiber, A the cross-sectional area of the fiber, E is the stress fiber elastic modulus, I the stress fiber second moment of inertia, γ the stress fiber internal (material) viscosity, V_P the restoring force by prestress, V_E the restoring force by flexural rigidity EI, V_{vis} the internal damping force by flexural material viscosity γI, C_v the cytosolic resistance coefficient for transverse motion, μ the viscosity of the cytosol, $\sigma_{f,v}$ the applied stress in transverse direction, and $\delta(x)$ the Dirac delta function. Similarly, the equation for axial motion is given by:

$$\rho A \frac{\partial^2 w_l}{\partial t^2} = \frac{\partial}{\partial x}\left(T_E + T_{vis}\right) - C_l \mu \frac{\partial w_l}{\partial t} + \sigma_{f,l} A \delta\left(x - x_0\right) \tag{3a}$$

with

$$T_E = EA \frac{\partial w_l}{\partial x}, \quad T_{vis} = \gamma A \frac{\partial^2 w_l}{\partial x \partial t}, \tag{3b}$$

where w_l is the axial displacement of the filament, T_E the restoring force by elastic modulus E, T_{vis} the internal damping force by the material viscosity γ, $\sigma_{f,l}$ the applied stress in axial direction, and C_l the cytosolic resistance coefficient for axial motion. It should be noted that the effect of prestress is present only for transverse motion because it acts as a restoring force only in this case but not in the case of axial motion. Prestress can change the stress fiber's elasticity and material viscosity; however, this only occurs when the axial deformation due to prestress is sufficiently large. In most cases, elasticity and material viscosity remain constant under physiological tensions (Deguchi et al. 2005); therefore, we assume that it is valid to neglect the effect of prestress on axial dynamics of the stress fiber.

At the nucleus, we impose displacement-free boundary conditions for both Eqs. (2a) and (3a) because the nucleus is much stiffer than the stress fiber. We also assume a zero moment condition in order to allow force transfer. At the integrin, we consider force- and stress-free boundary conditions in order to allow deformation of the stress fiber at this location when the forcing is applied near the integrin. Note that this situation roughly mimics experiments where force is applied directly using magnetic or optimal tweezers to microbeads bound to integrins on the cell membrane. Finally, it should be pointed out that the use of Eqs. (2a) and (3a) is not necessarily limited to the integrin-nucleus link because they can describe any situation in which a stress fiber links a receptor on the cell surface to a relatively stiff subcellular structure such as focal adhesion sites, cell-cell adhesion proteins, etc.

Using this model, we studied the transmission dynamics of either a constant or a purely oscillatory force applied to an actin stress fiber initially in static equilibrium. In the simulations, the amplitude of the applied stress was chosen as 20 Pa, which experiments have shown to be sufficient to elicit a biochemical response (Na et al. 2008). It should be noted, however, that a change in the forcing amplitude does not change the conclusions presented here because Eqs. (2a) and (3a) are linear. It has been suggested that conversion of mechanical forces into biochemical signals in cells occurs through mechanical deformation of proteins that bind signaling molecules (Na et al. 2008); therefore, we quantify mechanical force transmission through the stress fiber in the model by evaluating the stress due to deformation (strain) at the nucleus. For transverse motion, this deformation-related stress has both prestress and elasticity contributions; thus, it is given by:

$$\sigma_{def,v} \equiv \sigma_p \frac{\partial w_v}{\partial x}\Big|_{x=L} - \frac{EI}{A} \frac{\partial^3 w_v}{\partial x^3}\Big|_{x=L}. \tag{4}$$

On the other hand, for axial motion, only elasticity contributes to deformation, leading to:

$$\sigma_{def,l} \equiv E \frac{\partial w_l}{\partial x}\Big|_{x=L}. \tag{5}$$

We examined mechanical signal transmission dynamics by solving Eqs. (2) and (3) numerically for both a constant and an oscillatory applied force and for representative stress fiber geometric parameters and viscoelastic properties (for details, see Hwang and Barakat 2012). Figures. 3B and C depict the temporal evolution of deformation-related stress at the nucleus for transverse and axial motions, respectively. For both transverse and axial motions, the deformation-related stress at the nucleus gradually increases in time due to the applied forcing; however, the transverse motion exhibits a much faster response. For transverse motion, the deformation-related stress at the nucleus attains the applied stress (20 Pa) within a few milliseconds, whereas axial motion requires several seconds for transmission of the same applied stress to the nucleus.

In the case of oscillatory forcing, the deformation-related stress, as in the case of constant forcing, exhibits an initial transient before it eventually saturates to a time-periodic steady state response. Importantly, the saturation amplitude of the deformation-related stress at the nucleus is found to be strongly dependent on the forcing frequency. Figures 3D and E illustrate the saturation amplitude of deformation-related stress at the nucleus as a function of forcing frequency for transverse and axial forcing, respectively. In both the axial and transverse directions, a low-frequency (<0.1 Hz) mechanical stimulus is transmitted to the nucleus without decay in its amplitude, whereas a high-frequency mechanical stimulus undergoes significant decay in amplitude. This implies that individual stress fibers act as low-pass filters of mechanical forcing. Interestingly, transverse motion exhibits a much broader filter width than axial motion: the filter width for transverse motion extends to $f \sim 1000$ Hz whereas the one for axial motion extends to only $f \sim 1$ Hz.

Dimensional analysis of Eqs. (2) and (3) provides important physical insight into the force transmission dynamics. It can be readily shown that the contribution of inertial forces for both transverse and axial motion (the left-hand sides of Eqs. (2a) and (3a)) is negligibly small compared to the restoring forces (prestress and elasticity) and the damping forces (material and cytosolic viscosity). This implies that the time scales for force transmission can be simply obtained by balancing the restoring force terms with the damping force terms for both transverse and axial motion. Such a balance provides the following time scale for transverse motion of the stress fiber:

$$\tau_v \sim \frac{\gamma}{\sigma_p} \left(\frac{R}{L} \right)^2 \sim 1 \, m\sec, \tag{6}$$

and the following time scale for axial motion:

$$\tau_l \sim \frac{\gamma}{E} \sim 4 \sec, \tag{7}$$

where R is the stress fiber radius. These time scales obtained from a simple dimensional analysis are consistent with the numerical simulation results (Figs. 3B and C). A parametric study on elasticity, prestress, material and cytosolic viscosities also confirms the same conclusions (Hwang and Barakat 2012). These analyses suggest that the dynamics of force transmission through transverse and axial motions are significantly different from one other: a transverse mechanical force is mainly mediated by prestress while axial force transmission is propelled by elasticity of the stress fiber. These different propelling processes are counteracted by material viscosity which acts as a damping force. It is interesting to note that mechanical signal transmission through transverse motion is found to be strongly dependent on stress fiber aspect ratio (R/L) (see Eq. (6)). The fact that this aspect ratio is small is an essential reason why mechanical stimulus transmission through transverse motion of the stress fiber occurs so rapidly. Interestingly, the time scale for transverse motion in Eq. (6) is inversely proportional to the square of the length, implying that for the same stress fiber radius, a longer stress fiber would transmit mechanical signals more rapidly than a shorter one.

The results of the model provide valuable insight into the mechanisms of mechanical force transmission via stress fibers. It had previously been conjectured that mechanical stress transmission via stress fibers occurs through the propagation of elastic waves (Na et al. 2008, Wang et al. 2009). However, the time scales for transverse and axial motion (τ_v and τ_l) obtained in our model are several orders of magnitude larger than the time scale obtained from the elastic wave conjecture ($\tau \sim \sqrt{\rho L^2 / E} \sim 0.3\,\mu\mathrm{sec}$; obtained by dimensional analysis after neglecting material and cytosolic damping). Thus, our analysis suggests that viscous damping (primarily stress fiber material viscosity) plays an integral role in the dynamics of mechanical signal transmission. Considering this important role of the damping force, only mechanical stimulus transmission through transverse motion of the fiber provides a time scale (i.e., τ_v) that is consistent with the experimental observation that mechanical deformations travel a distance of 50 μm in less than 300 ms (Na et al. 2008). In contrast, the time scale for axial motion (τ_l) is significantly longer than this experimental observation. It should be noted that the time scale for axial motion τ_l is consistent with the experimental observation that viscoelastic retraction of a stress fiber in the axial direction occurs with a time scale of 4–6 s (Kumar et al. 2006). Taken together, the present results suggest that mechanical force transmission through transverse stress fiber motion is the only pathway for very rapid mechanotransduction and that stress fiber prestress is likely its main propeller, while material viscosity represents the primary limitation for the speed of its propulsion.

The results for oscillatory forcing can also be analyzed by dimensional analysis in a similar manner. Since the time scale is an inherent feature of a given system, its inverse provides the characteristic frequency of the system. Thus, the characteristic frequencies for transverse and axial motion f_v and f_l are as follows:

$$f_v \sim \frac{\sigma_p}{\gamma}\left(\frac{L}{R}\right)^2 \sim 1000 \; Hz, \tag{8}$$

$$f_l \sim \frac{E}{\gamma} \sim 0.25 \; Hz. \tag{9}$$

Frequencies of physiological time-periodic mechanical stimuli often fall in the range 0.1–10 Hz (e.g., cardiac and respiratory frequencies). The frequency scaling in Eqs. (8) and (9) implies that only transverse stress fiber motion allows such a time-periodic physiological mechanical signal to travel a long distance within a cell. In contrast, axial motion of stress fibers will not allow a physiologically relevant time-periodic mechanical signal to propagate deeply into a cell because material viscosity rapidly dampens the amplitude of the mechanical signal. When prestress (σ_p) is reduced, the critical frequency for transverse stress fiber motion is reduced accordingly. This implies that dissipating prestress prevents long-distance mechanical signal transmission of physiological time-periodic mechanical stimuli. This prediction is in qualitative agreement with the experimental finding that dissipation of stress fiber prestress by pharmacological agents inhibits long-distance mechanical stimulus transmission (Hu et al. 2003, 2005, Wang and Suo 2005).

Conclusions and Future Directions

Mechanical forces on EC surfaces activate intracellular signaling cascades that play an important role in regulating cell structure and function. Propagation of signaling molecules within cells occurs via diffusion-reaction processes, and these processes are relatively slow (time constant of at least several seconds). In light of experimental observations that mechanical forces induce very rapid responses within cells (response time of order 100 ms) at distances far away from the point of force application (>10 μm) (Na et al. 2008), there is a need to invoke alternative mechanisms for rapid long-distance intracellular transmission of mechanical signals. It has been proposed that the cytoskeleton provides such a mechanism although how mechanical signal transmission via the cytoskeleton occurs remains poorly understood. The mathematical models described in this chapter aim to provide insight into the dynamics of force transmission via the cytoskeleton and to shed light onto the physical processes that govern this transmission.

It has been conjectured in the literature that cytoskeleton-mediated mechanical signal transmission in cells occurs via the propagation of elastic stress waves in actin stress fibers; however, stress fibers exhibit prominent viscoelastic behavior. Our model suggests that the elastic wave conjecture is unlikely to be the mechanism for mechanical force transmission in cells. Rather, force transmission appears to be propelled by prestress in the cytoskeleton provided by molecular motor proteins, and the velocity of this propulsion is limited by internal damping provided primarily by the material viscosity of the viscoelastic cytoskeletal elements.

Our current models make a number of simplifying assumptions. For instance, we assume that the viscoelastic properties of the various intracellular structures involved in force transmission including the cytoskeleton remain constant. There is evidence that exposure of ECs to mechanical stimulation alters the mechanical properties of both the nucleus and the cytoskeleton (Deguchi et al. 2005, Sato et al. 1987, 1996); however, these changes appear to require hours to occur. Because sensing of mechanical forces at the EC surface and transmission of these forces to the intracellular space via the cytoskeleton occur over time scales that are considerably shorter than that, the assumption of constant mechanical properties appears justified. It should also be noted that the representation of the cytoskeleton in our present models focuses exclusively on actin stress fibers in the spatiotemporal model and on stress fibers and microtubules in the temporal model. Linker proteins among different cytoskeletal filament networks and the intermediate filament cytoskeleton have not yet been included in either model, even though they are likely to be actively involved in the EC response to flow (Helmke et al. 2003, Mott and Helmke 2007). Therefore, an important future direction is to incorporate these proteins into the models.

Our spatiotemporal model results for a single stress fiber indicate that only transverse motion of the stress fiber allows very rapid long-distance force transmission. For a network of linked stress fibers, one can envision particular linking patterns that preserve rapid long-distance force transmission with dynamics that are consistent with those observed in experiments. A representative example is the case of several stress fibers aligned in parallel and linked together at their ends. In this case, a force applied in the transverse direction will lead to transverse motion in each of the stress fibers, and force transmission will not be slowed down due to the absence of axial motion in any of the linked stress fibers (Hwang et al. 2012). In contrast, other links that allow the axial motion of one or more stress fibers to interfere with the transverse motion of other stress fibers would be expected to slow down force transmission dynamics. The simplest representative situation is the case where two stress fibers are aligned perpendicular to one another with one of them constrained to move only in the axial direction. In this case, the network of two stress fibers would be heavily dependent on the viscoelastic dynamics of axial motion because the material damping force for axial motion is much larger than that for transverse motion. Therefore, force transmission through this simple network would be significantly delayed (Hwang et al. 2012). This notion suggests that the topology of a cytoskeletal network is an important determinant of force transmission dynamics. Therefore, future modeling efforts need to quantitatively describe the effect of cytoskeletal network topology on cytoskeleton-mediated force transmission dynamics.

The cytoskeleton in living cells is highly dynamic and active. Actin filaments and microtubules continuously polymerize and depolymerize. The presence of molecular motors also leads to a highly dynamic prestress environment which is critical in determining cytoskeletal organization and intracellular transport. Moreover, these processes interact very actively with the extracellular matrix via focal adhesions. For example, formation of focal adhesions has been shown to be force-dependent (Balaban et al. 2001, Choquet et al. 1997) and to involve changes in the cytoskeleton. In all these very dynamic processes, multiple signaling pathways are tightly coupled together, rendering modeling of the cytoskeleton very challenging. However, understanding these very complex processes plays an essential role in regulating cell migration, organization, and adaptation. Therefore, developing relevant mathematical models for these processes promises to improve our understanding not only of cellular mechanotransduction but also of important processes such as vascular wound healing, morphogenesis, and angiogenesis.

Acknowledgment

This work was supported in part by a permanent endowment in Cardiovascular Cellular Engineering from the AXA Research Fund.

List of Abbreviations

EC	endothelial cell
TPMM	three parameter Maxwell model
CCAP	cell-cell adhesion protein
FAS	focal adhesion site
N	nucleus
A	actin
M	microtubule
SNR	signal-to-noise ratio
MS	mechanosensor
F	force
u	deformation
k	spring constant

μ	viscosity
σ_p	prestress
L	stress fiber length
f	force applied to stress fiber
x_0	point of force application on stress fiber
f	frequency
ρ	density of stress fiber
w	stress fiber displacement
A	cross-sectional area of stress fiber
V_p	restoring transverse force by prestress
V_E	restoring transverse force by flexural rigidity
V_{vis}	internal transverse damping force by flexural material viscosity
C	cytosolic resistance coefficient
σ_f	applied stress
δ	Dirac delta function
T_E	restoring longitudinal force by elastic modulus
T_{vis}	internal longitudinal damping force by material viscosity
E	stress fiber elastic modulus
I	stress fiber second moment of inertia
γ	stress fiber internal (material) viscosity
R	stress fiber radius
τ	characteristic time scale
Subscripts	
v	vertical
l	longitudinal

References

Asakura, T. and T. Karino. 1990. Flow patterns and spatial distribution of atherosclerotic lesions in human coronary arteries. Circ Res 66: 1045–1066.

Balaban, N.Q., U.S. Schwarz, D. Riveline et al. 2001. Force and focal adhesion assembly: a close relationship studied using elastic micropatterned substrates. Nat Cell Bio 3: 466–472.

Barakat, A.I., E.V. Leaver, P.A. Pappone et al. 1999. A flow-activated chloride-selective membrane current in vascular endothelial cells. Circ Res 85: 820–828.

Boal, D.H. 2002. Mechanics of the Cell. Cambridge University Press, Cambridge, UK.

Brangwynne, C.P., F.C. MacKintosh, S. Kumar et al. 2006. Microtubules can bear enhanced compressive loads in living cells because of lateral reinforcement. J Cell Biol 173: 733–741.

Chappell, D.C., S.E. Varner, R.M. Nerem et al. 1998. Oscillatory shear stress stimulates adhesion molecule expression in cultured human endothelium. Circ Res 82: 532–539.

Chien, S. 2007. Mechanotransduction and endothelial cell homeostasis: the wisdom of the cell (review). Am J Physiol 292: H1209–H1224.

Choquet, D., D.P. Felsenfeld and M.P. Sheetz. 1997. Extracellular matrix rigidity causes strengthening of integrin-cytoskeleton linkages. Cell 88: 39–48.

Davies, P.F. 1995. Flow-mediated endothelial mechanotransduction. Physiol Rev 75: 519–560.

Davies, P.F., J.A. Spaan and R. Krams. 2005. Shear stress biology of the endothelium. Ann Biomed Eng 33: 1714–1718.

Deguchi, S., K. Maeda, T. Ohashi et al. 2005. Flow-induced hardening of endothelial nucleus as an intracellular stress-bearing organelle. J Biomech 38: 1751–1759.

Fabry, B., G.N. Maksym, J.P. Butler et al. 2001. Scaling the microrheology of living cells. Phys Rev Lett 87: 148102.

Fung, Y.C. 1981. Biomechanics: Mechanical Properties of Living Tissues. Springer-Verlag, New York, USA.

Gautam, M., Y. Shen, T.L. Thirkill et al. 2006. Flow-activated chloride channels in vascular endothelium: shear stress sensitivity, desensitization dynamics, and physiological implications. J Biol Chem 281: 36492–36500.

Gittes, F. and F.C. MacKintosh. 1998. Dynamic shear modulus of a semiflexible polymer network. Phys Rev E 58: R1241–R1244.

Hahn, C. and M.A. Schwartz. 2009. Mechanotransduction in vascular physiology and atherogenesis. Nat Rev Mol Cell Bio 10: 53–62.

Helmke, B.P. 2005. Molecular control of cytoskeletal mechanics by hemodynamic forces. Physiology 20: 43–53.

Helmke, B.P., R.D. Goldman and P.F. Davies. 2000. Rapid displacement of vimentin intermediate filaments in living endothelial cells exposed to flow. Circ Res 86: 745–752.

Helmke, B.P., D.B. Thakker, R.D. Goldman et al. 2001. Spatiotemporal analysis of flow-induced intermediate filament displacement in living endothelial cells. Biophys J 80: 184–194.

Helmke, B.P., A.B. Rosen and P.F. Davies. 2003. Mapping mechanical strain of an endogenous cytoskeletal network in living endothelial cells. Biophys J 84: 2691–2699.

Hsu, H.J., C.F. Lee, A. Locke et al. 2010. Stretch-induced stress fiber remodeling and the activations of JNK and ERK depend on mechanical strain rate, but not FAK. PLoS One 5: e12470.

Hu, S., J. Chen, B. Fabry et al. 2003. Intracellular stress tomography reveals stress focusing and structural anisotropy in cytoskeleton of living cells. Am J Physiol 285: C1082–C1090.

Hu, S., J. Chen, J.P. Butler et al. 2005. Prestress mediates force propagation into the nucleus. Biochem Biophy Res Commun 329: 423–428.

Hwang, Y. and A.I. Barakat. 2012. Dynamics of mechanical signal transmission through prestessed actin stress fibers. PLoS One 7: e35343.

Hwang, Y., C.L.M. Gouget and A.I. Barakat. 2012. Mechanisms of cytoskeleton-mediated mechanical signal transmission in cells. Commun Integrat Biol 5: 1–5.

Ingber, D.E. 1997. Tensegrity: the architectural basis of cellular mechanotransduction. Annu Rev Physiol 59: 575–599.

Ingber, D.E. 2008. Tensegrity-based mechanosensing from macro to micro. Prog Biophys Mol Biol 97: 163–179.

Kim, T., W. Hwang, H. Lee et al. 2009. Computational analysis of viscoelastic properties of crosslinked actin networks. PLoS Comp Biol 5: e1000439.

Kollmansberger, P. and B. Fabry. 2011. Linear and nonlinear rheology of living cells. Annu Rev Mater Res 41: 75–97.

Ku, D.N., D.P. Giddens, C.K. Zarins et al. 1985. Pulsatile flow and atherosclerosis in the human carotid bifurcation: positive correlation between plaque location and low oscillating shear stress. Arteriosclerosis 5: 293–302.

Kumar, S., I.Z. Maxwell, A. Heisterkamp et al. 2006. Viscoelastic retraction of single living stress fibers and its impact on cell shape, cytoskeletal organization and extracellular matrix mechanics. Biophys J 90: 3762–3773.

Langille, B.L. and F. O'Donnell. 1986. Reductions in arterial diameter produced by chronic decreases in blood flow are endothelium-dependent. Science 231: 405–407.

Li, S., B.P. Chen, N. Azuma et al. 1999. Distinct roles for the small GTPases Cdc42 and Rho in endothelial responses to shear stress. J Clin Invest 103: 1141–1150.

Li, S., P.J. Butler, Y. Wang et al. 2002. The role of the dynamics of focal adhesion kinase in the mechanotaxis of endothelial cells. Proc Natl Acad Sci USA 99: 3546–3551.

Libby, P. 2002. Inflammation in atherosclerosis. Nature 420: 868–874.

Lieu, D.K., P.A. Pappone and A.I. Barakat. 2004. Differential membrane potential and ion current responses to different types of shear stress in vascular endothelial cells. Am J Physiol 286: C1367–C1375.

Lim, C.T., E.H. Zhou and S.T. Quekb. 2006. Mechanical models for living cells. J Biomech 39: 195–216.

Malek, A.M. and S. Izumo. 1996. Mechanism of endothelial cell shape change and cytoskeletal remodeling in response to fluid shear stress. J Cell Sci 109: 713–726.

Maniotis, A.J., C.S. Chen and D.E. Ingber. 1997. Demonstrations of mechanical connections between integrins, cytoskeletal filaments, and nucleoplasm that stabilize nuclear structure. Proc Natl Acad Sci USA 94: 849–854.

Mazzag, B. and A.I. Barakat. 2011. The effect of noisy flow on endothelial cell mechanotransduction: a computational study. Ann Biomed Eng 39: 911–921.

Mazzag, B., J.S. Tamaresis and A.I. Barakat. 2003. A model for shear stress sensing and transmission in vascular endothelial cells. Biophys J 84: 4087–4101.

Meyer, C.J., F.J. Alenghat, P. Rim et al. 2000. Mechanical control of cyclic AMP signalling and gene transcription through integrins. Nat Cell Biol 2: 666–668.

Mofrad, M.R.K. 2009. Rheology of the Cytoskeleton. Annu Rev Fluid Mech 41: 433–453.

Mott, R.E. and B.P. Helmke. 2007. Mapping the dynamics of shear stress-induced structural changes in endothelial cells. Am J Physiol 293: C1616–C1626.

Na, S., O. Collin, F. Chowdhury et al. 2008. Rapid signal transduction in living cells is a unique feature of mechanotransduction. Proc Natl Acad Sci USA 105: 6626–6631.

Nerem, R.M., M.J. Levesque and J.F. Cornhill. 1981. Vascular endothelial morphology as an indicator of the pattern of blood flow. J Biomech Eng 103: 172–176.

Olesen, S.P., D.E. Clapham and P.F. Davies. 1988. Hemodynamic shear stress activates a K^+ current in vascular endothelial cells. Nature 331: 168–170.

Ookawa, K., M. Sato and N. Ohshima. 1992. Changes in the microstructure of cultured porcine aortic endothelial cells in the early stage after applying a fluid-imposed shear stress. J Biomech 25: 1321–1328.

Pohl, U., J. Holtz, R. Busse et al. 1986. Crucial role of endothelium in the vasodilator response to increased flow *in vivo*. Hypertension 8: 37–44.

Salwen, S.A., D.H. Szarowski, J.N. Turner et al. 1998. Three-dimensional changes of the cytoskeleton of vascular endothelial cells exposed to sustained hydrostatic pressure. Med Biol Eng Comput 36: 520–527.

Sato, M., M.J. Levesque and R.M. Nerem. 1987. Micropipette aspiration of cultured bovine aortic endothelial cells exposed to shear stress. Arterioscler Thromb Vasc Biol 7: 276–286.

Sato, M., N. Ohshima and R.M. Nerem. 1996. Viscoelastic properties of cultured porcine aortic endothelial cells exposed to shear stress. J Biomech 29: 461–467.

Semmrich, C., T. Storz, J. Glaser et al. 2007. Glass transition and rheological redundancy in F-actin solutions. Proc Natl Acad Sci USA 104: 20199–20203.

Shyy, J.Y. and S. Chien. 2002. Role of integrins in endothelial mechanosensing of shear stress. Circ Res 91: 769–775.

Silberberg, Y.R., A.E. Pelling, G.E. Yakubov et al. 2008. Mitochondrial displacements in response to nanomechanical forces. J Mol Recognit 21: 30–36.

Tarbell, J.M. and M.Y. Pahakis. 2006. Mechanotransduction and the glycocalyx. J Intern Med 259: 339–350.

Tedgui, A. and Z. Mallat. 2001. Anti-inflammatory mechanisms in the vascular wall. Circ Res 88: 877–887.

Tschoegl, N.W. 1989. The Phenomenological Theory of Linear Viscoelastic Behavior: An Introduction. Springer-Verlag, Berlin, Germany.

Tzima, E., M.A. Del Pozo, W.B. Kiosses et al. 2002. Activation of Rac1 by shear stress in endothelial cells mediates both cytoskeletal reorganization and effects on gene expression. EMBO J 21: 6791–6800.

Tzima, E., M. Irani-Tehrani, W.B. Kiosses et al. 2005. A mechanosensory complex that mediates the endothelial cell response to fluid shear stress. Nature 437: 426–431.

Wang, N. and Z. Suo. 2005. Long-distance propagation of forces in a cell. Biochem Biophy Res Commun 328: 1133–1138.

Wang, N., J.P. Butler and D.E. Ingber. 1993. Mechanotransduction across the cell surface and through the cytoskeleton. Science 260: 1124–1127.

Wang, N., J.D. Tytell and D.E. Ingber. 2009. Mechanotransduction at a distance: mechanically coupling the extracellular matrix with the nucleus. Nat Rev Mol Cell Biol 10: 75–82.

Wechezak, A.R., R.F. Viggers and L.R. Sauvage. 1985. Fibronectin and F-actin redistribution in cultured endothelial cells exposed to shear stress. Lab Invest 53: 639–647.

Weinbaum, S., J.M. Tarbell and E.R. Damiano. 2007. The structure and function of the endothelial glycocalyx layer. Annu Rev Biomed Eng 9: 121–167.

Yamada, H. and H. Ando. 2007. Orientation of apical and basal actin stress fibers in isolated and subconfluent endothelial cells as an early response to cyclic stretching. Mol Cell Biomech 4: 1–12.

6

The Functional Role of the Microtubule/Microfilament Cytoskeleton in the Regulation of Pulmonary Vascular Endothelial Barrier

*Irina B. Alieva[1,2,3] and Alexander D. Verin[3,4,a,]**

Introduction

The endothelial cells (EC) lining the vessels are in close contact with each other, rendering the vascular wall into a tight barrier, which control such diverse processes as vascular tone, homeostasis, adhesion of platelets and leukocytes to the vascular wall and permeability of vascular wall for cells and fluids (Bazzoni and Dejana 2004, Dudek and Garcia 2001, Komarova

[1]A.N. Belozersky Institute of Physical and Chemical Biology, Moscow State University, Moscow, Russia.
[2]Department of Histology, Cytology and Embryology, Medical Faculty People's Friendship University of Russia, Moscow, Russia.
[3]Vascular Biology Center, Georgia Health Science University, Augusta, GA 30912, USA.
[4]Pulmonary and Critical Care Medicine, Georgia Health Science University, Augusta, GA 30912, USA.
[a]Email: AVERIN@georgiahealth.edu
*Corresponding author

and Malik 2010, Ware and Matthay 2000). Lung endothelium regulates movement of fluid, macromolecules, and leukocytes into the interstitium and subsequently into the alveolar air spaces. The integrity of the pulmonary EC monolayer, therefore, is a critical requirement for preservation of pulmonary function. This barrier is dynamic and highly susceptible to the regulation, by various stimuli, of physiological and pathological origin. Any breach in the EC barrier results in leakage of fluid from the lumen of the vessels into the interstitial tissue and/or alveolar lumen, severely impairing gas exchange. Disruption of endothelial barrier occurs during inflammatory disease states such as acute lung injury (ALI) and its more severe form, acute respiratory distress syndrome (ARDS), which remains a major cause of morbidity and mortality with an overall mortality rate of 30–40% (Ware and Matthay 2000), results in the uncontrolled movement of fluid and macromolecules into the interstitium and pulmonary air spaces causing pulmonary edema (Ermert et al. 1995). Data of literature have proved that normal functioning of the endothelial barrier is provided by the balance between contracting and stretching forces generated by EC cytoskeleton (Bogatcheva and Verin 2008, Dudek and Garcia 2001, Komarova et al. 2007). In this review, we will analyze the cytoskeletal elements whose reorganization affects endothelial permeability, and emphasize the role of microtubules/microfilament crosstalk in lung EC barrier regulation.

Endothelial Permeability and EC Cytoskeleton

Reorganization of the endothelial cytoskeleton, which is composed of actin filaments, microtubules (MT) and intermediate filaments, leads to alteration in cell shape and provides a structural basis for increase of vascular permeability, which has been implicated in the pathogenesis of many diseases including asthma, sepsis, and acute lung injury (Dudek and Garcia 2001, Lee and Gotlieb 2003a, b, Lum and Malik 1996). The majority of transendothelial trafficking of fluids and leukocytes occurs via the paracellular pathway (Dudek and Garcia 2001). Paracellular gap formation is regulated by the balance of competing contractile forces, which generate centripetal tension, and adhesive cell-cell and cell-matrix tethering forces, which together regulate cell shape changes (Dudek and Garcia 2001). Both competing forces in this model are linked through actin microfilaments, which are connected to multiple membrane adhesive proteins of the zona occludens and zona adherens, functional intercellular proteins and focal adhesion complex proteins (Bogatcheva and Verin 2008).

F-actin microfilaments are dynamic structures and the time of their exchange near the cell margin is close to a few seconds (Amann and Pollard 2000); nevertheless, they effectively regulate the shape of the cells *in vitro* and provide permanent maintenance of the cells in the sprawled state.

In endothelial cells, dynamic actin polymerization/depolymerization allows for the rapid reorganization of actin structures and the transition from the quiescent phenotype, characterized by thick cortical actin ring and the decreasing or absence of stress fibers, to the activated contractile cell phenotype with thin or no cortical actin and abundant stress fibers. Cell adhesions—focal adhesion and intracellular contacts (Adams et al. 1998, Birukova et al. 2004d)—are also able to undergo quick assembly and disassembly within a few minutes.

Disruption of actin cytoskeleton dramatically increases EC permeability *in vitro* (Garcia et al. 2001). In perfused rabbit lungs, selective disruption of actin microfilaments leads to a significant increase in vascular permeability and marked interstitial edema formation, implicating direct involvement of microfilaments in the regulation of lung vascular permeability *in vivo* (Ermert et al. 1995).

While information is limited about involvement of microtubule component in the vascular barrier maintenance *in vivo*, intravenous administration of anti-cancer drugs and MT inhibitors (such as the vinca alkaloids) can lead to the sudden development of pulmonary edema in breast cancer patients suggesting the involvement of MT network in the regulation of lung permeability (Cattan and Oberg 1999). Published data (Bogatchev et al. 2007, Gorshkov et al. 2012) indicate that disruption of microtubule structure causes vascular leak in mice and triggers Rho and p38 MAPK activation, which potentially can cause contraction and barrier dysfunction via alterations in the actomyosin cytoskeleton. However, the precise linkage between the microtubule network and contractile cytoskeleton is not fully explored.

Rho GTPases and Endothelial Cytoskeleton/permeability Regulation

The members of the Rho family Ras homology—small GTPases, Rho, Rac, and Cdc42—act as molecular switches, cycling between an active GTP-bound and inactive GDP-bound state and interact with a number of downstream targets that in turn trigger various intracellular processes, such as proliferation, cell motility, gene expression, and actin remodeling (Bishop and Hall 2000, Hall 1998, Narumiya et al. 1997). In particular, they are responsible for transducing extracellular signals to control of the actin cytoskeleton in most cell types (Tapon and Hall 1997). It is interesting that in the endothelium Rho GTPases regulate endothelial cell barrier function in different ways (Wojciak-Stothard and Ridley 2002, Wojciak-Stothard et al. 2001) while Rac and Cdc42 induce lamellipodia and filopodia formation (Nakahara et al. 2003, Nobes and Hall 1995, Wojciak-Stothard and Ridley

2002) and have EC barrier-protective effects *in vitro* (Birukova et al. 2007a, b, 2012, Jacobson et al. 2006, Kouklis et al. 2004), Rho proteins (RhoA, RhoB, RhoC), and RhoA in particular, induce F-actin stress fiber and focal adhesion formation stimulating contractile mechanisms, leading to barrier dysfunction (Birukova et al. 2004, Bogatcheva et al. 2007). However, some studies have suggested that Rho inhibition leads to endothelial barrier compromise with minimal effect on endothelial contractile properties (Vouret-Craviari et al. 1998).

Interestingly overexpression of the wild-type RhoA causes disruption of microtubules in cancer cells (Son et al. 2000), and RhoA activation induced phosphorylation of the microtubule-associated protein tau, its dissociation from microtubules and microtubule destabilization (Sayas et al. 1999). These findings support the critical role of Rho-dependent pathway in the regulation of microtubule integrity.

Rho family GTPases are activated by the Dbl family of guanosine nucleotide exchange factors (GEFs), which play a major role in Rho regulation by a variety of external stimuli (Bishop and Hall 2000, Takuwa 2002, Wettschureck and Offermanns 2002, Zheng 2001). Identification of p115RhoGEF established the direct link between heterotrimeric G-proteins and the monomeric small GTPase Rho (Hart et al. 1998, Kozasa et al. 1998). Binding of p115RhoGEF through the N-terminal RGS domain to activated G12/13 α subunit induces p115RhoGEF membrane translocation and stimulates its GEF activity (Bhattacharyya and Wedegaertner 2003). G12/13-p115RhoGEF signaling cascade is directly involved in thrombin-induced EC permeability changes (Birukova et al. 2004).

GEF-H1 has been characterized as a Rho-specific GEF, which localizes on microtubules and exhibits Rho-specific activity (Ren et al. 1998). In its MT-bound state, the guanosine-exchange activity of GEF-H1 is suppressed, whereas GEF-H1 release caused by MT disassembly stimulates Rho-specific GEF activity (Krendel et al. 2002). It was shown that GEF-H1 is involved in MT-mediated Rho activation and barrier dysfunction (Kratzer et al. 2012).

Among several Rho targets, Rho-kinase (p160 ROCK) is directly involved in actin stress fiber formation and increasing of myosin light chains (MLC) phosphorylation thus initiating actomyosin contraction (Amano et al. 1996, 1997, Birukova et al. 2004a, 2004c, 2004d, Essler et al. 1998, Parizi et al. 2000). ROCK increases MLC phosphorylation by two potential mechanisms: direct phosphorylation of MLC at Ser-19 and indirectly via phosphorylation of the myosin-associated phosphatase type 1 (PP1), myosin-binding subunit (MYPT1) at Thr-695, which leads to PP1 inactivation and dissociation from myosin (Amano et al. 1996, Csortos et al. 2007, Fukata et al. 2001, Kimura et al. 1996). RhoA activation also induces actin polymerization and stress fiber formation by inhibiting the actin binding protein cofilin, independent of the ROCK-induced increase in MLC phosphorylation, but dependent on LIM kinase activity (Gorovoy et al. 2005, Maekawa et al. 1999).

The Role of MLC Phosphorylation in EC Barrier Regulation

A key EC contractile event in several models of agonist-induced barrier dysfunction is the phosphorylation of MLC (Garcia et al. 1995b, Moy et al. 1993, Sheldon et al. 1993, Wysolmerski and Lagunoff 1991). The inflammatory agonists such as thrombin and histamine produce rapid increases in MLC phosphorylation, actomyosin interaction, and endothelial permeability (Garcia et al. 1995b, Moy et al. 1993). Thrombin increased EC centripetal tension and MLC phosphorylation, which peaked at 2 min and returned to nearly control levels by 60 min indicating involvement of both MLC kinase and MLC phosphatase (MLCP) activity in the thrombin response (Garcia et al. 1995a, b, Verin et al. 1995, 1998). MLC kinases, which are able to phosphorylate MLC *in vitro* and *in vivo*, include Ca^{2+}-calmodulin dependent MLC kinase (MLCK) and ROCK (Amano et al. 1996, Goeckeler and Wysolmerski 1995, Wysolmerski and Lagunoff 1990, 1991). Both kinases phosphorylate MLC at Ser-19 resulting in the initiation of contractile activity which finally leads to movement of actin and myosin filaments past one another.

Dephosphorylation of MLC is provided by MLCP. Smooth muscle (SM) MLCP is a holoenzyme composed of three subunits: a catalytic subunit (CS1) of 38 kDa and two non-catalytic subunits of 21 and 110–130 kDa (Alessi et al. 1992, Hartshorne 1998, Shimizu et al. 1994, Shirazi et al. 1994). The larger one, MYPT1, binds to the catalytic subunit and targets it to MLC insuring substrate specificity (Alessi et al. 1992, Shimizu et al. 1994). The function of the smaller subunit is unknown. ROCK has been proposed to mediate the inhibition of MLCP in response to various agonists (Hartshome et al. 1998, Kimura et al. 1996). Phosphorylation of the regulatory site (Thr-695 in chicken MYPT1) by ROCK induces inhibition of MLCP activity (Csortos et al. 2007, Essler et al. 1998, 1999, Verin et al. 2001). Inhibition of MLCP activity leads to a net increase in MLC phosphorylation induced by the MLCK and ROCK; increased MLC phosphorylation increases actomyosin interaction, causing F-actin to bundle into stress fibers and increase endothelial permeability (Birukova et al. 2004d, Csortos et al. 2007, Verin et al. 1995, 2000a). Thus, the level of MLC phosphorylation is subject to regulation by two major intracellular factors, Ca^{2+} level and the activity of a small G protein Rho, regulatory protein for actin filaments and microtubules.

In addition, several other factors are likely involved in modulation of MLC phosphorylation level. Protein kinase C (PKC) seems to be able to phosphorylate MLC in bovine pulmonary endothelium; this phosphorylation occurs at the sites, distinct from MLCK-specific sites, and is associated with the formation of actin network, distinct from classical stress fiber pattern (Bogatcheva et al. 2003). Although not able to produce contractile response, this network formation is concomitant with the increase in EC permeability (Bogatcheva et al. 2003, Verin et al. 2000).

MLC-independent Contractile Mechanisms of EC Barrier Regulation

Phorbol esters, such as PMA, cause a slow developing, but sustained increase in tension, without a further increase in MLC phosphorylation in smooth muscles (Sato et al. 1992, Whitney et al. 1995). This sustained contraction suggests an important role for additional proteins in the regulation of the contractile apparatus.

Caldesmon (CaD) contains distinct binding sites for actin and myosin, thereby potentially regulating actomyosin interactions and promoting actin filament formation in the absence of MLC phosphorylation (Marston and Redwood 1991, Sobue and Sellers 1991). *In vitro* studies indicate that in the absence of MLC phosphorylation and in the presence of low [Ca^{2+}], CaD binding to actin filaments inhibits myosin Mg^{2+}-ATPase activity and actin/myosin binding (Marston and Redwood 1991, Pritchard and Marston 1993, Sobue and Sellers 1991). PMA-induced phosphorylation of SM CaD correlates with contraction and has been postulated as the on/off switch regulating actin-myosin interactions (Marston and Redwood 1991, Sobue and Sellers 1991). In SM, the Ser/Thr ERK MAP kinases are responsible for phorbol ester-stimulated phosphorylation of CaD (Adam et al. 1989). The responsible kinases and the functional significance of CaD phosphorylation have not been fully investigated in EC. However, there are two potentially key consequences of this event: 1) removal of inhibition of actin-activated myosin ATPase, thus increasing actomyosin interaction 2) remodeling of actin filaments (Marston and Redwood 1991, Sobue and Sellers 1991, Yamashiro et al. 1995). In addition to the ERK MAP kinases, p38 MAPK may also be involved in cytoskeletal protein rearrangement. It is able to phosphorylate CaD at the same sites as ERKs *in vitro* (Hedges et al. 1998). Both p38 and ERK can phosphorylate CaD in activated endothelium (Borbiev et al. 2004, Bogatcheva and Verin 2008, Bogatcheva et al. 2006). In thrombin-challenged EC, CaD was shown to dissociate from myosin (Borbiev et al. 2003), allowing for possible contraction. Inhibition of p38 strengthens CaD-myosin complex formation and attenuates thrombin-induced stress fiber formation and permeability increase (Borbiev et al. 2004). Hereby, CaD phosphorylation represents another potential mechanism by which actomyosin contraction involves in increased transendothelial permeability.

Small heat shock actin-capping protein, HSP-27, phosphorylates by MAP kinase-activated protein kinase 2 (MAPKAP kinase 2), that is in turn phosphorylated by p38 MAP kinase (Guay et al. 1997, Rogalla et al. 1999, Schafer et al. 1998, 1999). Phosphorylation of HSP-27 promotes F-actin formation, membrane blebbing and mediates actin reorganization and cell migration in human endothelium (Guay et al. 1997, Huot et al. 1998, Piotrowicz and Levin 1997, Rousseau et al. 1997). In addition, HSP-27 is phosphorylated *in vivo* in smooth muscle in response to contractile agonists, and contraction of permeabilized cells is inhibited with anti-HSP-27 antibody (Bitar et al. 1991, Larsen et al. 1997). An accumulating body of evidence indicates that HSP27 phosphorylation in p38/MAPKAP-2 pathway is important for the regulation of endothelial cytoskeleton and permeability. In ATP-depleted cells, elevation of HSP-27 phosphorylation induced by heat shock pre-conditioning is associated with the increased stability of F-actin (Loktionova and Kabakov 1998). HSP-27 phosphorylation is also elevated in endothelial cells, treated with several edemagenic and stress-fiber inducing agonists, like thrombin (Tar et al. 2006), hydrogen peroxide (Huot et al. 1997), LPS (Hirano et al. 2004, Kratzer et al. 2012), and TGFβ (Antonov et al. 2012). In hydrogen peroxide-treated cells, the phosphorylation of HSP-27 seems to be a prerequisite of contractile response, as the response depends on both level of HSP-27 expression and p38 activation (Huot et al. 1997). However, the exact mechanism of the contractile response and permeability induction by phospho-HSP-27 remains unclear. For example, in LPS-treated EC, HSP-27 was shown to dissociate from actin upon phosphorylation (Hirano et al. 2004). Consistent with this data an early publication demonstrated that only unphosphorylated HSP-27 binds to microfilaments *in vitro* (Benndorf et al. 1994). Hirano and co-authors speculated that dissociation of HSP-27 from F-actin stress fibers is necessary to promote cell contraction (Hirano et al. 2004), as interaction of dephospho-HSP-27 with actin inhibits its polymerization (Benndorf et al. 1994). However, another study demonstrated that HSP-27 has no effect on actin polymerization *in vitro* and HSP-27-actin interaction depends on their conformation (Panasenko et al. 2003). Further, depletion of HSP-27 did not prevent TGFβ-induced barrier dysfunction in human umbilical vein endothelium (Lu et al. 2006). Whether or not phospho-HSP-27 is directly associated with F-actin and promotes stress fibers formation directly or via interaction with other cytoskeletal proteins, p38 MAPK-mediated HSP-27 phosphorylation is tightly correlated with contractile response and endothelial barrier disruption in most studies.

Role of p38 MAPK Pathway in EC Permeability

P38 MAPK is a mammalian homologue of the HOG1 kinase in Saccharomyces which is activated by osmotic stress (Paul et al. 1997). P38 MAPK belongs to a multigene family of MAP (mitogen-activated protein) Ser/Thr kinases that currently includes more than 12 cloned members (Bogatcheva et al. 2003b, Herlaar and Brown 1999, Paul et al. 1997). All members of MAPK family (often referred as "proline-directed" protein kinases) recognized minimally required consensus sequence Ser/Thr-Pro (Davis 1993) in their protein targets. Upstream activation of p38 MAPK is thought to be via MAP kinase kinases (MAPKK), MKK3, MKK4 and MKK6, which directly phosphorylate the dual site Thr-Gly-Tyr in p38 MAPK (Garrington and Johnson 1999). The MAPKK themselves are regulated by upstream MAPK kinases, such as p21-activated kinase (PAK), TGF-β-activated kinase-1 (TAK1), which can respond to receptor activation and/or small GTPases, such as Rac1 and CDC42 (Garrington and Johnson 1999). P38 MAPK activation regulates a variety of cellular processes including inflammation, apoptosis, differentiation, cell cycle progression, cell migration and contraction (Bogatcheva et al. 2003b, Graves et al. 1995, Herlaar and Brown 1999). Data of literature demonstrated the direct involvement of p38 MAPK in endothelial cytoskeletal remodeling and permeability (Becker et al. 2001, Bogatcheva et al. 2011, Clauss et al. 2001, Ferrero et al. 2001, Garcia et al. 2002, Kevil et al. 2001, Kiemer et al. 2002, Petrache et al. 2001). P38 MAPK downstream targets contain several cytoskeletal proteins such as HSP-27 (via MAPKAP kinase II), caldesmon and tau (Guay et al. 1997, Hedges et al. 1998, Reynolds et al. 1997). It was shown that pertussis toxin- and thrombin-induced EC permeability was temporally linked to p38 MAPK activation and phosphorylation of HSP-27 and caldesmon (Bogatcheva et al. 2007, Borbiev et al. 2004, Garcia et al. 2002). However, exact cytoskeletal targets of p38 MAPK in endothelium remain undetermined.

Information is controversial about link between p38 MAPK and microtubule cytoskeleton. For example, it was shown that microtubule disruption activates MAP kinase(s) through phosphorylation in rat fibroblasts (Shinohara-Gotoh et al. 1991). P38, but not ERK or JNK MAPK, was activated in 3T3 fibroblasts when the cells were arrested in M phase by disruption of the spindle with nocodazole (Takenaka et al. 1998). In contrast, MT-active drugs do not activate p38 MAPK pathway in leukemia cells (Blagosklonny et al. 1999). Microtubule stabilizer, taxol, activates p38 MAPK in human ovarian and breast carcinoma cells, but reduced basal p38 MAPK activity in human epidermal carcinoma cells (Bacus et al. 2001, Seidman et al. 2001, Stone and Chambers 2000). P38 MAPK stimulated MT-mediated adenoviral motility towards nucleus (Suomalainen et al. 2001). The information about the role of p38 MAPK activity in EC microtubule arrangement is limited. However, published data indicated that pharmacological inhibition of p38 MAPK partially protected MT structure from microtubule damage induced by MT inhibitors and stabilization of microtubules attenuated EC permeability induced by edemagenic agonists (Bogatcheva et al. 2007, Petrache et al. 2003) indicating a tight link between p38 MAPK activity and microtubule remodeling.

Effects of cAMP Elevation on EC Barrier Properties

The second messenger cAMP is an important mediator of SM relaxation and cAMP/protein kinase A (PKA) pathway promotes barrier integrity in the endothelium (Garcia et al. 1995b, Moy et al. 1993, Patterson and Garcia 1994, Patterson et al. 2000). Elevation of cAMP level followed by activation of PKA significantly attenuates thrombin-, LPS-, PMA- and pertussis toxin-induced EC barrier dysfunction (Bogatcheva et al. 2009, Patterson and Garcia 1994, Patterson et al. 1994, 2000). Inhibition of either cAMP (by Rp-cAMPS) or PKA (by PKI) significantly increases amount of stress fibers and formation of paracellular gaps indicating barrier compromise (Liu et al. 2001). cAMP-mediated attenuation of thrombin-induced permeability is strongly correlated with a decrease of the level of thrombin-induced MLC phosphorylation and MLCK activity (Garcia et al. 1995b, Verin et al. 1998b) suggesting a barrier protective effect of cAMP targeted on MLC-dependent contractile mechanism of EC barrier dysfunction. EC MLCK sequence includes highly conserved potential phosphorylation sites for PKA in CaM-binding region (Garcia et al. 1997). Augmentation of intracellular cAMP levels markedly enhanced MLCK phosphorylation and reduced kinase activity in EC MLCK immunoprecipitates, suggesting that cAMP affects MLCK activity via PKA-dependent MLCK phosphorylation (Verin et al. 1998). In addition, PKA activation can also facilitate MLCK association with myosin and MLC dephosphorylation (Bindewald et al. 2004).

PKA-mediated modulation of Rho GTPase activity is another potentially important mechanism for regulation of actin cytoskeletal organization (Dong et al. 1998, Lang et al. 1996). Elevation of intracellular cAMP and increased PKA activity attenuates Rho activation via RhoA phosphorylation at Ser-188 (Lang et al. 1996), which decreased Rho association with ROCK (Dong et al. 1998). PKA activation also increases interaction of Rho with Rho-GDP dissociation inhibitor (Rho-GDI) and translocation of Rho from the membrane to the cytosol (Lang et al. 1996, Qiao et al. 2003, Tamma et al. 2003). Thus, the overall effect of PKA on Rho is inhibition of RhoA activity and stabilization of cortical actin cytoskeleton, which promotes EC barrier enhancement. Overall, activation of cAMP/PKA pathway can be considered a perspective strategy to counteract endothelial barrier dysfunction.

Increase in cAMP level represents a classical way of PKA activation. However, it was also shown that PKA activation may be independent of cAMP elevation, but depends upon coupling of specific trimeric G proteins with a member of the PKA-anchoring protein (AKAP) family, followed by release of the PKA catalytic subunit from its regulatory subunit (Niu et al. 2001, Zieger et al. 2001). Recent studies demonstrate that AKAPs are involved in the regulation of EC permeability (Sehrawat et al. 2011). Another mechanism may utilize activation of PKA via stimulation of transcription nuclear factor kappa B (NF-κB) (Zieger et al. 2001). Further, it was demonstrated that ATP-induced EC barrier enhancement is PKA-dependent, but cAMP-independent (Kolosova et al. 2005).

Importantly, that cAMP elevation activates not only PKA, but several other targets including the cAMP-dependent GEF, Epac1, and the small GTPase, Rap1 (de Rooij et al. 1998, Kawasaki et al. 1998). Elevation of intracellular cAMP levels may lead to Rac1 activation (Birukova et al. 2008), which is ultimately involved in endothelial barrier enhancement (Birukova et al. 2007b). The stimulation of both PKA and Epac-1 results in activation of the Rac1-specific GEFs, Tiam1 and Vav2 and Rac 1-mediated EC barrier protective responses (Birukova et al. 2008, 2010) although PKA does not require the activation of Rap1. Further, pharmacological activation of PKA and Epac 1 pathways attenuates LPS-induced EC barrier compromise with combined PKA/Epac 1 activation having an additive effect (Bogatcheva et al. 2009) suggesting that both pathways are independently involved in EC barrier preservation.

Elevation of cAMP level attenuates endothelial hyperpermeability induced by MT inhibitors suggesting an involvement of cAMP-dependent mechanisms in barrier protection after microtubule dissolution. Importantly, treatment of cells with forskolin also significantly attenuates the increase in MLC phosphorylation and stress fiber formation produced by MT inhibitors (Birukova et al. 2004d), implicating the involvement of cAMP in EC barrier protection against MT inhibitors-induced barrier compromise, at least in part, via inhibition of contractile MLC-dependent mechanism. However, it does not preclude the possibility that cAMP/PKA exerts control over additional MLC-independent biochemical reactions that contribute to barrier regulation.

For example, PKA activation stimulates phosphorylation of several actin-binding and focal adhesion proteins, including vasodilator-stimulated phospho-protein VASP. VASP is highly expressed in endothelium (Markert et al. 1996). It is located in cell-cell adherens junctions and cell-matrix focal adhesions where it simultaneously binds to F-actin microfilaments and other cell junction proteins including vinculin, zyxin, ZO1 and α-catenin (Holt et al. 1998, Lawrence et al. 2002, Reinhard et al. 2001). Phosphorylation of VASP by either PKA or protein kinase G (PKG) interferes with actin bundling and stress fiber formation (Harbeck et al. 2000, Krause et al. 2003, Price and Brindle 2000). Depletion of VASP leads to attenuation of ATP-dependent EC barrier enhancement and baseline permeability is significantly increased in VASP-deficient EC implicating the involvement of VASP in the EC barrier regulation (Bogatcheva et al. 2009, Kolosova et al. 2005, Schlegel et al. 2008). However, participation of VASP phosphorylation in PKA/PKG-mediated EC barrier protection remains unclear. Although the level of phospho-VASP seems to correlate with barrier enhancement in adenosine and ATP-treated endothelial cells (Comerford et al. 2002, Kolosova et al. 2005), other studies fail to show that elevated VASP phosphorylation is critical for PKA/PKG-dependent barrier enhancement (Rentsendorj et al. 2008, Schlegel et al. 2008).

Microtubule Network and its Role in Endothelial Permeability

One of the main components of cytoskeleton, microtubules, have important functions in various cellular processes, such as cell shape formation, cell polarization and motility (Bershadsky et al. 2003, 2006a, b, Broussard et al. 2008, Efimov et al. 2007, Prager-Khoutorsky et al. 2007, Schober et al. 2007, Small and Kaverina 2003, Small et al. 2002). Microtubules are composed of α,β-tubulin heterodimers, which assemble in a head-to-tail manner to form a linear protofilament. Protofilaments are subsequently joined through lateral interactions to form the wall of the cylindrical microtubule (Nogales 2000). In many cell types, microtubules are organized in a radial array with their minus-ends anchored at the centrosome and their plus-ends extending toward the cell periphery (Akhmanova and Hoogenraad 2005, Howard and Hyman 2003, Moritz and Agard 2001). Microtubules are highly dynamic, constantly switching between phases of growth and shrinkage mediated by the addition or loss of tubulin dimers from the ends of the microtubule. They are constantly growing or shortening, even when the region of the cytoplasm they occupy is not visibly changed. This property of microtubules is based on binding and hydrolysis of GTP by tubulin subunits (Desai and Mitchison 1997). The ends of individual microtubules undergo growth and disassembly at the distance of a few microns (Mitchison and Kirschner 1984a, b), and all the system continuously interchanges with the pool of dissolved tubulin in the cytoplasm with the characteristic time of exchange of 5–20 min (Vorobjev et al. 1999, Vorobjev et al. 2003). This behavior is called "dynamic instability" (Mitchison and Kirschner 1984a, b, Walker et al. 1988) since each individual microtubule, in contrast to the whole system, is not in a steady-stable condition. The discovery of the microtubule dynamic instability (Mitchison and Kirschner 1984a) has raised a question about their biological feasibility. "Search-and-Capture" model (Kirschner and Mitchison 1986) predicted that during transitions between the growth and shortening of the plus ends, microtubules rapidly explore 3D intracellular space and search for targets to interact with and capture. Despite the dynamic behavior of individual microtubules, changes of the whole microtubule system occur slowly enough to preserve microtubule structure. Even in normally moving cells, as well as in immobile or slow-moving cells, the microtubule system is stored in space and is almost constant in time. The assembly and stability of microtubules is regulated not only by the nucleotide state of tubulin, but also by interaction with cellular factors like microtubule-associated proteins (MAPs) and other regulatory cytoskeletal proteins (MacRae 1992).

Microtubules are known to interact with the cell-cell adhesion machinery in fibroblasts (Mary et al. 2002) and in epithelial cells (Akhmanova et al. 2009, Carramusa et al. 2007, Meng et al. 2008, Shewan et al. 2005, Vogelmann and Nelson 2005). Microtubules are extended throughout cell cytoplasm, stabilizing cell shape, and mediate intracellular transport and localization of organelles. Many signaling pathways involved in EC barrier regulation relied on microtubules; in particular, elements of Rho pathway, Rho family GTPases and their regulatory proteins, were shown to be tightly associated with polymerized tubulin (Lee and Gotlieb 2003a, b). Rho activation and Rho-dependent barrier compromise were shown to occur in response to agents that induce microtubule breakdown (Birukova et al. 2004d, Bogatcheva et al. 2007, Verin et al. 2001). Furthermore, thrombin-, TGFβ- TNF- and ROS-induced hyperpermeability is associated with both destabilization of the peripheral microtubule network and Rho activation (Birukova et al. 2004a, c, 2005b, Kratzer et al. 2012, Petrache et al. 2003). The mechanism of MT-dependent-Rho activation likely involves the release of GTP exchange factor GEF-H1 from degrading microtubules. Indeed, GEF-H1 depletion attenuates Rho activation and the increase in vascular permeability (Birukova et al. 2006).

Another small GTPase GEF was recently linked to the regulation of the state of microtubules. Rap1 GEF Epac Iis a cAMP-responsive protein, which provides PKA-independent regulation of cytoskeleton organization. Epac localizes with microtubules; its specific activation with cAMP analogue o-Me-cAMP results in microtubule elongation and, more importantly, reverses microtubule-dependent increases in vascular permeability induced by TNFα and TGFβ (Sehrawat et al. 2008). Surprisingly, this effect of Epac was found to be Rap1-independent (Sehrawat et al. 2008).

Collectively, published data demonstrate the critical role of microtubules in the maintaining of endothelial barrier *in vitro* (Birukova et al. 2004a, d, 2005b, Bogatcheva et al. 2007, Kratzer et al. 2012, Petrache et al. 2003, Verin et al. 2001) and vascular barrier integrity *in vivo* (Gorshkov et al. 2012,

Kolosova et al. 2008, Kratzer et al. 2012). The microtubules apparently are the first target in the circuit of reactions leading to the pulmonary EC barrier compromise. Further, we and others (Alieva et al. 2010, Komarova et al. 2012, Sehrawat et al. 2011, Smurova et al. 2008) have shown that dynamic microtubules play an essential role in the barrier function *in vitro*; peripheral microtubules depolymerization is a necessary and sufficient condition for initiation of endothelial barrier dysfunction (Smurova et al. 2008).

Physiological Role of Microtubule Dynamic in the Endothelial Cell

The microtubule cytoskeleton is a major determinant of cortical dynamics and microtubules can interact with the cortices of animal cells in a variety of ways. One such interaction involves microtubule plus-ends, which are commonly oriented towards the cell periphery (Akhmanova and Hoogenraad 2005, Gundersen et al. 2004). Dynamic MT instability allows these plus ends to grow outwards and potentially explore peripheral structures (Howard and Hyman 2003) including integrin-based focal adhesions (Small and Kaverina 2003), and organize vesicular transport to the cell surface (Watanabe et al. 2005), as well as the delivery of regulatory molecules to the cortex (Rodriguez et al. 2003). Further, due to dynamic instability of microtubule ends, their ability to frequent exchange of assembly-disassembly phase, microtubules are able to locally modulate the dynamics of cell contacts (in moving fibroblasts by direct interaction of microtubule plus-ends with the focal contacts ("targeting" contacts) (Efimov et al. 2007, Kaverina et al. 1999), as well as to regulate the dynamics of cell-cell contacts (Akhmanova et al. 2009, Meng et al. 2008, Shewan et al. 2005, Stehbens et al. 2006). In addition, microtubules can control the organization of the actin system of the cell, locally altering the contractility of actomyosin at the ends of stress fibers (Kaverina and Straube 2011, Schober et al. 2007, Small and Kaverina 2003). Microtubules are known to interact with the cell-cell adhesion machinery in fibroblasts (Mary et al. 2002) and in epithelial cells (Akhmanova et al. 2009, Carramusa et al. 2007, Meng et al. 2008, Shewan et al. 2005, Vogelmann and Nelson 2005) and microtubule dynamics is an important factor in regulation of cell-cell contacts (Akhmanova et al. 2009, Meng et al. 2008, Stehbens et al. 2006). Dynamic instability behavior of microtubules is cell-type-specific (Akhmanova et al. 2001, Cassimeris 1993, Sammak and Borisy 1988, Schulze and Kirschner 1986, 1987, Shelden and Wadsworth 1993, Vaughan et al. 2002) and growth rates are generally dissimilar in different cell types and in different cell areas (Stepanova et al. 2003). Our data of direct quantification of microtubule dynamics in EC (Alieva et al. 2010) demonstrated that the growth rates in EC are comparable with those obtained for epithelial cells and fibroblasts (Komarova et al. 2002, Salaycik et al. 2005). The instantaneous rate of microtubule growth was similar for microtubules extending toward the front and rear of the cell (Alieva et al. 2010) with high growth rate comparable to microtubule growth rate in fibroblasts in the cell interior and lower growth rate near the cell boundary in the area of cell contacts. Microtubule plus-ends are highly dynamic in the internal cytoplasm of EC and their growth rate is comparable with microtubule growth rate in fibroblasts or in functionally active cytoplasm areas in polarized and migrating cells (Komarova et al. 2002, Salaycik et al. 2005). These dynamic microtubules are capable of adjusting existing contacts and can also adjust endothelial permeability (Alieva et al. 2010).

Microtubule population in EC is heterogeneous and can be divided into two subpopulations: stable, modified (acetylated) and dynamic (tyrosinated) microtubules which is consistent with the dynamic characteristics of its constituent microtubules (Birukova et al. 2004a, d, Alieva et al. 2010). Acetylated microtubules are less dynamic than intact tyrosinated microtubules and more resistant to the effects of external factors. Therefore, under conditions compromising vascular endothelium integrity, they may confer stability on the endothelial microtubule network. Further, some EC barrier-enhancing factors may shift the ratio in favor of stable microtubule subpopulation and increase overall stability of the EC cytoskeleton (Alieva et al. 2010, Smurova et al. 2008).

Microtubule disassembly is critically involved in EC barrier dysfunction induced by edemagenic agents *in vitro* (Birukova et al. 2004a, b, Smurova et al. 2008) and in murine models of acute lung injury (Kolosova et al. 2008, Kratzer et al. 2012). Further, published data suggest that changes in microtubule structure are an early event in the circuit of the reactions leading to the changes in pulmonary EC barrier permeability (Smurova et al. 2008, 2011). Barrier-disruptive treatment leads not only to the disappearance of peripheral microtubules. The entire system of microtubules has been changed: microtubules growing out of centrosome become significantly shorter (Alieva et al. 2010). Such shortening is possible if microtubules do not reach the cell periphery due to decrease of either plus end growth rate or shortening of the progressive growth phase. This may also happen if microtubules rapidly shorten reaching the cell periphery (due to the absence of pauses and increased frequency of catastrophes at the cell periphery). The calculations of the plus end growth rates indicate that at least one of the explanations is correct: upon the development of the barrier dysfunction induced by nanomolar concentrations of nocodazole, the plus end growth rates were dropped markedly. As a result, the plus ends, apparently, cannot reach the cell periphery (Alieva et al. 2010). Cell-cell contacts in lung epithelial cells were reported to stabilize the dynamic behavior of microtubule plus-ends (Waterman-Storer et al. 2000). Microtubule plus-end growth rate was reduced from the cell center to the cell periphery, indicating that the microtubule dynamics varied in different regions of human EC. On the other hand, microtubule growth rate was lower in EC cultivated in the monolayer than in single cells (Alieva et al. 2010), and it may well be that microtubule plus-ends growth rate decrease with EC monolayer formation (Alieva et al. 2010).

Study of microtubule dynamics in living newt lung epithelial cells demonstrated that microtubules in the extending lamellae at the leading edge are dynamic, whereas microtubules in lamellae that contact neighboring cells can be either dynamic or stable (Waterman-Storer et al. 2000). We suppose that in the EC monolayer, where VE-cadherin adherent junctions (AJs) are well-structured, dynamic microtubules may interact with AJs. This interaction may lead to MT stabilization in the area of the contact.

In contrast, it was shown that in polarized, motile cells, microtubules extended into newly formed protrusions at the leading edge. These "pioneering" microtubules (Waterman-Storer and Salmon 1997) demonstrated different behavior when compared with microtubules in non-leading lateral edges indicating the region-specific differences in microtubule dynamics and suggesting the existence of different mechanisms of local regulation of microtubule plus ends elongation in EC.

We suggest that high microtubule dynamics and local distinctions in microtubule growth rate underlie the specific function of EC where fast delivery of molecular signals to the cell edge (to the area of cell-cell junctions) is essential for their active and fast local regulation in response to external and internal signals. Rapid growth allows nascent microtubules to elongate from the centrosome area to the cell boundary in a very short time. In the EC monolayer, where AJ contacts are organized, microtubules may interact with AJs and this interaction may lead to their stabilization in the area of contact. Cell-cell contact in lung epithelial cells was reported to stabilize the dynamic behavior of microtubule plus-ends (Stehbens et al. 2006, Waterman-Storer et al. 2000). According to our results (Alieva et al. 2010), dynamic microtubules are capable of adjusting existing contacts and can adjust endothelial permeability along with the other regulatory mechanisms.

Cross-talk between Microtubules and Microfilaments in EC Barrier Regulation

Interplay between microtubule and actin cytoskeleton is essential for a wide variety of processes, including vesicle and organelle transport, contractile responses, tissue cell migration and wound repair (Lee and Gotlieb 2003a, b, Signor and Scholey 2000, Wehrle-Haller and Imhof 2003). Microtubules and the actin cytoskeleton function in concert in normal endothelium as well as during endothelial injury. Microtubule stabilization strengthens EC barrier function (Kratzer et al. 2012, Petrache et al. 2003, Suzuki et al. 2004d) whereas microtubule depolymerization is associated with dissolution of the cortical actin cytoskeleton, MLC phosphorylation, increased stress fiber formation, contraction, and EC barrier dysfunction (Birukova et al. 2004d, Bogatcheva et al. 2007, Verin et al. 2001). These effects were linked to the activation of small GTPase Rho and its effector Rho kinase and can be attenuated by cell pretreatment with paclitaxel (taxol), which promotes microtubule assembly (Birukova et al. 2004d, Bogatcheva et al. 2007, Verin et al. 2001). Microtubule disruption releases Rho activators, RhoGEFs, from bound tubulin, causing RhoA activation and stress fiber formation (Krendel et al. 2002, van Horck et al. 2001).

Nocodazole-induced isometric contraction in fibroblasts correlated well with the level of nocodazole-induced MLC phosphorylation (Kolodney and Elson 1995). Based on these data, it was proposed that MLC phosphorylation is a common mechanism for MT inhibitors- and thrombin-mediated contraction (Kolodney and Elson 1995). It was also demonstrated that thrombin and MT inhibitors induced p38 MAPK activation. Further, p38 MAPK inhibition significantly attenuates thrombin- and MT inhibitors-induced EC barrier compromise (Birukova et al. 2005a, Bogatcheva et al. 2007, Borbiev et al. 2004). Collectively, these data indicate that the disruption of microtubule structure triggers Rho and p38 MAPK activation, which potentially can cause contraction and barrier dysfunction via alterations in the actomyosin cytoskeleton. In contrast, microtubule polymerization sequesters LIM kinase and limits its access to the actin cytoskeleton, preventing LIM kinase-mediated reorganization of the cortical actin rim into stress fibers by phosphorylating and inhibiting activity of the actin depolymerizing factor cofilin (Gorovoy et al. 2005, Maekawa et al. 1999). Thus, the interaction between microtubules and actin critically affects the nature of cytoskeletal interactions that control EC barrier integrity. Overall, these findings reveal an important role for MT-actin cross-talk in regulation of endothelial barrier function.

MT/actin filaments coordination is accomplished at several levels by cross-linker proteins that join microtubules and actin filaments (Bershadsky et al. 2006a, b, Fuchs and Karakesisoglou 2001, Rodriguez et al. 2003), as well as by small GTPases of the Rho family that mediate dynamics and organization in both the actin filaments and microtubule network (Cook et al. 1998, Daub et al. 2001, Fukata et al. 2002, Ishizaki et al. 2001, Jaffe and Hall 2005, Palazzo et al. 2001).

Cross-linker proteins by which microtubules and actin filaments are connected directly were described (Bershadsky et al. 2006, Fuchs and Karakesisoglou 2001, Rodriguez et al. 2003). However, a more flexible temporal and spatial coordination of cytoskeletal dynamic fibrils requires a different system of regulation, rather than just their mechanical connection. It was found that Rho GTPases can regulate the dynamics of actin and the organization as a system, and microtubules (Cook et al. 1998, Daub et al. 2001, Fukata et al. 2002, Ishizaki et al. 2001, Jaffe and Hall 2005, Palazzo et al. 2001), as well as focal and cell-cell contacts (Mehta and Malik 2006). One of the mechanisms underlying such a parallel regulation is possibly related to the properties of direct downstream effector of Rho GTPase-mDia (Alberts 2002, Jaffe and Hall 2005, Wallar and Alberts 2003) which has been shown to mediate the effect of Rho on the adhesive structure (Bershadsky et al. 2006a, b, Carramusa et al. 2007, Watanabe et al. 2004, 2005).

Active mDia stimulates actin polymerization and at the same time affects the dynamics of microtubules, providing concentration of MT plus ends near growing focal contacts. Although microtubules inhibit myosin-dependent contraction, their accumulation near the focal contacts prevents their growth and even induces their disassembly. Thus, there is a system with a negative control, which regulates the growth of focal contacts. The effect of mDia on microtubules may be a result of its not yet identified interaction with the molecular complexes that directly regulate microtubule dynamics, such as proteins associated with the ends of microtubules—EB1,

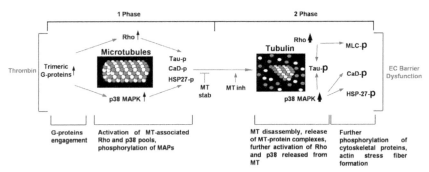

Figure 1. Schematic representation of thrombin-induced MT-mediated EC barrier compromise. Thrombin can induce cytoskeletal reorganization and EC barrier dysfunction in two stages. On the phase 1 thrombin-induced engagement of trimeric G-proteins activates Rho and p38 MAPK pools associated with microtubules leading to phosphorylation of microtubule-associated cytoskeletal targets and microtubule disassembly. On the second phase microtubule dissolution releases microtubule-associated protein complexes, further activates Rho and p38 MAPK pathways, increases phosphorylation of cytoskeletal targets and leads to stress fiber formation and EC barrier compromise. Abbreviations: MT stab: MT stabilization; MT inh: MT inhibitors; MAPs: MT–associated proteins.

Color image of this figure appears in the color plate section at the end of the book.

CLIP-170 and APC (Bershadsky et al. 2006a, b, Heald and Nogales 2002, Jaffe and Hall 2005). EB1 and APC bind to mDia, and this fact may indicate the participation of Rho in microtubules capping (Wen et al. 2004). Thus, mDia (a direct effector of Rho) is a factor governing the system of microtubules, actin filaments and cell contacts in general. It is also possible that mDia modulates microtubule dynamics indirectly. For example, there is evidence that mDia may cause the activation of Rac 1 (Tsuji et al. 2002), including a pathway, dependent of Rac-effector IQGAP, which interacts with CLIP-170 (Fukata et al. 2002, Gundersen 2002) and binds to the APC (Watanabe et al. 2004). Importantly, it was recently shown that IQGAP/Rac signaling is involved in the regulation of EC permeability *in vitro* and in murine model (Bhattacharya et al. 2012, Yamaoka-Tojo et al. 2004). The data of the past few years demonstrate that CLIPs interact with IQGAP1, possibly functioning as links between the MT plus ends and cortical actin network downstream of Rac1 and Cdc42 (Fukata et al. 2002, Lansbergen and Akhmanova 2006). On the other hand, CLIP-170 interacts with IQGAP and may be involved in the regulation of cell-cell contact (Lansbergen and Akhmanova 2006, Small and Kaverina 2003). CLIPs and APC are also involved in the interaction of microtubules with intercellular contacts (Komarova et al. 2007, Mehta and Malik 2006), and IQGAP can bind to β-catenin at adherens junctions (Akiyama and Kawasaki 2006, Gundersen 2002, Gundersen et al. 2004, Komarova et al. 2007) being a key factor involved in the regulation of cytoskeletal and adhesion structures of cells.

Conclusion

Understanding remains limited as to the role of cytoskeletal components, particularly microtubules and microfilaments in the regulation of EC barrier function. We proposed a complex mechanism by which edemagenic agents like thrombin can induce cytoskeletal and barrier dysfunction (Fig. 1). This includes two stages. In the initial stage, thrombin-induced engagement of heterotrimeric G-proteins activates Rho and p38 MAPK pools associated with microtubules, which leads to phosphorylation of microtubule-associated cytoskeletal targets and microtubule disassembly. In the second stage, microtubule dissolution releases microtubule-associated protein complexes, further activates Rho and p38 MAPK pathways, increases phosphorylation of cytoskeletal targets and leads to stress fiber formation and EC barrier compromise.

References

Adam, L.P., J.R. Haeberle and D.R. Hathaway. 1989. Phosphorylation of caldesmon in arterial smooth muscle. J Biol Chem 264: 7698–7703.

Adams, J.A., G.M. Omann and J.J. Linderman. 1998. A mathematical model for ligand/receptor/G-protein dynamics and actin polymerization in human neutrophils. J Theor Biol 193: 543–560.

Akhmanova, A. and C.C. Hoogenraad. 2005. Microtubule plus-end-tracking proteins: mechanisms and functions. Curr Opin Cell Biol 17: 47–54.

Akhmanova, A., C.C. Hoogenraad, K. Drabek et al. 2001. Clasps are CLIP-115 and -170 associating proteins involved in the regional regulation of microtubule dynamics in motile fibroblasts. Cell 104: 923–935.

Akhmanova, A., S.J. Stehbens and A.S. Yap. 2009. Touch, grasp, deliver and control: functional cross-talk between microtubules and cell adhesions. Traffic 10: 268–274.

Akiyama, T. and Y. Kawasaki. 2006. Wnt signalling and the actin cytoskeleton. Oncogene 25: 7538–7544.

Alberts, A.S. 2002. Diaphanous-related Formin homology proteins. Current Biol: CB 12: R796.

Alessi, D., L.K. MacDougall, M.M. Sola et al. 1992. The control of protein phosphatase-1 by targetting subunits. The major myosin phosphatase in avian smooth muscle is a novel form of protein phosphatase-1. Eur J Biochem/FEBS 210: 1023–1035.

Alieva, I.B., E.A. Zemskov, I.I. Kireev et al. 2010. Microtubules growth rate alteration in human endothelial cells. J Biomed and Biotech 2010: 671536.

Amann, K.J. and T.D. Pollard. 2000. Cellular regulation of actin network assembly. Current biology: CB 10: R728–730.

Amano, M., M. Ito, K. Kimura et al. 1996. Phosphorylation and activation of myosin by Rho-associated kinase (Rho-kinase). J Biol Chem 271: 20246–20249.

Amano, M., K. Chihara, K. Kimura et al. 1997. Formation of actin stress fibers and focal adhesions enhanced by Rho-kinase. Science 275: 1308–1311.

Antonov, A.S., G.N. Antonova, M. Fujii et al. 2012. Regulation of endothelial barrier function by TGF-beta type I receptor ALK5: potential role of contractile mechanisms and heat shock protein 90. J Cell Physiol 227: 759–771.

Bacus, S.S., A.V. Gudkov, M. Lowe et al. 2001. Taxol-induced apoptosis depends on MAP kinase pathways (ERK and p38) and is independent of p53. Oncogene 20: 147–155.

Bazzoni, G. and E. Dejana. 2004. Endothelial cell-to-cell junctions: molecular organization and role in vascular homeostasis. Physiol Rev 84: 869–901.

Becker, P.M., A.D. Verin, M.A. Booth et al. 2001. Differential regulation of diverse physiological responses to VEGF in pulmonary endothelial cells. American journal of physiology. Lung Cell and Mol Physiol 281: L1500–1511.

Benndorf, R., K. Hayess, S. Ryazantsev et al. 1994. Phosphorylation and supramolecular organization of murine small heat shock protein HSP25 abolish its actin polymerization-inhibiting activity. J Biol Chem 269: 20780–20784.

Bershadsky, A.D., N.Q. Balaban and B. Geiger. 2003. Adhesion-dependent cell mechanosensitivity. Annu Rev Cell Dev Biol 19: 677–695.

Bershadsky, A.D., C. Ballestrem, L. Carramusa et al. 2006a. Assembly and mechanosensory function of focal adhesions: experiments and models. Eur J Cell Biol 85: 165–173.

Bershadsky, A.D., M. Kozlov and B. Geiger. 2006b. Adhesion-mediated mechanosensitivity: a time to experiment, and a time to theorize. Curr. Opinion. Cell Biol 18: 472–481.

Bhattacharya, M., G. Su, X. Su et al. 2012. IQGAP1 is necessary for pulmonary vascular barrier protection in murine acute lung injury and pneumonia. American journal of physiology. Lung Cell and Mol Physiol 303: L12–19.

Bhattacharyya, R. and P.B. Wedegaertner. 2003. Characterization of G alpha 13-dependent plasma membrane recruitment of p115RhoGEF. Biochem J 371: 709–720.

Bindewald, K., D. Gunduz, F. Hartel et al. 2004. Opposite effect of cAMP signaling in endothelial barriers of different origin. Am J Physiol Cell physiology 287: C1246–1255.

Birukova, A.A., K.G. Birukov, K. Smurova et al. 2004a. Novel role of microtubules in thrombin-induced endothelial barrier dysfunction. FASEB J 18: 1879–1890.

Birukova, A.A., F. Liu, J.G. Garcia et al. 2004b. Protein kinase A attenuates endothelial cell barrier dysfunction induced by microtubule disassembly. Am J Cell Physiol. Lung Cell and Mol Physiol 287: L86–93.

Birukova, A.A., K. Smurova, K.G. Birukov et al. 2004c. Role of Rho GTPases in thrombin-induced lung vascular endothelial cells barrier dysfunction. Microvascular Res 67: 64–77.

Birukova, A.A., K. Smurova, K.G. Birukov et al. 2004d. Microtubule disassembly induces cytoskeletal remodeling and lung vascular barrier dysfunction: role of Rho-dependent mechanisms. J Cell Physiol 201: 55–70.

Birukova, A.A., K.G. Birukov, B. Gorshkov et al. 2005a. MAP kinases in lung endothelial permeability induced by microtubule disassembly. Am J Cell Physiol. Lung Cell and Mol Physiol 289: L75–84.

Birukova, A.A., K.G. Birukov, D. Adyshev et al. 2005b. Involvement of microtubules and Rho pathway in TGF-beta1-induced lung vascular barrier dysfunction. J Cell Physiol 204: 934–947.

Birukova, A.A., D. Adyshev, B. Gorshkov et al. 2006. GEF-H1 is involved in agonist-induced human pulmonary endothelial barrier dysfunction. Am J Physiol Lung Cell and Mol Physiol 290: L540–548.

Birukova, A.A., E. Alekseeva, A. Mikaelyan et al. 2007a. HGF attenuates thrombin-induced endothelial permeability by Tiam1-mediated activation of the Rac pathway and by Tiam1/Rac-dependent inhibition of the Rho pathway. FASEB J: official publication of the Federation of American Societies for Experimental Biology 21: 2776–2786.

Birukova, A.A., T. Zagranichnaya, P. Fu et al. 2007b. Prostaglandins PGE(2) and PGI(2) promote endothelial barrier enhancement via PKA- and Epac1/Rap1-dependent Rac activation. Exp Cell Res 313: 2504–2520.

Birukova, A.A., T. Zagranichnaya, E. Alekseeva et al. 2008. Epac/Rap and PKA are novel mechanisms of ANP-induced Rac-mediated pulmonary endothelial barrier protection. Journal of cellular physiology 215: 715–724.

Birukova, A.A., D. Burdette, N. Moldobaeva et al. 2010. Rac GTPase is a hub for protein kinase A and Epac signaling in endothelial barrier protection by cAMP. Microvasc Res 79: 128–138.

Birukova, A.A., T. Wu, Y. Tian et al. 2012. Iloprost improves endothelial barrier function in LPS-induced lung injury. Eur. Respiratory J. official journal of the European Society for Clinical Respiratory Physiology. 2013 jan: 41(1): 165–176.

Bishop, A.L. and A. Hall. 2000. Rho GTPases and their effector proteins. The Biochemical Journal 348(Pt 2): 241–255.

Bitar, K.N., M.S. Kaminski, N. Hailat et al. 1991. Hsp27 is a mediator of sustained smooth muscle contraction in response to bombesin. Biochem and Biophys Res Commun 181: 1192–1200.

Blagosklonny, M.V., Y. Chuman, R.C. Bergan et al. 1999. Mitogen-activated protein kinase pathway is dispensable for microtubule-active drug-induced Raf-1/Bcl-2 phosphorylation and apoptosis in leukemia cells. Leukemia: official journal of the Leukemia Society of America, Leukemia Research Fund, UK 13: 1028–1036.

Bogatcheva, N.V. and A.D. Verin. 2008. The role of cytoskeleton in the regulation of vascular endothelial barrier function. Microvascular Research 76: 202–207.

Bogatcheva, N.V., A.D. Verin, P. Wang et al. 2003a. Phorbol esters increase MLC phosphorylation and actin remodeling in bovine lung endothelium without increased contraction. Am J Physiol Lung Cell and Mol Physiol 285: L415–426.

Bogatcheva, N.V., S.M. Dudek, J.G. Garcia et al. 2003b. Mitogen-activated protein kinases in endothelial pathophysiology. J. Inves. Med. the official publication of the American Federation for Clinical Research 51: 341–352.

Bogatcheva, N.V., A. Birukova, T. Borbiev et al. 2006. Caldesmon is a cytoskeletal target for PKC in endothelium. J Cell Biochem 99: 1593–1605.

Bogatcheva, N.V., D. Adyshev, B. Mambetsariev et al. 2007. Involvement of microtubules, p38, and Rho kinases pathway in 2-methoxyestradiol-induced lung vascular barrier dysfunction. American journal of physiology. Lung Cell and Mol Physiol 292: L487–499.

Bogatcheva, N.V., M.A. Zemskova, Y. Kovalenkov et al. 2009. Molecular mechanisms mediating protective effect of cAMP on lipopolysaccharide (LPS)-induced human lung microvascular endothelial cells (HLMVEC) hyperpermeability. J Cell Physiol 221: 750–759.

Bogatcheva, N.V., M.A. Zemskova, B.A. Gorshkov et al. 2011. Ezrin, radixin, and moesin are phosphorylated in response to 2-methoxyestradiol and modulate endothelial hyperpermeability. Am J Resp and Mol Biol 45: 1185–1194.

Borbiev, T., A.D. Verin, A. Birukova et al. 2003. Role of CaM kinase II and ERK activation in thrombin-induced endothelial cell barrier dysfunction. Am J Physiol Lung Cell and Mol Physiol 285: L43–54.

Borbiev, T., A. Birukova, F. Liu et al. 2004. p38 MAP kinase-dependent regulation of endothelial cell permeability. American journal of physiology. Lung Cell and Mol Physiol 287: L911–918.

Broussard, J.A., D.J. Webb and I. Kaverina. 2008. Asymmetric focal adhesion disassembly in motile cells. Curr Opin Cell Biol 20: 85–90.

Carramusa, L., C. Ballestrem, Y. Zilberman et al. 2007. Mammalian diaphanous-related formin Dia1 controls the organization of E-cadherin-mediated cell-cell junctions. J Cell Sci 120: 3870–3882.

Cassimeris, L. 1993. Regulation of microtubule dynamic instability. Cell Motil Cytoskeleton 26: 275–281.

Cattan, C.E. and K.C. Oberg. 1999. Vinorelbine tartrate-induced pulmonary edema confirmed on rechallenge. Pharmacotherapy 19: 992–994.

Clauss, M., C. Sunderkotter, B. Sveinbjornsson et al. 2001. A permissive role for tumor necrosis factor in vascular endothelial growth factor-induced vascular permeability. Blood 97: 1321–1329.

Comerford, K.M., D.W. Lawrence, K. Synnestvedt et al. 2002. Role of vasodilator-stimulated phosphoprotein in PKA-induced changes in endothelial junctional permeability. FASEB J: official publication of the Federation of American Societies for Experimental Biology 16: 583–585.

Cook, T.A., T. Nagasaki and G.G. Gundersen. 1998. Rho guanosine triphosphatase mediates the selective stabilization of microtubules induced by lysophosphatidic acid. J Cell Biol 141: 175–185.

Csortos, C., I. Kolosova and A.D. Verin. 2007. Regulation of vascular endothelial cell barrier function and cytoskeleton structure by protein phosphatases of the PPP family. Am J Physiol: lung Cell and Mol Physiol 293: L843–854.

Daub, H., K. Gevaert, J. Vandekerckhove et al. 2001. Rac/Cdc42 and p65PAK regulate the microtubule-destabilizing protein stathmin through phosphorylation at serine 16. J Biol Chem 276: 1677–1680.

Davis, R.J. 1993. The mitogen-activated protein kinase signal transduction pathway. The Journal of biological chemistry 268: 14553–14556.

de Rooij, J., F.J. Zwartkruis, M.H. Verheijen et al. 1998. Epac is a Rap1 guanine-nucleotide-exchange factor directly activated by cyclic AMP. Nature 396: 474–477.

Desai, A. and T.J. Mitchison. 1997. Microtubule polymerization dynamics. An Rev Cell and Dev Biol 13: 83–117.

Dong, J.M., T. Leung, E. Manser et al. 1998. cAMP-induced morphological changes are counteracted by the activated RhoA small GTPase and the Rho kinase ROKalpha. J Biol Chem 273: 22554–22562.

Dudek, S.M. and J.G. Garcia. 2001. Cytoskeletal regulation of pulmonary vascular permeability. J Appl Physiol 91: 1487–1500.

Efimov, A., A. Kharitonov, N. Efimova et al. 2007. Asymmetric CLASP-dependent nucleation of noncentrosomal microtubules at the trans-Golgi network. Dev Cell 12: 917–930.

Ermert, L., H. Bruckner, D. Walmrath et al. 1995. Role of endothelial cytoskeleton in high-permeability edema due to botulinum C2 toxin in perfused rabbit lungs. Am J Physiol: Cell Physiol 268: L753–761.

Essler, M., M. Amano, H.J. Kruse et al. 1998. Thrombin inactivates myosin light chain phosphatase via Rho and its target Rho kinase in human endothelial cells. J Biol Chem 273: 21867–21874.

Essler, M., M. Retzer, M. Bauer et al. 1999. Mildly oxidized low density lipoprotein induces contraction of human endothelial cells through activation of Rho/Rho kinase and inhibition of myosin light chain phosphatase. J Biol Chem 274: 30361–30364.

Ferrero, E., M.R. Zocchi, E. Magni et al. 2001. Roles of tumor necrosis factor p55 and p75 receptors in TNF-alpha-induced vascular permeability. Am J Physiol: Cell Physiol 281: C1173–1179.

Fuchs, E. and I. Karakesisoglou. 2001. Bridging cytoskeletal intersections. Genes & development 15: 1–14.

Fukata, M., T. Watanabe, J. Noritake et al. 2002. Rac1 and Cdc42 capture microtubules through IQGAP1 and CLIP-170. Cell 109: 873–885.

Fukata, Y., M. Amano and K. Kaibuchi. 2001. Rho-Rho-kinase pathway in smooth muscle contraction and cytoskeletal reorganization of non-muscle cells. Trends Pharmacol Sci 22: 32–39.

Garcia, J.G., F.M. Pavalko and C.E. Patterson. 1995a. Vascular endothelial cell activation and permeability responses to thrombin. Blood coagulation & fibrinolysis: an international J Haemost and Thromb 6: 609–626.

Garcia, J.G., H.W. Davis and C.E. Patterson. 1995b. Regulation of endothelial cell gap formation and barrier dysfunction: role of myosin light chain phosphorylation. J Cell Physiol 163: 510–522.

Garcia, J.G., V. Lazar, L.I. Gilbert-McClain et al. 1997. Myosin light chain kinase in endothelium: molecular cloning and regulation. Am J Resp Cell and Mol Biol 16: 489–494.

Garcia, J.G., F. Liu, A.D. Verin et al. 2001. Sphingosine 1-phosphate promotes endothelial cell barrier integrity by Edg-dependent cytoskeletal rearrangement. J Clinic Invest 108: 689–701.

Garcia, J.G., P. Wang, K.L. Schaphorst et al. 2002. Critical involvement of p38 MAP kinase in pertussis toxin-induced cytoskeletal reorganization and lung permeability. FASEB journal: official publication of the FASEB 16: 1064–1076.

Garrington, T.P. and G.L. Johnson. 1999. Organization and regulation of mitogen-activated protein kinase signaling pathways. Curr Opi Cell Biol 11: 211–218.

Goeckeler, Z.M. and R.B. Wysolmerski. 1995. Myosin light chain kinase-regulated endothelial cell contraction: the relationship between isometric tension, actin polymerization, and myosin phosphorylation. J Cell Biol 130: 613–627.

Gorovoy, M., J. Niu, O. Bernard et al. 2005. LIM kinase 1 coordinates microtubule stability and actin polymerization in human endothelial cells. J Biol Chem 280: 26533–26542.

Gorshkov, B.A., M.A. Zemskova, A.D. Verin et al. 2012. Taxol alleviates 2-methoxyestradiol-induced endothelial permeability. Vasc Pharmacol 56: 56–63.

Graves, J.D., J.S. Campbell and E.G. Krebs. 1995. Protein serine/threonine kinases of the MAPK cascade. Ann New York Acad Sci 766: 320–343.

Guay, J., H. Lambert, G. Gingras-Breton et al. 1997. Regulation of actin filament dynamics by p38 map kinase-mediated phosphorylation of heat shock protein 27. J Cell Sci 110(Pt 3): 357–368.

Gundersen, G.G. 2002. Microtubule capture: IQGAP and CLIP-170 expand the repertoire. Current biology: CB 12: R645–647.

Gundersen, G.G., E.R. Gomes and Y. Wen. 2004. Cortical control of microtubule stability and polarization. Curr Opin Cell Biol 16: 106–112.

Hall, A. 1998. Rho GTPases and the actin cytoskeleton. Science 279: 509–514.

Harbeck, B., S. Huttelmaier, K. Schluter et al. 2000. Phosphorylation of the vasodilator-stimulated phosphoprotein regulates its interaction with actin. J Biol Cell 275: 30817–30825.

Hart, M.J., X. Jiang, T. Kozasa et al. 1998. Direct stimulation of the guanine nucleotide exchange activity of p115 RhoGEF by Galpha13. Science 280: 2112–2114.

Hartshorne, D.J. 1998. Myosin phosphatase: subunits and interactions. Acta Physiol Scv 164: 483–493.

Hartshorne, D.J., M. Ito and F. Erdodi. 1998. Myosin light chain phosphatase: subunit composition, interactions and regulation. J Muscle Res Cell Motil 19: 325–341.

Heald, R. and E. Nogales. 2002. Microtubule dynamics. Journal of Cell Science 115: 3–4.

Hedges, J.C., I.A. Yamboliev, M. Ngo et al. 1998. p38 mitogen-activated protein kinase expression and activation in smooth muscle. Am. J. Physiol: Cell Physiol 275: C527–534.

Herlaar, E. and Z. Brown. 1999. p38 MAPK signalling cascades in inflammatory disease. Mol Med Today 5: 439–447.

Hirano, S., R.S. Rees, S.L.Yancy et al. 2004. Endothelial barrier dysfunction caused by LPS correlates with phosphorylation of HSP27 *in vivo*. Cell Biol Toxicology 20: 1–14.

Holt, M.R., D.R. Critchley and N.P. Brindle. 1998. The focal adhesion phosphoprotein, VASP. Int J Biochem and Biol 30: 307–311.

Howard, J. and A.A. Hyman. 2003. Dynamics and mechanics of the microtubule plus end. Nature 422: 753–758.

Huot, J., F. Houle, F. Marceau et al. 1997. Oxidative stress-induced actin reorganization mediated by the p38 mitogen-activated protein kinase/heat shock protein 27 pathway in vascular endothelial cells. Circulation Res 80: 383–392.

Huot, J., F. Houle, S. Rousseau et al. 1998. SAPK2/p38-dependent F-actin reorganization regulates early membrane blebbing during stress-induced apoptosis. J Cell Biol 143: 1361–1373.

Ishizaki, T., Y. Morishima, M. Okamoto et al. 2001. Coordination of microtubules and the actin cytoskeleton by the Rho effector mDia1. Nat Cell Biol 3: 8–14.

Jacobson, J.R., S.M. Dudek, P.A. Singleton et al. 2006. Endothelial cell barrier enhancement by ATP is mediated by the small GTPase Rac and cortactin. Am J Physiol: Lung Cell and Mol Physiol 291: L289–295.

Jaffe, A.B. and A. Hall. 2005. Rho GTPases: biochemistry and biology. Annu. Rev. and Dev. Biol. 21: 247–269.

Kaverina, I. and A. Straube. 2011. Regulation of cell migration by dynamic microtubules. Semm Cell and Dev Biol 22: 968–974.

Kaverina, I., O. Krylyshkina and J.V. Small. 1999. Microtubule targeting of substrate contacts promotes their relaxation and dissociation. J Cell Biol 146: 1033–1044.

Kawasaki, H., G.M. Springett, N. Mochizuki et al. 1998. A family of cAMP-binding proteins that directly activate Rap1. Science 282: 2275–2279.

Kevil, C.G., T. Oshima and J.S. Alexander. 2001. The role of p38 MAP kinase in hydrogen peroxide mediated endothelial solute permeability. Endothelium: J Endoth Cell Res 8: 107–116.

Kiemer, A.K., N.C. Weber, R. Furst et al. 2002. Inhibition of p38 MAPK activation via induction of MKP-1: atrial natriuretic peptide reduces TNF-alpha-induced actin polymerization and endothelial permeability. Circulation Research 90: 874–881.

Kimura, K., M. Ito, M. Amano et al. 1996. Regulation of myosin phosphatase by Rho and Rho-associated kinase (Rho-kinase). Science 273: 245–248.

Kirschner, M.W. and T. Mitchison. 1986. Microtubule dynamics. Nature 324: 621.

Kolodney, M.S. and E.L. Elson. 1995. Contraction due to microtubule disruption is associated with increased phosphorylation of myosin regulatory light chain. PNAS of the United States of America 92: 10252–10256.

Kolosova, I.A., T. Mirzapoiazova, D. Adyshev et al. 2005. Signaling pathways involved in adenosine triphosphate-induced endothelial cell barrier enhancement. Circ Res 97: 115–124.

Kolosova, I.A., T. Mirzapoiazova, L. Moreno-Vinasco et al. 2008. Protective effect of purinergic agonist ATPgammaS against acute lung injury. Am J Physiol Lung Cell Mol Physiol 294: L319–324.

Komarova, Y. and A.B. Malik. 2010. Regulation of endothelial permeability via paracellular and transcellular transport pathways. Ann Rev Physiol 72: 463–493.

Komarova, Y.A., I.A. Vorobjev and G.G. Borisy. 2002. Life cycle of MTs: persistent growth in the cell interior, asymmetric transition frequencies and effects of the cell boundary. J Cell Sci 115: 3527–3539.

Komarova, Y.A., D. Mehta and A.B. Malik. 2007. Dual regulation of endothelial junctional permeability. Science's STKE: signal transduction knowledge environment 2007: re8.

Komarova, Y.A., F. Huang, M. Geyer et al. 2012. VE-Cadherin signaling induces EB3 phosphorylation to suppress microtubule growth and assemble adherens junctions. Molecular cell.

Kouklis, P., M. Konstantoulaki, S. Vogel et al. 2004. Cdc42 regulates the restoration of endothelial barrier function. Circulation Research 94: 159–166.

Kozasa, T., X. Jiang, M.J. Hart et al. 1998. p115 RhoGEF, a GTPase activating protein for Galpha12 and Galpha13. Science 280: 2109–2111.

Kratzer, E., Y. Tian, N. Sarich et al. 2012. Oxidative stress contributes to lung injury and barrier dysfunction via microtubule destabilization. Am J Resp Cell and Mol Biol 47: 688–697.

Krause, M., E.W. Dent, J.E. Bear et al. 2003. Ena/VASP proteins: regulators of the actin cytoskeleton and cell migration. Ann Rev Cell and Dev Biol 19: 541–564.

Krendel, M., F.T. Zenke and G.M. Bokoch. 2002. Nucleotide exchange factor GEF-H1 mediates cross-talk between microtubules and the actin cytoskeleton. Nature Cell Biology 4: 294–301.

Lang, P., F. Gesbert, M. Delespine-Carmagnat et al. 1996. Protein kinase A phosphorylation of RhoA mediates the morphological and functional effects of cyclic AMP in cytotoxic lymphocytes. The EMBO Journal 15: 510–519.

Lansbergen, G. and A. Akhmanova. 2006. Microtubule plus end: a hub of cellular activities. Traffic 7: 499–507.

Larsen, J. K., I.A. Yamboliev, L.A.Weber et al. 1997. Phosphorylation of the 27-kDa heat shock protein via p38 MAP kinase and MAPKAP kinase in smooth muscle. Am J Physiol: Lung Cell and Mol Physiol 273: L930–940.

Lawrence, D.W., K.M. Comerford and S.P. Colgan. 2002. Role of VASP in reestablishment of epithelial tight junction assembly after Ca^{2+} switch. Am J Physiol: Cell Physiol 282: C1235–1245.

Lee, J.S. and A.I. Gotlieb. 2003a. Understanding the role of the cytoskeleton in the complex regulation of the endothelial repair. Histol and Histopathol 18: 879–887.

Lee, T.Y. and A.I. Gotlieb. 2003b. Microfilaments and microtubules maintain endothelial integrity. Microscopy Research and Technique 60: 115–127.

Liu, F., A.D.Verin, T. Borbiev et al. 2001. Role of cAMP-dependent protein kinase A activity in endothelial cell cytoskeleton rearrangement. Am J Physiol: Lung Cell and Mol Physiol 280: L1309–1317.

Loktionova, S.A. and A.E. Kabakov. 1998. Protein phosphatase inhibitors and heat preconditioning prevent Hsp27 dephosphorylation, F-actin disruption and deterioration of morphology in ATP-depleted endothelial cells. FEBS Letters 433: 294–300.

Lu, Q., E.O. Harrington, H. Jackson et al. 2006. Transforming growth factor-beta1-induced endothelial barrier dysfunction involves Smad2-dependent p38 activation and subsequent RhoA activation. J Apply Physiol 101: 375–384.

Lum, H. and A.B. Malik. 1996. Mechanisms of increased endothelial permeability. Can J Physiol Pharmacol 74: 787–800.

MacRae, T.H. 1992. Microtubule organization by cross-linking and bundling proteins. Biochem and Biophysic Acta 1160: 145–155.

Maekawa, M., T. Ishizaki, S. Boku et al. 1999. Signaling from Rho to the actin cytoskeleton through protein kinases ROCK and LIM-kinase. Science 285: 895–898.

Markert, T., V. Krenn, J. Leebmann et al. 1996. High expression of the focal adhesion- and microfilament-associated protein VASP in vascular smooth muscle and endothelial cells of the intact human vessel wall. Basic Research in Cardiology 91: 337–343.

Marston, S.B. and C.S. Redwood. 1991. The molecular anatomy of caldesmon. Biochem J 279(Pt 1): 1–16.

Mary, S., S. Charrasse, M. Meriane et al. 2002. Biogenesis of N-cadherin-dependent cell-cell contacts in living fibroblasts is a microtubule-dependent kinesin-driven mechanism. Mol Biol Cell 13: 285–301.

Mehta, D. and A.B. Malik. 2006. Signaling mechanisms regulating endothelial permeability. Physiol Rev 86: 279–367.

Meng, W., Y. Mushika, T. Ichii et al. 2008. Anchorage of microtubule minus ends to adherens junctions regulates epithelial cell-cell contacts. Cell 135: 948–959.

Mitchison, T. and M. Kirschner. 1984a. Dynamic instability of microtubule growth. Nature 312: 237–242.

Mitchison, T. and M. Kirschner. 1984b. Microtubule assembly nucleated by isolated centrosomes. Nature 312: 232–237.

Moritz, M. and D.A. Agard. 2001. Gamma-tubulin complexes and microtubule nucleation. Curr Opin Struct Biol. 11: 174–181.

Moy, A.B., S.S. Shasby, B.D. Scott et al. 1993. The effect of histamine and cyclic adenosine monophosphate on myosin light chain phosphorylation in human umbilical vein endothelial cells. J Clin Invest 92: 1198–1206.

Nakahara, H., T. Otani, T. Sasaki et al. 2003. Involvement of Cdc42 and Rac small G proteins in invadopodia formation of RPMI7951 cells. Genes to cells: devoted to molecular & cellular mechanisms 8: 1019–1027.

Narumiya, S., T. Ishizaki and N. Watanabe. 1997. Rho effectors and reorganization of actin cytoskeleton. FEBS Letters 410: 68–72.

Niu, J., R. Vaiskunaite, N. Suzuki et al. 2001. Interaction of heterotrimeric G13 protein with an A-kinase-anchoring protein 110 (AKAP110) mediates cAMP-independent PKA activation. Current biology: CB 11: 1686–1690.

Nobes, C.D. and A. Hall. 1995. Rho, rac, and cdc42 GTPases regulate the assembly of multimolecular focal complexes associated with actin stress fibers, lamellipodia, and filopodia. Cell 81: 53–62.

Nogales, E. 2000. Structural insights into microtubule function. Ann Rev Biochem 69: 277–302.

Palazzo, A.F., T.A. Cook, A.S. Alberts et al. 2001. mDia mediates Rho-regulated formation and orientation of stable microtubules. Nature Cell Biology 3: 723–729.

Panasenko, O.O., M.V. Kim, S.B. Marston et al. 2003. Interaction of the small heat shock protein with molecular mass 25 kDa (hsp25) with actin. European Journal of Biochemistry/FEBS 270: 892–901.

Parizi, M., E.W. Howard and J.J. Tomasek. 2000. Regulation of LPA-promoted myofibroblast contraction: role of Rho, myosin light chain kinase, and myosin light chain phosphatase. Exp Cell Res 254: 210–220.

Patterson, C.E. and J.G. Garcia. 1994. Regulation of thrombin-induced endothelial cell activation by bacterial toxins. Blood Coagul Fibrinolysis 5: 63–72.

Patterson, C.E., H.W. Davis, K.L. Schaphorst et al. 1994. Mechanisms of cholera toxin prevention of thrombin- and PMA-induced endothelial cell barrier dysfunction. Microvascular Research 48: 212–235.

Patterson, C.E., H. Lum, K.L. Schaphorst et al. 2000. Regulation of endothelial barrier function by the cAMP-dependent protein kinase. Endothelium: J Endoth Cell Res 7: 287–308.

Paul, A., S. Wilson, C.M. Belham et al. 1997. Stress-activated protein kinases: activation, regulation and function. Cell Signal 9: 403–410.

Petrache, I., A.D. Verin, M.T. Crow et al. 2001. Differential effect of MLC kinase in TNF-alpha-induced endothelial cell apoptosis and barrier dysfunction. Am J Physiol: Lung Cell and Mol Physiol 280: L1168–1178.

Petrache, I., A. Birukova, S.I. Ramirez et al. 2003. The role of the microtubules in tumor necrosis factor-alpha-induced endothelial cell permeability. Am J Resp Cell and Mol Biol 28: 574–581.

Piotrowicz, R.S. and E.G. Levin. 1997. Basolateral membrane-associated 27-kDa heat shock protein and microfilament polymerization. J Biol Chem 272: 25920–25927.

Prager-Khoutorsky, M., I. Goncharov, A. Rabinkov et al. 2007. Allicin inhibits cell polarization, migration and division via its direct effect on microtubules. Cell Motil. Cytoskeleton 64: 321–337.

Price, C.J. and N.P. Brindle. 2000. Vasodilator-stimulated phosphoprotein is involved in stress-fiber and membrane ruffle formation in endothelial cells. Art Thromb and Vasc Biol 20: 2051–2056.

Pritchard, K. and S.B. Marston. 1993. The Ca(2+)-sensitizing component of smooth muscle thin filaments: properties of regulatory factors that interact with caldesmon. Biochem Biophys Res Commu 190: 668–673.

Qiao, J., F. Huang and H. Lum. 2003. PKA inhibits RhoA activation: a protection mechanism against endothelial barrier dysfunction. Am J Physiol: Lung Cell and Mol Cell 284: L972–980.

Reinhard, M., T. Jarchau and U. Walter. 2001. Actin-based motility: stop and go with Ena/VASP proteins. Trends in biochemical sciences 26: 243–249.

Ren, Y., R. Li, Y. Zheng et al. 1998. Cloning and characterization of GEF-H1, a microtubule-associated guanine nucleotide exchange factor for Rac and Rho GTPases. J Biol Chem 273: 34954–34960.

Rentsendorj, O., T. Mirzapoiazova, D. Adyshev et al. 2008. Role of vasodilator-stimulated phosphoprotein in cGMP-mediated protection of human pulmonary artery endothelial barrier function. Am J Physiol: Lung Cell and Mol Physiol 294: L686–697.

Reynolds, C.H., A.R. Nebreda, G.M. Gibb et al. 1997. Reactivating kinase/p38 phosphorylates tau protein *in vitro*. J Neurochem 69: 191–198.

Rodriguez, O.C., A.W. Schaefer, C.A. Mandato et al. 2003. Conserved microtubule-actin interactions in cell movement and morphogenesis. Nat Cell Biol 5: 599–609.

Rogalla, T., M. Ehrnsperger, X. Preville et al. 1999. Regulation of Hsp27 oligomerization, chaperone function, and protective activity against oxidative stress/tumor necrosis factor alpha by phosphorylation. J Biol Chem 274: 18947–18956.

Rousseau, S., F. Houle, J. Landry et al. 1997. p38 MAP kinase activation by vascular endothelial growth factor mediates actin reorganization and cell migration in human endothelial cells. Oncogene 15: 2169–2177.

Salaycik, K.J., C.J. Fagerstrom, K. Murthy et al. 2005. Quantification of microtubule nucleation, growth and dynamics in wound-edge cells. J Cell Sci 118: 4113–4122.

Sammak, P.J. and G.G. Borisy. 1988. Direct observation of microtubule dynamics in living cells. Nature 332: 724–726.

Sato, K., M. Hori, H. Ozaki et al. 1992. Myosin phosphorylation-independent contraction induced by phorbol ester in vascular smooth muscle. J Phamacol and Exp Therap 261: 497–505.

Sayas, C.L., M.T. Moreno-Flores, J. Avila et al. 1999. The neurite retraction induced by lysophosphatidic acid increases Alzheimer's disease-like Tau phosphorylation. J Biol Chem 274: 37046–37052.

Schafer, C., S.E. Ross, M.J. Bragado et al. 1998. A role for the p38 mitogen-activated protein kinase/Hsp 27 pathway in cholecystokinin-induced changes in the actin cytoskeleton in rat pancreatic acini. J Biol Chem 273: 24173–24180.

Schafer, C., P. Clapp, M.J. Welsh et al. 1999. HSP27 expression regulates CCK-induced changes of the actin cytoskeleton in CHO-CCK-A cells. The American journal of physiology 277: C1032–1043.

Schlegel, N., S. Burger, N. Golenhofen et al. 2008. The role of VASP in regulation of cAMP- and Rac 1-mediated endothelial barrier stabilization. Am J Physiol: Cell Physiol 294: C178–188.

Schober, J.M., Y.A. Komarova, O.Y. Chaga et al. 2007. Microtubule-targeting-dependent reorganization of filopodia. J Cell Sci 120: 1235–1244.

Schulze, E. and M. Kirschner. 1986. Microtubule dynamics in interphase cells. J Cell Biol 102: 1020–1031.

Schulze, E. and M. Kirschner. 1987. Dynamic and stable populations of microtubules in cells. J Cell Biol 104: 277–288.

Sehrawat, S., X. Cullere, S. Patel et al. 2008. Role of Epac1, an exchange factor for Rap GTPases, in endothelial microtubule dynamics and barrier function. Mol Biol Cell 19: 1261–1270.

Sehrawat, S., T. Ernandez, X. Cullere et al. 2011. AKAP9 regulation of microtubule dynamics promotes Epac1-induced endothelial barrier properties. Blood 117: 708–718.

Seidman, R., I. Gitelman, O. Sagi et al. 2001. The role of ERK 1/2 and p38 MAP-kinase pathways in taxol-induced apoptosis in human ovarian carcinoma cells. Exp Cell Res 268: 84–92.

Shelden, E. and P. Wadsworth. 1993. Observation and quantification of individual microtubule behavior *in vivo*: microtubule dynamics are cell-type specific. J Cell Biol 120: 935–945.

Sheldon, R., A. Moy, K. Lindsley et al. 1993. Role of myosin light-chain phosphorylation in endothelial cell retraction. Am J Physiol: Lung Cell and Mol Physiol 265: L606–612.

Shewan, A.M., M. Maddugoda, A. Kraemer et al. 2005. Myosin 2 is a key Rho kinase target necessary for the local concentration of E-cadherin at cell-cell contacts. Mol Biol Cell 16: 4531–4542.

Shimizu, H., M. Ito, M. Miyahara et al. 1994. Characterization of the myosin-binding subunit of smooth muscle myosin phosphatase. J Biol Chem 269: 30407–30411.

Shinohara-Gotoh, Y., E. Nishida, M. Hoshi et al. 1991. Activation of microtubule-associated protein kinase by microtubule disruption in quiescent rat 3Y1 cells. Exp Cell Res 193: 161–166.

Shirazi, A., K. Iizuka, P. Fadden et al. 1994. Purification and characterization of the mammalian myosin light chain phosphatase holoenzyme. The differential effects of the holoenzyme and its subunits on smooth muscle. J Biol Chem 269: 31598–31606.

Signor, D. and J.M. Scholey. 2000. Microtubule-based transport along axons, dendrites and axonemes. Essays in Biochemistry 35: 89–102.

Small, J.V. and I. Kaverina. 2003. Microtubules meet substrate adhesions to arrange cell polarity. Curr Opin Cell Biol 15: 40–47.

Small, J.V., B. Geiger, I. Kaverina et al. 2002. How do microtubules guide migrating cells? Nat Rev Mol Cell Biol 3: 957–964.

Smurova, K.M., A.A. Biriukova, A.D. Verin et al. 2008. The microtubule system in endothelial barrier dysfunction: disassembly of peripheral microtubules and microtubules reorganization in internal cytoplasm. Tsitologiia 50: 49–55.

Smurova, K.M., A.D. Verin and I.B. Alieva. 2011. The effect of Rho-kinase inhibition depends on the nature of factors that modify endothelial permeability. Tsitologiia 53: 359–366.

Sobue, K. and J.R. Sellers. 1991. Caldesmon, a novel regulatory protein in smooth muscle and nonmuscle actomyosin systems. J Biol Chem 266: 12115–12118.

Song, Y., C. Wong and D.D. Chang. 2000. Overexpression of wild-type RhoA produces growth arrest by disrupting actin cytoskeleton and microtubules. J Cell Biochem 80: 229–240.

Stehbens, S.J., A.D. Paterson, M.S. Crampton et al. 2006. Dynamic microtubules regulate the local concentration of E-cadherin at cell-cell contacts. J Cell Sci 119: 1801–1811.

Stepanova, T., J. Slemmer, C.C. Hoogenraad et al. 2003. Visualization of microtubule growth in cultured neurons via the use of EB3-GFP (end-binding protein 3-green fluorescent protein). J Neurosci 23: 2655–2664.

Stone, A.A. and T.C. Chambers. 2000. Microtubule inhibitors elicit differential effects on MAP kinase (JNK, ERK, and p38) signaling pathways in human KB-3 carcinoma cells. Exp Cell Res 254: 110–119.

Suomalainen, M., M.Y. Nakano, K. Boucke et al. 2001. Adenovirus-activated PKA and p38/MAPK pathways boost microtubule-mediated nuclear targeting of virus. The EMBO Journal 20: 1310–1319.

Suzuki, S., H. Bing, T. Sugawara et al. 2004. Paclitaxel prevents loss of pulmonary endothelial barrier integrity during cold preservation. Transplantation 78: 524–529.

Takenaka, K., T. Moriguchi and E. Nishida. 1998. Activation of the protein kinase p38 in the spindle assembly checkpoint and mitotic arrest. Science 280: 599–602.

Takuwa, Y. 2002. Subtype-specific differential regulation of Rho family G proteins and cell migration by the Edg family sphingosine-1-phosphate receptors. Biochim Biophy Acta 1582: 112–120.

Tamma, G., E. Klussmann, G. Procino et al. 2003. cAMP-induced AQP2 translocation is associated with RhoA inhibition through RhoA phosphorylation and interaction with RhoGDI. J Cell Sci 116: 1519–1525.

Tapon, N. and A. Hall. 1997. Rho, Rac and Cdc42 GTPases regulate the organization of the actin cytoskeleton. Curr Opin Cell Biol 9: 86–92.

Tar, K., C. Csortos, I. Czikora et al. 2006. Role of protein phosphatase 2A in the regulation of endothelial cell cytoskeleton structure. Journal of Cellular Biochemistry 98: 931–953.

Tsuji, T., T. Ishizaki, M. Okamoto et al. 2002. ROCK and mDia1 antagonize in Rho-dependent Rac activation in Swiss 3T3 fibroblasts. J Cell Biol 157: 819–830.

van Horck, F.P., M.R. Ahmadian, L.C. Haeusler et al. 2001. Characterization of p190RhoGEF, a RhoA-specific guanine nucleotide exchange factor that interacts with microtubules. J Biol Chem 276: 4948–4956.

Vaughan, P.S., P. Miura, M. Henderson et al. 2002. A role for regulated binding of p150(Glued) to microtubule plus ends in organelle transport. J Cell Biol 158: 305–319.

Verin, A.D., C.E. Patterson, M.A. Day et al. 1995. Regulation of endothelial cell gap formation and barrier function by myosin-associated phosphatase activities. Am J Physiology: Lung Cell and Mol Physiol 269: L99–108.

Verin, A.D., C. Cooke, M. Herenyiova et al. 1998a. Role of Ca²⁺/calmodulin-dependent phosphatase 2B in thrombin-induced endothelial cell contractile responses. The American Journal of Physiology 275: L788–799.

Verin, A.D., L.I. Gilbert-McClain, C.E. Patterson et al. 1998b. Biochemical regulation of the nonmuscle myosin light chain kinase isoform in bovine endothelium. Am J Resp Cell and Mol Biol 19: 767–776.

Verin, A.D., C. Csortos, S.D. Durbin et al. 2000a. Characterization of the protein phosphatase 1 catalytic subunit in endothelium: involvement in contractile responses. J Cell Biochem 79: 113–125.

Verin, A.D., F. Liu, N. Bogatcheva et al. 2000b. Role of ras-dependent ERK activation in phorbol ester-induced endothelial cell barrier dysfunction. Am J Physiol: Lung Cell and Mol Physiol 279: L360–370.

Verin, A.D., A. Birukova, P. Wang et al. 2001. Microtubule disassembly increases endothelial cell barrier dysfunction: role of MLC phosphorylation. Am J Physiol: Lung Physiol Cell and Mol Physiol 281: L565–574.

Vogelmann, R. and W.J. Nelson. 2005. Fractionation of the epithelial apical junctional complex: reassessment of protein distributions in different substructures. Mol Biol Cell 16: 701–716.

Vorobjev, I.A., V.I. Rodionov, I.V. Maly et al. 1999. Contribution of plus and minus end pathways to microtubule turnover. J Cell Sci 112(Pt 14): 2277–2289.

Vorobjev, I.A., I.B. Alieva, I.S. Grigoriev et al. 2003. Microtubule dynamics in living cells: direct analysis in the internal cytoplasm. Cell Biol Int 27: 293–294.

Vouret-Craviari, V., P. Boquet, J. Pouyssegur et al. 1998. Regulation of the actin cytoskeleton by thrombin in human endothelial cells: role of Rho proteins in endothelial barrier function. Mol Biol Cell 9: 2639–2653.

Walker, R.A., E.T. O'Brien, N.K. Pryer et al. 1988. Dynamic instability of individual microtubules analyzed by video light microscopy: rate constants and transition frequencies. J Cell Biol 107: 1437–1448.

Wallar, B.J. and A.S. Alberts. 2003. The formins: active scaffolds that remodel the cytoskeleton. Trends in Cell Biology 13: 435–446.

Ware, L.B. and M.A. Matthay. 2000. The acute respiratory distress syndrome. The New England Journal of Medicine 342: 1334–1349.

Watanabe, T., S. Wang, J. Noritake et al. 2004. Interaction with IQGAP1 links APC to Rac1, Cdc42, and actin filaments during cell polarization and migration. Developmental Cell 7: 871–883.

Watanabe, T., J. Noritake and K. Kaibuchi. 2005. Regulation of microtubules in cell migration. Trends Cell Biol 15: 76–83.

Waterman-Storer, C.M. and E.D. Salmon. 1997. Actomyosin-based retrograde flow of microtubules in the lamella of migrating epithelial cells influences microtubule dynamic instability and turnover and is associated with microtubule breakage and treadmilling. J Cell Biol 139: 417–434.

Waterman-Storer, C.M., W.C. Salmon and E.D. Salmon. 2000. Feedback interactions between cell-cell adherens junctions and cytoskeletal dynamics in newt lung epithelial cells. Mol Biol Cell 11: 2471–2483.

Wehrle-Haller, B. and B.A. Imhof. 2003. Actin, microtubules and focal adhesion dynamics during cell migration. Int J Biochem and Cell Biol 35: 39–50.

Wen, Y., C.H. Eng, J. Schmoranzer et al. 2004. EB1 and APC bind to mDia to stabilize microtubules downstream of Rho and promote cell migration. Nat Cell Biol 6: 820–830.

Wettschureck, N. and S. Offermanns. 2002. Rho/Rho-kinase mediated signaling in physiology and pathophysiology. J Mol Med 80: 629–638.

Whitney, G., D. Throckmorton, C. Isales et al. 1995. Kinase activation and smooth muscle contraction in the presence and absence of calcium. J Vasc Surgery 22: 37–44.

Wojciak-Stothard, B. and A.J. Ridley. 2002. Rho GTPases and the regulation of endothelial permeability. Vasc Pharmacol 39: 187–199.

Wojciak-Stothard, B., S. Potempa, T. Eichholtz et al. 2001. Rho and Rac but not Cdc42 regulate endothelial cell permeability. J Cell Sci 114: 1343–1355.

Wysolmerski, R.B. and D. Lagunoff. 1990. Involvement of myosin light-chain kinase in endothelial cell retraction. PNAS of the United States of America 87: 16–20.

Wysolmerski, R.B. and D. Lagunoff. 1991. Regulation of permeabilized endothelial cell retraction by myosin phosphorylation. Am J Physiol: Cell Physiol 261: C32–40.

Yamaoka-Tojo, M., M. Ushio-Fukai, L. Hilenski et al. 2004. IQGAP1, a novel vascular endothelial growth factor receptor binding protein, is involved in reactive oxygen species—dependent endothelial migration and proliferation. Circulation Research 95: 276–283.

Yamashiro, S., Y. Yamakita, K. Yoshida et al. 1995. Characterization of the COOH terminus of non-muscle caldesmon mutants lacking mitosis-specific phosphorylation sites. J Biol Chem 270: 4023–4030.

Zheng, Y. 2001. Dbl family guanine nucleotide exchange factors. Trends Biochem Sci 26: 724–732.

Zieger, M., S. Tausch, P. Henklein et al. 2001. A novel PAR-1-type thrombin receptor signaling pathway: cyclic AMP-independent activation of PKA in SNB-19 glioblastoma cells. Biochem Biophy Res Commun 282: 952–957.

7

Membrane-Cytoskeleton Interactions and Control of Vesicle Trafficking in Endothelial Cells

Felicia Antohe

Introduction

Endothelial cells are specialized epithelial cells that cover all blood vessels of the cardiovascular system in the body (Simionescu and Simionescu 1988). The vascular network is responsible for the transport of blood from the heart to the peripheral tissues, assuring rapid exchange of nutrients and waste products. The study of macromolecule trafficking in endothelial cells and the transmigration of cellular elements and pathogenic microorganisms clearly demonstrate that there are complex temporal and structural interactions between the intracellular cytoskeleton and associated components and the membrane proteins, governed by controlled energy production and involvement of different small protein kinases that mediate various signaling pathways. The transcellular transport of blood circulating macromolecules, cells or pathogens is severely restricted and

Institute of Cellular Biology and Pathology Nicolae Simionescu, Bucharest, Romania.
Email: felicia.antohe@icbp.ro

requires specific processes and highly regulated pathways (Simionescu et al. 2002). To maintain the permeability and the non-thrombogenic surface of the endothelium, the integrity of the monolayer is crucial in order to prevent vascular endothelial cell dysfunctions. Thus, the proper structure and function of the endothelial cytoskeleton governs the complex interactions between the protein complexes during cell motility, cell-cell and cell-substratum adhesion, cell signaling and wound repair (Lee and Gotlieb 2003). The endothelial cell membranes consist of heterogeneous but regulated domains formed by proteins and lipids that form specialized membrane structures (lipid rafts, coated pits and vesicles, plasmalemmal pits and vesicles or caveolae and channels) in which specific receptors are located (Simionescu and Simionescu 1991). In addition, the endothelial cells are endowed with apparently less organized cytoskeleton, containing as major components the actin, myosin II, tropomyosin, α-actinin and actin-binding proteins, such as fodrin, gelsolin, girdin, protein 4.1, filamin, vinculin, talin, vimentin and many others (Lee Tsu-Yee and Gotlieb 2003). The cytoskeleton in endothelial cells is a highly dynamic structure, and its role in establishing the protein composition of membrane microdomains to activate or inhibit various membrane functions in response to both internal and external stimulations has come under intensive investigations (Chichili and Rodgers 2009, Manneville et al. 2003, Waschke et al. 2005).

The connections between the plasma membrane proteins and the cytoskeletal elements confer on the endothelial monolayer, specific characteristics, namely cellular polarity, phenotypic heterogeneity and the existence of differentiated functional membrane microdomains that make the endothelium a key player in the regulation of all vascular functions.

The Cytoskeleton Forms Endothelial Cell Polarity

Cell polarity, as a fundamental biological property, is very well illustrated by endothelial cell functions that assure the separation of the intra- and extra-vascular compartment, and the vectorial exchange of substances. These properties coexist and maintain the asymmetric accumulation of mobile regulatory molecules between opposite fronts of the cell and the intracellular oriented components of the cytoskeleton elements, mainly actin and microtubules. Rapid polymerization or de-polymerization of globular actin (G-actin) sub-units or α- and β-tubulin heterodimeric sub-units takes place by ATP and GTP hydrolysis, respectively. The intrinsic polarity and dynamics of actin filaments and microtubules through a large number of cytoskeleton-associated proteins generate structural asymmetry and well oriented functions with rapid turnover and local reorganization in response to polarity signals. The polarity of the endothelium also has to be taken into account when developing different therapeutic strategies, since not only the efficient expression of the desired product, but also its correct cellular location or secretion into the correct extracellular compartment will assure the success of the applied therapeutic strategy.

Pioneering experiments performed *in situ* on mouse endothelial cells using cationized ferritin demonstrated microdomains of different charges on the luminal and abluminal front of the cells accompanied by preferential distribution of specific lectins (Simionescu et al. 1981).

Endothelial cells are connected through intercellular junctions (Dejana et al. 2009) that separate the functionally and structurally distinct luminal and abluminal cell surfaces. The luminal plasma membrane is in contact with the blood and takes part in the homeostasis of circulating macromolecules while the abluminal cell membrane connects the endothelial cell with the basement membrane and modulates blood flow, monitoring the selective transport of nutrients and adequate release of vasoactive agents, hormones or altered substances. The polarized cell phenotype is generated by specific protein sorting and regulated protein trafficking between the trans-Golgi network and the cell surface. The polarized distribution of cell membrane proteins is maintained by anchorage to the cytoskeleton and limitation of lateral diffusion by tight junctions. Immunological erythrocyte related proteins seem to be responsible for the characteristic cobblestone morphology of the endothelial cell monolayer. Indeed, indirect immunofluorescence on endothelial cells, both *in situ* and *in vitro*, showed the expression of proteins that cross-reacted with antibodies against spectrin, fodrin (Heltianu et al. 1986) and protein 4.1 (Constantinescu et al. 1986). These findings indicate that endothelial cells express proteins antigenically related to the spectrin family (Fig. 1a and b) that support the shape, thickness and lateral integrity of the cells. Disturbances of spatial and temporal interactions between these partners may contribute to the dysfunctions in disease states, including altered permeability, ischemia, inflammation, and carcinogenesis.

Endothelial cell polarity is defined as a functional adaptation for the specific exchanges between blood and interstitial fluids. The cells display a different composition of plasma membrane constituents on the luminal and basal front in terms of proteins, receptors, specific enzymes and others (Simionescu and Antohe 2006). Control of cell polarity is crucial during tissue morphogenesis and renewal. The proper function or dysfunction of the endothelium depends on spatial cues provided by the extracellular environment. Mechanical forces and cyclic stretching supported by endothelial cells were different on apical versus basal surface, generating adequate responses in the endothelial monolayer. Blood flow generates several types of hemodynamic forces (hydrostatic pressure, cyclic strains, and wall shear stresses) that modulates endothelial cell motility and polarity. In human umbilical vein endothelial cells (HUVEC), the Ca^{2+}-dependent

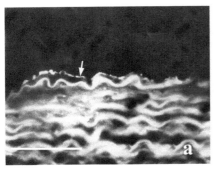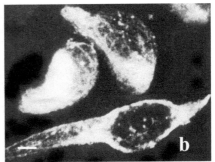

Figure 1. Spectrin and band 4.1 protein in endothelial cells. a. Frozen section of rabbit aorta incubated with affinity-purified anti-spectrin IgG followed by FITC-labeled secondary antibody. The fluorescence was restricted to endothelial cells (arrow). Underneath endothelial layer, the elastic lamellae show autofluorescence; b. Endothelial cells freshly isolated from human umbilical vein, attached to poly-L-lysine coated coverslips and incubated with anti-protein 4.1 antibody, followed by FITC-labeled secondary antibody. The fluorescence is displayed over the entire cytoplasm. bar = 10 µm. (By permission from E. Constantinescu).

intracellular cysteine protease calpain, which is localized to stress-induced focal adhesion sites, induces cell motility through cleavage of the integrin-β3 sub-unit, as well as talin, focal adhesion kinase (FAK) and vinculin, primarily through the PI3K pathway (Miyazaki et al. 2007). Yamada and Ando (2007) demonstrated in porcine endothelial cells that the differential formation of apical and basal actin stress fibers may be due to differences in the transmission of mechanical stretching to the central and apical regions of the cell through the actin fibers network.

Cell-to-cell contact might change the strength, orientation, and anchorage of apical actin filaments playing a critical role in mechanical signal transduction. Experimental model on human aortic endothelial cells (HAEC) in culture that were axially and cyclically stretched on silicone rubber membranes coated with various concentrations of fibronectin, collagen type IV and laminin, showed that the extracellular stress produces different amounts of adhesiveness. The authors claimed that the orientation response to cyclic stretching is not a spontaneous motile response, but determined in large part by factors that affect the cell substratum adhesiveness and actin-myosin intracellular contractile level (Ngu et al. 2008). The intermediate filaments composed of vimentin were highly abundant compared to stress fibers, mainly in the basal part of the endothelial cells (Uehara and Uehara 2010). The authors showed that some of the intermediate filaments were closely mixed with actin filaments in stress fibers, and were associated with coated vesicles. Plectin, a scaffold protein, was predominantly localized in the layers of vimentin and stress fibers, thus contributing to the mechanical stabilization of endothelial cells.

Using micro-patterned substrates to impose reproducible cell-cell interactions, Dupin et al. showed that in the absence of other polarizing cues, cell-cell contacts are the main regulator of nucleus and centrosome positioning and intracellular polarized organization (Dupin et al. 2009). Nucleus and centrosome location is controlled by N-cadherin through the regulation of cell interactions with the extracellular matrix, whereas the orientation of the nucleus-centrosome axis is determined by the geometry of N-cadherin-mediated contacts. Important participants in defining the endothelium polarity are claudins, proteins that represent the major constituents of endothelial tight junctions. The claudins are connected via a C-terminal PDZ-binding motif with several junction associated proteins that provide a link to the actin cytoskeleton, at the apical tight junction. This indicates that crucial determinants for stable tight junction incorporation of claudins reside in a cytoplasmic C-terminal sequence implicated in specific protein-protein interactions with the elements of the cytoskeleton (Rüffer and Gerke 2004).

The endothelial cell planar polarity in blood vessels was considered to be age and vessel-specific, being caused by shear-related regulation of glycogen synthase kinase-3β (GSK-3β), known to be involved in endothelial microtubule stability. Moreover, when the activity of GSK-3β is inhibited, endothelial cell polarity is reversed, demonstrating that the vascular endothelium displays a specific adaptive mode of mechano-sensitive response to the local blood flow (McCue et al. 2006).

Cell polarity is an essential feature of cell movement, cellular regeneration, and angiogenesis. Cell motility is characterized by anterior-posterior polarization of multiple cell structures such as integrin-cytoskeleton complexes, actin filaments, actin binding proteins and signaling proteins that are preferentially localized in lamellipodial extensions. Experiments performed on endothelial cells in culture showed that plasma membrane micro-viscosity is increased at the cell leading edge due to a reduced cholesterol concentration, which induces the increases of actin-mediated vesicle deformation (Vasanji et al. 2004).

Specific inhibitors of endothelial cell movement may reduce angiogenesis and tumor growth. Indeed, the matrix component endostatin (the C-terminal fragment of collagen XVIII) is a potent inhibitor of angiogenesis. Wickström et al., showed that endostatin induced the recruitment of $\alpha_5\beta_1$ integrin into the raft fraction via a heparin sulfate proteoglycan-dependent mechanism (Wickström et al. 2003). The close vicinity of $\alpha_5\beta_1$ integrin and heparan sulfate in lipid raft-mediated interactions triggered Src-dependent activation of p190RhoGAP (p190), a Rho family GTPase-activating protein, with concomitant decrease in RhoA activity, and disassembly of actin stress fibers and focal adhesions. Contrary to this, when wound repair or increased proliferation are needed, Wickström et al. claimed that the low concentration of membrane cholesterol could be beneficial.

In the human umbilical vein, endothelial cells infected by adenoviral vectors expressing dominant-negative or constitutively active mutants of RhoA (N19RhoA/RhoA63), ROCK (RB/PH(TT)/CAT) and Rac1 (N17RAC), the binding of T-cadherin (T-cad) to the T-cad receptors promotes cell elongation and polarization while adhesion to the matrix was inhibited. The reported data demonstrated that the RhoA/ROCK pathway is necessary for cell contraction, stress fiber assembly, and inhibition of spreading, whereas Rac is required for formation of actin-rich lamellipodia at the leading edges of polarized cells (Philippova et al. 2005).

Caveolin-1 was shown to undergo phosphorylation of Tyr^{14} to promote interaction with intermediate filaments during the anterior polarization of transmigrating endothelial cells. It was shown by immuno-electron microscopy that caveolin-1 is distributed along cytoskeleton structures and co-localizes with intermediate filaments, thus facilitating optimal transmigration (Santilman et al. 2007).

Endothelial cell angiogenesis is a crucial process both in vasculogenesis and in vascular repair associated with vascular inflammatory disease and cancer. The cells suffer profound changes in shape and polarity accompanied by cytoskeleton remodeling and by cyclic interactions of the cytoskeleton with plasma membrane elements. The elongation of endothelial cells during angiogenesis is accompanied by stabilization of microtubules and their alignment into parallel arrays directed along the growing cellular tip. This is regulated by members of the formin family (FMNL). The novel, recently described FMNL3 regulator seems to be responsible for the conversion of quiescent endothelial cells into their elongated angiogenic morphology that allows the proliferation of endothelial cells under appropriate stimulation (Hetheridge et al. 2012).

Highly regulated ligand-dependent mechanisms accompanied by post-translational modifications of intracellular proteins maintain the preferential distributions of different receptors on the apical or basolateral surface of endothelial cells and differentiate between the healthy versus diseased conditions. Indeed, biochemical assays performed on endothelial cells grown on semi-permeable filters indicated that, upon ligand binding, the LDL receptors are translocated preferentially from the apical to the basal cell membrane. This ligand-induced redistribution of LDL receptors may explain, in part, the lipid accumulation in the vessel walls in hyperlipidemia (Antohe et al. 1997, 1999). The functional polarity of endothelial cells was also demonstrated by the selective transport mechanism of IgG from the placenta to the lumen of the fetal blood vessels. FcRn receptors expressed in human placental endothelial cells underwent endocytosis, recycling, or transcytosis upon IgG binding. The FcRn receptor is the key element that discriminates and monitors the pathway of native and modified IgG, maintaining the asymmetric transcellular transport of IgG (Antohe et al. 2001, 2004). However, the interactions of the cytoskeleton with components of the pathways that regulate IgG transcellular transport are largely unknown and should be addressed in the future.

One important intracellular player involved both in the transfer of information as second messenger and in the polymerization and, or depolymerization dynamics of cytoskeleton elements, are Ca^{+2} ions generated on the plasma membrane by receptor-ligand interactions. Published data showed that in human umbilical vein endothelial cells, alignment and cytoskeleton remodeling was accompanied by calpain-mediated cleavage of vinculin and talin, activated by the increase of Ca^{+2} influx that was elicited by prolonged application of shear stress. This activity seems to be facilitated primarily through the PI3K pathway, which mediates shear-induced endothelial cell polarity (Miyazaki et al. 2007).

All the reported data support the concept that local tuning of the actin cytoskeleton at adherens junction could allow monolayers of endothelial cells to behave as a functional unit that can withstand the stress of the continuous changes in blood flow that take place in the organism. The same local tuning regulates the junctional and actin rearrangements that occur during inflammation, leukocyte transmigration and angiogenic responses.

Cytoskeleton Modulators and Endothelial Membrane Heterogeneity Maintain the Selectivity of the Endothelial Barrier

The phenotypic heterogeneity of endothelial cells, conferred by differences in the frequency and occurrence of caveolae, trans-endothelial channels, fenestrae and inter-endothelial junctions, results in the structural attributes that underlie the various functions of the vascular endothelium. Under the influence of local physiological conditions, endothelial cells undergo specific segmental differentiation that involves mainly the frequency and occurrence of their main cellular constituents. Modulation of these constituents by the dynamics of the cortical actin cytoskeleton leads to large phenotypic heterogeneity, expressed in distinct characteristics of the endothelial cells of large vessels versus microvessels, and in significant differences between the arteriolar, capillary and venular endothelia among the latter (Simionescu 2000, Simionescu and Antohe 2006).

However specific protein-protein interactions in endothelial cells, including membrane protein-cytoskeleton, and protein-lipid interactions related to a specific location or associated with disease are not fully known.

The endothelial monolayer is the first barrier that makes contact with the nutrients, growth factors, hormones and other bioactive molecules, as well as metabolites, modified circulating macromolecules and pathogens. All of these components form specific or non-specific contact with components of the plasma membrane that trigger a response. Different stress factors, such as hyperglycemia, induce the formation of advanced glycation end-products (AGEs), tightly correlated with the severity of diabetic complications. Recently published data showed that in the chemically or genetically-induced diabetes in animal model, the expression of caveolin-1 is increased in endothelial cells. The cholesterol content and ACE enzymatic activity in detergent-resistant membrane domains of these cells were linked to structural and biochemical modifications. These alterations were even more pronounced in genetically-induced diabetes. Based on these data, the authors predicted that in the pathogenesis of diabetes, upregulation of caveolin-1 and increased ACE enzymatic activity may result in an inhibition of eNOS and, consequently, a decrease in NO production leading to the endothelial cell dysfunction (Uyy et al. 2010). Under this condition, moesin, a protein linker between actin filaments and the plasma membrane, is phosphorylated on Thr[558] by p38 kinase and Rho kinase (ROCK) in response to advanced glycation end-products (AGE). The level of moesin phosphorylation, was shown to increase in human dermal and cerebral microvessels, along with an increase in the permeability of the blood-brain barrier. Inhibition of p38 kinase and ROCK attenuated these responses (Li et al. 2011). These observations are suggestive examples demonstrating the critical role of cytoskeleton in endothelial dysfunction associated with diabetes.

The Rho GTPases of the Rho, Rac, and Cdc42 families were initially characterized as key regulators of actin cytoskeleton remodeling induced by extracellular signals (Song et al. 1996, Ridley 2001). Rho is required for endothelial cell migration across transwell filters in response to sphingosine-1-phosphate (Liu et al. 2001, Zeng et al. 2002), while Rac is required for both sphingosine-1-phosphate and VEGF-induced migration of endothelial cells (Ryu et al. 2002, Soga et al. 2001, Zeng et al. 2002). Rho GTPases have also been implicated in endothelial responses to shear stress (Li et al. 1999, Tzima et al. 2001), where Rho and Cdc42 translocate to the membrane and mediate the activation of the transcription factor AP-1 through the c-Jun NH2-terminal kinases (Li et al. 1999). Another group (Wojciak-Stothard and Ridley 2003) demonstrated that RhoA, Rac1, and Cdc42 are all activated during the early stages of shear stress-induced remodeling of the actin cytoskeleton in endothelial cells. RhoA is required for initial cell contraction and depolarization, and subsequently the coordinated action of RhoA and Rac1 is required for the alignment and movement of endothelial cells in the direction of the flow. It seems that rearrangements of the actin cytoskeleton associated with early phases of cell spreading in human microvascular endothelial (HMVE) cells suppress Rho activity by promoting accumulation of p190RhoGAP in the lipid rafts where it inhibits Rho. The results suggested that cell rounding fails to prevent accumulation of p190RhoGAP in lipid rafts and to increase Rho activity in cells that lack the cytoskeletal protein filamin. During spreading, filamin is degraded and cells that express a calpain-resistant form of filamin have high Rho activity even when spread. Filamin may therefore represent the link that connects cytoskeleton-dependent changes of cell shape to Rho inactivation during the earliest phases of cell spreading by virtue of its ability to promote accumulation of p190RhoGAP in the lipid rafts (Mammoto et al. 2007).

Other mechanisms involved in the control of actin dynamics that occurs at the plasma membrane employ PIP2, PIP3, small GTPases and G proteins. PIP3 turns on the polymerization of actin filaments and PIP2 ensures that the filaments grow in the desired directions. This mechanism has not yet been demonstrated in endothelial cells. The timing and location of actin polymerization are triggered by stimuli at the plasma membrane (Insall and Weiner 2001). The connections between endothelial cell response and various signals is still an open ground for new discoveries.

Lipid Raft-Cytoskeleton Interactions Assure the Structural Basis of the Signaling Pathways in Endothelial Cell

The existence of differentiated microdomains (Palade et al. 1982) was structurally and chemically defined by the variable distribution of specialized membrane structures such as rafts, caveolae, coated pits and vesicles; these microdomains represent the functional significance of membrane-cytoskeleton coupling. Distinct microdomains are maintained within the fluid bilayer either by interaction between the membrane proteins themselves or between membrane proteins and cytoskeleton elements located on the cytoplasmic side of the membrane. In addition, the sterol ring that surrounds the vesicle neck is thought to be generated by lateral phase separation between the transiently fused vesicle membrane and the plasmalemma proper (Simionescu et al. 1983). The temporally and topologically dynamic interplay between these complex factors and structures modulated by the metabolic requirements of the host tissue generates the structural and functional heterogeneity of the endothelium, which reflects its local differentiation (Simionescu et al. 2002).

Complex mechanisms involving controlled macromolecule interactions correspond to functionally discrete microdomains. The concept of plasma membrane microdomains was validated and elaborated when it was established that membrane domains form through associations between cholesterol and other membrane lipids, generating a lipid phase named *lipid rafts* with which specific membrane proteins associate (Simons and Ikonen 1997, Smart et al. 1999). These specialized structures on the plasma membrane have specific lipid-protein compositions that establish links to the cytoskeleton. It has been proposed that these structures are membrane domains in which signaling mechanisms might occur through a clustering of receptors and components of receptor-activated signaling cascades. The localization of these proteins in lipid rafts, which is regulated by the cytoskeleton, influences the potency and efficacy of different receptors and transporters both in normal and pathological conditions.

Membrane heterogeneity includes discrete membrane domains that could vary from nano- to micrometer size, formed through a combination of protein and lipid interactions. These domains are important for cell viability and any changes in the structure and composition of these membrane platforms could induce severe cellular dysfunctions (Banfi et al. 2006, Brown and London 2000, Jury et al. 2004, Koenig et al. 2008). Moreover, some pathogens utilize membrane domains to gain entry or exit from the target cell using the energy and driving forces of cytoskeleton components (Bhattacharya and Roy 2008, Cambi et al. 2004, Manes et al. 2003, Marsh and Helenius 2006). The functional properties of membrane domains underscore the importance of membrane structure and the interactions with the cortical actin cytoskeleton, which is composed of a lattice network of filaments that underline and attach to the plasma membrane. Proteomic studies on different cell types (mainly T cells, neutrophils, smooth muscle cells) show that isolated detergent resistant membrane (DRM) domains are particularly enriched in cytoskeleton proteins, indicative of interactions between the actin cytoskeleton and membrane rafts that could be important in forming and maintaining the rafts. The cytoskeletal proteins that are enriched in rafts include actin, tubulin, myosin, actinin, and supervillin. An important regulator of membrane–cytoskeleton interactions is the phosphoinositide PIP2, which serves as a co-factor for many of the proteins that anchor actin filaments to the plasma membrane or to G-protein coupled receptors (GPCRs), to extracellular-signal-regulated kinases (ERKs), and to Src family kinases (SFKs), (reviewed by Chichili and Rodgers 2009).

Membrane-cytoskeleton coupling was demonstrated to play major roles in many cellular responses, such as cell growth, differentiation, polarization, motility, biomechanical transduction, and others. A growing amount of evidence indicating that membrane-cytoskeleton interactions are regulated by the composition of the plasma membrane, suggests that the cholesterol-rich and caveolin-containing lipid rafts, are essential for membrane-cytoskeleton coupling. Several models for raft-cytoskeleton interactions were described in an excellent review published by Levitan and Gooch (Levitan and Gooch 2007). The authors evaluated the role of rafts in membrane-cytoskeleton interactions and cytoskeleton-dependent processes suggested by two groups of experimental data. The first documented the association of rafts with cytoskeleton modulators such as phosphatidylinositol 4.5 bisphophate (PIP2), Rho-GTPases, and integrins. The second showed direct associations of several cytoskeleton proteins such as vimentin, fodrin, myosin IIA, α-actinin, and F-actin with signaling molecules localized in the lipid rafts. Most data were generated in non-endothelial cells such as lymphocytes, neuronal cells, or smooth muscle cells. The mechanisms by which the rafts can influence the structure of the underlying cytoskeleton in different cell types remain controversial for the moment. However the main characteristics and the mechanisms described in other cell types systems are probably active in endothelial cells as well. Their specific functions in the endothelial monolayer remain to be validated.

Elegant experiments performed on endothelial cells demonstrated that the exposure of cells to oxLDL (considered to be one of the major risk factors for the development of atherosclerosis), disrupts lipid rafts and caveolae by cholesterol depletion and hydrolysis of sphingomyelin, another major component of the lipid rafts (Levitan and Gooch 2007). Membrane raft or caveolae disruption results in partial disappearance of individual stress fibers and clustering of F-actin in the peripheral region of the cells by activation of RhoA and of the Rho kinase pathway (Essler et al. 1999, Zhao et al. 1997). Endothelial cells exposed in culture to hyperlipidemic condition become gradually loaded with modified lipids, becoming in late stages of atherosclerosis, endothelial derived foam cells. These activated cells express adhesion molecules (VCAM-1, VLA-4), have enhance intracellular Ca^{2+} release, and demonstrate a high level of heat shock proteins such as Hsp27, Hsp70, and Hsp90 (Ivan and Antohe 2010). Endothelial-derived foam cells showed a modified pattern of actin and vinculin localization (Constantinescu et al. 1997). Actin appeared as a weakly-stained network around the nucleus, whereas vinculin was distributed as small granules throughout the cell cytoplasm (Fig. 2a and b). These experimental data suggest that in advanced atherosclerosis, some of the endothelial cytoskeleton proteins undergo modifications that could represent one of the important factors involved in further development of the atherosclerotic plaque. Exposure of endothelial cells in culture to oxLDL also induced low plasma membrane expression of the lipid raft marker G_{M1} (Byfield et al. 2006) and internalization of endothelial caveolin (Walton et al. 2003).

In summary, the data documented that the lipid rafts and caveolae disruption and internalization, accompanied by profound reorganization of the cytoskeleton, affect the latter's polymerization, stabilization, cross-linking, and membrane association, and induce changes in endothelial

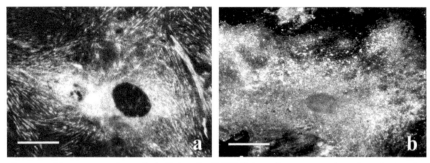

Figure 2. Hamster aortic endothelial cells (HAEC) in culture stained with anti-vinculin monoclonal antibody. a. HAEC after 24 hr incubation with the normal control serum; b. HAEC after 24 hr incubation with the hyperlipidemic serum. Bar = 10 μm. (By permission from E. Constantinescu).

cell function. The ability of the cells to sense and respond to mechanical stimuli involving changes in cell morphology, spreading, proliferation, and differentiation confers their capacity to form new blood vessels under different pathological conditions, including the development of atherosclerosis (Sieminski et al. 2004).

In late stages of atherosclerosis, intracellular accumulation of lipid droplets may disrupt membrane-cytoskeleton interaction with severe consequences for endothelial function (increased permeability and shift toward a secretor phenotype). Profound structural modifications accompanied by significant up-regulation of caveolin-1 expression, correlated with augmented ACE activity and cellular cholesterol content were observed in diabetic lungs (Uyy et al. 2010). However direct demonstration that endothelial dysfunction is correlated with cytoskeleton reorganization is missing.

A large amount of published results suggests that transmembrane proteins and the actin cortical cytoskeleton can organize lipids into short-lived nano-scale assemblies that can form larger domains under certain conditions. This supports the concept that interactions between lipids, cholesterol, and proteins create and maintain functional lateral heterogeneity in the endothelial cell membrane. The experimental data are supported by in silico theoretical studies. The relationship between membrane micro-heterogeneity and anomalous diffusion through cell membranes was evaluated by Monte Carlo simulations of two-component lipid membranes. The authors concluded that the near-critical fluctuations in the membrane lead to transient diffusion, while membrane-cytoskeleton interaction strongly affects phase separation, enhances diffusion, and eventually leads to diffusion of lipids through the opposite leaflet of the membrane (Ehrig et al. 2011). Lipid rafts facilitate lateral compartmentalization of molecules at the cell surface and facilitate the internalization of ligand-receptor complexes via raft-dependent endocytosis, both under physiological and pathological conditions.

Well known as a major risk factor of atherosclerosis, a high level of plasma cholesterol induces cell activation due to the altered composition of the plasma membrane. Modified phospholipids bind Annexin A6 (AnxA6) in a Ca^{2+} dependent manner, thus becoming closely associated with the plasma membrane, endosomes, caveolae and membrane rafts. This promotes interaction with signaling proteins, the endocytic machinery, and the actin cytoskeleton, that inhibit epidermal growth factor receptor and Ras signaling (Grewal et al. 2010). The authors claimed that AnxA6 may regulate transient membrane-actin interactions during endocytic transport, contributing to intracellular cholesterol homeostasis.

Metabolic dysfunction associated with elevated blood homocysteine induces clustering of lipid rafts platforms of activated NADPH oxidase sub-units, gp91(phox), and p47(phox) in the membrane of glomerular endothelial cells. Raft redox signaling platform formation is accompanied by enhancement of endothelial permeability by disruption of microtubule stability (Yi et al. 2009).

In addition, it was demonstrated that lipid rafts form a platform that mediates interaction of extracellular matrix components (i.e. endostatin, the C-terminal fragment of collagen XVIII) with actin-rich regions of endothelial cells, and coordinates the proteins involved in the regulation of angiogenic function of the cell. A model for endostatin signaling was proposed that might explain the inhibition of endothelial cell migration induced by disassembly of actin filaments. In the first step, endostatin binds to integrin α5β1 on the endothelial cell surface, where it interacts with a heparan sulfate proteoglycan (*HSPG*) and caveolin-1 (cav-1). Caveolin-associated Src kinase is then activated, and Src subsequently phosphorylates p190RhoGAP. RhoA is inactivated by p190RhoGAP, resulting in the disassembly of actin stress fibers and focal adhesions, with dramatic effect on the migratory capacity of endothelial cells. This cascade of events represents an efficient pathway by which a fragment of the extracellular matrix and an inhibitor of angiogenesis transmits signals to the cytoplasm and regulates cell adhesion and proliferation (Wickström et al. 2003).

Experiments performed on human brain microvascular endothelial cells (HBMEC) showed that lipid raft-dependent endocytosis mediates *C. neoformans* (a neurotropic fungal pathogen) internalization and that the lipid-raft localization of CD44, the cytoskeleton, and the intracellular kinase-DYRK3 are involved in this process (Huang et al. 2011).

Detailed scanning electron and fluorescent microscopy showed that the apical surface of endothelial cells constitutively forms small filopodia-like protrusions that are positive for ICAM-1 and freely move within the lateral plane of the membrane. Clustering of ICAM-1, using anti-ICAM-1 antibody-coated beads, efficiently and rapidly recruits ICAM, which makes connections with beta-actin, filamin and myosin-II. By inhibiting actin polymerization with cytochalasin B, it was shown that ICAM-1 clustering is regulated in an inside-out fashion through the actin cytoskeleton (van Buul et al. 2010).

In endothelial cell, numerous factors are involved in the complex system for the delivery and compartmentalization of proteins and lipids in order to achieve spatial and temporal coordination of signaling. Rafts or caveolae are active participants that regulate signaling pathways and processes that have multiple steps controlled by energy and by different proteins to ensure coordination of signaling cascades. Cytoskeleton elements such as actin bundles were shown to be closely associated with caveolae by electron microscopy (Rothberg et al. 1992) through filamin, which may act as a physical linker between actin and caveolae, as demonstrated by yeast two-hybrid strategy (Stahlhut and van Deurs 2000).

It is now becoming clear that lipid microdomains on the cell surface participate in signal transduction and that they may constitute a missing link between death and life receptor domains in endothelial cells. Findings that numerous receptors and signaling molecule sub-units are raft-associated proteins highlight the importance of these cholesterol rich domains in endothelial cells. Future challenges will be to identify the significance of lipid rafts as key players in signaling in the physiological regulation of vascular function, and in the pathophysiological effects of different stress factors associated with vascular disease. To identify the precise roles played by different components and compartments, we have to understand the molecular basis of diseases that exploit intracellular pathways. This may enable the design of specific successful therapies.

The Cytoskeleton Supports Intracellular Trafficking and Vesicular Transport in Endothelial Cells

The best-studied, raft-mediated endocytic route is controlled by caveolins. Recent data suggest that integrin-mediated cell adhesion is a key regulator of caveolar endocytosis.

Caveolae are caveolin-1-enriched smooth invaginations of the plasma membrane that form a sub-domain of lipid rafts. Endocytosis of rafts, including caveolar but also non-caveolar dynamin-dependent and dynamin-independent pathways, is characterized mainly by its cholesterol sensitivity and clathrin-independence (Lajoie and Nabi 2010). These flask-shaped invaginations are present in the plasma membrane of many cell types. They have long been implicated in endocytosis, transcytosis, and cell signaling. Recent work has confirmed that caveolae are directly involved in the internalization of membrane components (glycosphingolipids and glycosylphosphatidylinositol-anchored proteins), extracellular ligands (folic acid, albumin, autocrine motility factor), bacterial toxins (cholera toxin, tetanus toxin), and several non-enveloped viruses (Simian virus 40, Polyoma virus). Unlike clathrin-mediated endocytosis, internalization through caveolae is a triggered event that involves complex signaling. Details regarding the mechanism of endocytosis and the subsequent intracellular pathways that the internalized substances take have been emerging in the last decade (Pelkmans and Helenius 2002). Important contributions to the mapping of caveolae plasmalemma are coming from M&N Simionescu's group that create a new image of these highly dynamic, specialized structural, chemical and functional microdomains of the endothelium. Endocytosis and transcytosis via caveolae is an important route for the regulation of endothelial barrier function and may participate in different vascular pathologies (Simionescu and Antohe 2006, Simionescu et al. 2002).

The intracellular traffic of coated or uncoated vesicles, cargo molecules, and organelles is governed by motor proteins such as myosins, kinesins and dynein. These molecular motors generate force by hydrolyzing ATP. Actin contributes to the maintenance of cell polarity in a dynamic manner through endocytosis and transport of cytosolic and cortical components, whereas microtubules stabilize the initial asymmetry created by actin network, being important for long-lasting polarity (Numata et al. 2008).

The main function of endothelial cells is to sort and direct permeate molecules to their right destination. Caveolae, channels, coated pits, and vesicles are specialized membrane structures that take up and transport macromolecules within endothelial cell by endocytosis and by transcytosis. The final destinations of endocytosed molecules are tightly regulated and depend on the tissue location and the physiological or pathological state of the cells. In addition to serving as mechanical transducers and signaling hubs, caveolae are considered to be the main instrument involved both in endocytosis and transcytosis (Minshall et al. 2003, Simionescu and Simionescu 1991, Simionescu et al. 2002). Electron microscopy data suggests that caveolae are closely linked to intracellular actin filaments (Izumi et al. 1991, Morone et al. 2006, Richter et al. 2008); however, the role of cytoskeleton-mediated endocytosis and transcytosis is still poorly defined (Parkar et al. 2009). Caveolae-mediated endocytosis was shown to be dependent on the organization of the actin cytoskeleton (Mundy et al. 2002, Parton 1994, Pelkmans et al. 2002). The association of filamin-A with caveolin-1 promotes, via Src signaling, caveolae-mediated transport of macromolecules by regulating vesicle internalization, clustering, and intracellular trafficking (Sverdlov et al. 2009). The same pathway is used by certain pathogens. Simian virus 40 (SV40) utilizes endocytosis through caveolae for infectious entry into host cells. After binding to caveolae, virus particles induce transient breakdown of actin stress fibers. Actin is then recruited to virus-loaded caveolae as actin patches that serve as sites for actin "tail" formation. Dynamin II is recruited transiently to the actin tails. These events depend on the presence of cholesterol and the tyrosine phosphorylation of caveolar proteins, which are necessary for formation of caveolae-derived endocytic vesicles and for infection of the cell. Thus, caveolar endocytosis is ligand-triggered and involves extensive rearrangement of the actin cytoskeleton (Pelkmans et al. 2002). For example, E-selectin is located in lipid rafts and upon ligand binding (i.e., leukocytes or specific antibodies) it associates with the actin binding proteins actinin, vinculin, filamin, and focal adhesion kinase and its substrate paxillin. These associations result in specific E-selectin-dependent signaling (Yoshida et al. 1996, 1998). The redistribution of E-selectin in the lipid bi-layer leads to increased co-localization with caveolin-1, demonstrating the outside-in signaling events implicated in regulating transcellular transport, survival, and inflammatory reactions in endothelial cells (Kiely et al. 2003).

Little is known about the mechanism of caveolae internalization and intracellular traffic. Studies indicate that tyrosine phosphorylation of caveolin-1 after caveolae formation is required for this process (Aoki et al. 1999). Close association between the actin cytoskeleton and microtubules to caveolae were shown in different experimental systems, supporting the idea that cytoskeleton elements were active participants in caveolae dynamics and internalization. However more studies are needed to reveal the mechanisms implicated in the complex vesicular traffic, especially in endothelial cells. A yeast two-hybrid strategy showed that caveolin-1 interacts with filamin, which may act as a physical linker between filamentous actin and caveolae. Additional interactions of caveolin-1 with RhoA and with stress fibers determine the intracellular organization of caveolae (Stahlhut and van Deurs 2000). While actin fibers seem to restrict the movement and to control the internalization of caveolae (Parton et al. 1994), microtubules regulate the accumulation of intracellular caveolae and their long-range movements (Tagawa et al. 2005). However, the proteins that link caveolae and microtubules are currently unknown.

Conclusion

To summarize, in this chapter we tried to bring together some of the reported experimental evidences suggesting that membrane-cytoskeleton interactions in endothelial cells play key roles in several cytoskeleton-dependent processes, particularly in the regulation of cellular polarity, in maintaining membrane heterogeneity, in promoting signal transduction, and in the control of intracellular vesicular transport. However the complex molecular mechanisms that regulate the outside-in and inside-out transfer of information are still largely unknown. Since the well-known risk factors for vascular diseases may directly affect or disrupt the integrity and structure of the endothelial cytoskeleton, leading to endothelial dysfunction and impaired repair mechanisms, it is essential to unravel the structure and function of the endothelial cytoskeleton. The results could promote the understanding of the implications of membrane-cytoskeleton interactions in endothelial cell dysfunctions related to different vascular pathologies.

Acknowledgments

The author thanks Dr. Elena Constantinescu and Dr. Constantina Heltianu for carefully reading the manuscript and for their suggestions. I also appreciate the generosity of Dr. Elena Constantinescu in providing the unpublished figures included in this chapter.

This work was supported by Romanian Academy, National Ministry of Education and CNCSIS-UEFISCSU grants PN-II-PCCA code No. 135 and 153.

References

Antohe, F., G. Serban, L. Radulescu et al. 1997. Transcytosis of albumin in endothelial cells is brefeldin A-independent. Endothelium 5: 125–136.

Antohe, F., M.J. Poznansky and M. Simionescu. 1999. Low density lipoprotein binding induces asymmetric redistribution of the low density lipoprotein receptors in endothelial cells. Eur J Cell Biol 78: 407–415.

Antohe, F., L. Rădulescu, A. Gafencu et al. 2001. Expression of functionally active FcRn and the differentiated bidirectional transport of IgG in human placental endothelial cells. Human Immunol 62: 93–105.

Antohe, F., L. Lin, G.Y. Kao et al. 2004. Transendothelial movement of liposomes *in vitro* mediated by cancer cells, neutrophils or histamine. J Liposome Res 14: 1–25.

Aoki, T., R. Nomura and T. Fujimoto. 1999. Tyrosine phosphorylation of caveolin-1in the endothelium. Exp Cell Res 253: 629–36.

Banfi, C., M. Brioschi, R. Wait et al. 2006. Proteomic analysis of membrane microdomains derived from both failing and non-failing human hearts. Proteomics 6: 1976–1988.

Bhattacharya, B. and P. Roy. 2008. Bluetongue virus outer capsid protein VP5 interacts with membrane lipid rafts via a snare domain. J Virol 82: 10600–10612.

Brown, D.A. and E. London. 2000. Structure and function of sphingolipid- and cholesterol-rich membrane rafts. J Biol Chem 275: 17221–17224.

Byfield, F.J., S. Tikku, G.H. Rothblat et al. 2006. OxLDL increases endothelial stiffness, force generation and network formation. J Lipid Res 47: 715–723.

Cambi, A., F. de Lange, N.M. van Maarseveen et al. 2004. Microdomains of the C-type lectin DC-SIGN are portals for virus entry into dendritic cells. J Cell Biol 164: 145–155.

Chichili, G.R. and W. Rodgers. 2009. Cytoskeleton-membrane interactions in membrane raft structure. Cell Mol Life Sci 66: 2319–2328.

Constantinescu, E., C. Heltianu and M. Simionescu. 1986. Immunological detection of an analogue of the erythroid protein 4.1 in endothelial cells. Cell Biol Internatl Rep 10: 861–868.

Constantinescu, E., D. Alexandru, M. Raicu et al. 1997. Exposure to hypercholesterolemic serum modifies the expression of cytoskeletal proteins in cultured endothelia. J Submicrosc Cytol Pathol 29: 543–551.

Dejana, E., E. Tournier-Lasserve and B.M. Weinstein. 2009. The control of vascular integrity by endothelial cell junctions: molecular basis and pathological implications. Dev Cell 16: 209–221.

Dupin, I., E. Camand and S. Etienne-Manneville. 2009. Classical cadherins control nucleus and centrosome position and cell polarity. J Cell Biol 185: 779–786.

Ehrig, J., E.P. Petrov and P. Schwille. 2011. Near-critical fluctuations and cytoskeleton-assisted phase separation lead to subdiffusion in cell membranes. Biophys 100: 80–89.

Essler, M., M. Retzer, M. Bauer et al. 1999. Mildly oxidized low density lipoprotein induces contraction of human endothelial cells through activation of rho/rho kinase and inhibition of myosin light chain phosphatase. J Biol Chem 274: 30361–30364.

Grewal, T., M. Koese, C. Rentero et al. 2010. Annexin A6-regulator of the EGFR/Ras signaling pathway and cholesterol homeostasis. Int J Biochem Cell Biol 42: 580–584.

Heltianu, C., I. Bogdan, E. Constantinescu et al. 1986. Endothelial cells express a spectrin-like cytoskeletal protein. Circ Res 58: 605–610.

Hetheridge, C., A.N. Scott, R.K. Swain et al. 2012. The novel formin FMNL3 is a cytoskeletal regulator of angiogenesis. J Cell Sci., Jan 24. PMID: 22275430.

Huang, S.H., M. Long, C.H. Wu et al. 2011. Invasion of Cryptococcus neoformans into human brain microvascular endothelial cells is mediated through the lipid rafts-endocytic pathway via the dual specificity tyrosine phosphorylation-regulated kinase 3 (DYRK3). J Biol Chem 286: 34761–34769.

Insall, R.H. and O.D. Weiner. 2001. PIP3, PIP2, and cell movement-similar messages, different meanings? Dev Cell 1: 743–747.

Ivan, L. and F. Antohe. 2010. Hyperlipidemia induces endothelial-derived foam cells in culture. J Recept Signal Transduct Res 30: 106–114.

Izumi, T., Y. Shibata and T. Yamamoto. 1991. Quick-freeze, deep-etch studies of endothelial components, with special reference to cytoskeletons and vesicle structures. J Electron Microsc 19: 316–326.

Jury, E.C., P.S. Kabouridis, F. Flores-Borja et al. 2004. Altered lipid raft-associated signaling and ganglioside expression in T lymphocytes from patients with systemic lupus erythematosus. J Clin Invest 113: 1176–1187.

Kiely, J.M., Y. Hu, G. García-Cardeña et al. 2003. Lipid raft localization of cell surface E-selectin is required for ligation-induced activation of phospholipase C gamma. J Immunol 171: 3216–3224.

Koenig, A., J.Q. Russell, W.A. Rodgers et al. 2008. Spatial differences in active caspase-8 define its role in T cell activation versus cell death. Cell Death Differ 15: 1701–1711.

Lajoie, P. and I.R. Nabi. 2010. Lipid rafts, caveolae, and their endocytosis. Int Rev Cell Mol Biol 282: 135–163.

Lee, J.S. and A.I. Gotlieb. 2003. Understanding the role of the cytoskeleton in the complex regulation of the endothelial repair. Histol Histopathol 18: 879–887.

Lee Tsu-Yee, J. and A.I. Gotlieb. 2003. Microfilaments and microtubules maintain endothelial integrity. Microsc. Res. Tech. 60: 115–127.

Levitan, I. and K.J. Gooch. 2007. Lipid rafts in membrane-cytoskeleton interactions and control of cellular biomechanics: actions of oxLDL. Antioxid Redox Signal 9: 1519–1534.

Li, Q., H. Liu, J. Du et al. 2011. Advanced glycation end products induce moesin phosphorylation in murine brain endothelium. Journal Brain Research Vol. 1373. pp. 1–10.

Li, S., B.P. Chen, N. Azuma et al. 1999. Distinct roles for the small GTPases Cdc42 and Rho in endothelial responses to shear stress. J Clin Invest 103: 1141–1150.

Liu, F., A.D. Verin, P. Wang et al. 2001. Differential regulation of sphingosine-1-phosphate and VEGF-induced endothelial cell chemotaxis. Am J Respir Cell Mol Biol 24: 711–719.

Mammoto, A., S. Huang and D.E. Ingber. 2007. Filamin links cell shape and cytoskeleton structure to Rho regulation by controlling accumulation of p190RhoGAP in lipid rafts. J Cell Sci 120: 456–467.

Manes, S., G. del Real and A.C. Martinez. 2003. Pathogens: raft hijackers. Nat Rev Immunol 3: 557–568.

Manneville, J.B., S. Etienne-Manneville, P. Skehel et al. 2003. Interaction of the actin cytoskeleton with microtubules regulates secretory organelle movement near the plasma membrane in human endothelial cells. J Cell Sci 116: 3927–3938.

Marsh, M. and A. Helenius. 2006. Virus entry: open sesame. Cell 124: 729–740.

McCue, S., D. Dajnowiec, F. Xu et al. 2006. Shear stress regulates forward and reverse planar cell polarity of vascular endothelium *in vivo* and *in vitro*. Circ Res 98: 939–946.

Minshall, R.D., W.C. Sessa, R.V. Stan et al. 2003. Caveolin regulation of endothelial function. Am J Physiol Lung Cell Mol Physiol 285: L1179–L1183.

Miyazaki, T., K. Honda and H. Ohata. 2007. Requirement of Ca^{2+} influx- and phosphatidylinositol 3-kinase-mediated m-calpain activity for shear stress-induced endothelial cell polarity. Am J Physiol Cell Physiol 293: C1216–C1225.

Morone, N., T. Fujiwara, K. Murase et al. 2006. Three-dimensional reconstruction of the membrane skeleton at the plasma membrane interface by electron tomography. J Cell Biol 174: 851–862.

Mundy, D.I., T. Machleidt, Y.S. Ying et al. 2002. Dual control of caveolar membrane traffic by microtubules and the actin cytoskeleton. J Cell Sci 115: 4327–4339.

Ngu, H., L. Lu, S.J. Oswald et al. 2008. Strain-induced orientation response of endothelial cells: effect of substratum adhesiveness and actin-myosin contractile level. Mol Cell Biomech 5: 69–81.

Numata, N., T. Kon, T. Shima et al. 2008. Molecular mechanism of force generation by dynein, a molecular motor belonging to the AAA+ family. Biochem Soc Trans 36: 131–135.

Palade, G.E., N. Simionescu and M. Simionescu. 1982. Differentiated microdomains in the vascular endothelium. *In*: H.L. Nossel and H.J. Vogel [eds.]. Pathobiology of the Endothelial Cell. Academic Press, New York, USA pp. 23–33.

Parkar, N.S., B.S. Akpa, L.C. Nitsche et al. 2009. Vesicle formation and endocytosis: function, machinery, mechanisms, and modeling. Antioxid Redox Signal 11: 1301–1312.

Parton, R.G. 1994. Ultrastructural localization of gangliosides; GM1 is concentrated in caveolae. J Histochem Cytochem 42: 155–166.

Parton, R.G., B. Joggerst and K. Simons. 1994. Regulated internalization of caveolae. J Cell Biol 127: 1199–1215.

Pelkmans, L. and A. Helenius. 2002. Endocytosis via caveolae. Traffic 3: 311–320.

Pelkmans, L., D. Püntener and A. Helenius. 2002. Local actin polymerization and dynamin recruitment in SV40-induced internalization of caveolae. Science 296: 535–539.

Philippova, M., D. Ivanov, R. Allenspach et al. 2005. RhoA and Rac mediate endothelial cell polarization and detachment induced by T-cadherin. FASEB J 19: 588–590.

Richter, T., M. Floetenmeyer, C. Ferguson et al. 2008. High-resolution 3D quantitative analysis of caveolar ultrastructure and caveolae-cytoskeleton interactions. Traffic 9: 893–909.

Ridley, A.J. 2001. Rho family proteins: coordinating cell responses. Trends Cell Biol 11: 471–477.

Rothberg, K.G., J.E. Heuser, W.C. Donzell et al. 1992. Caveolin, a protein component of caveolae membrane coats. Cell 68: 673–682.

Rüffer, C. and V. Gerke. 2004. The C-terminal cytoplasmic tail of claudins 1 and 5 but not its PDZ-binding motif is required for apical localization at epithelial and endothelial tight junctions. Eur J Cell Biol 83: 135–144.

Ryu, Y., N. Takuwa, N. Sugimoto et al. 2002. Sphingosine-1-phosphate, a platelet-derived lysophospholipid mediator, negatively regulates cellular Rac activity and cell migration in vascular smooth muscle cells. Circ Res 90: 325–332.

Santilman, V., J. Baran, B. Anand-Apte et al. 2007. Caveolin-1 polarization in transmigrating endothelial cells requires binding to intermediate filaments. Mol Cell Biomech 4: 1–12.

Sieminski, A.L., R.P. Hebbel and K.J. Gooch. 2004. The relative magnitudes of endothelial force generation and matrix stiffness modulate capillary morphogenesis *in vitro*. Exp Cell Res 297: 574–584.

Simionescu, M. 2000. Structural, biochemical and functional differenciation of the vascular endothelium. *In*: W. Risau [ed.]. Morphogenesis of the Endothelium. Harwood Academic Publishers, United Kingdom, pp. 1–23.

Simionescu, M. and N. Simionescu. 1991. Endothelial transport of macromoelecules: transytosis and endocytosis. A look from cell biology. Cell Biology Reviews, Springer International, Berlin, Germany 25: 1–78.

Simionescu, M. and F. Antohe. 2006. Functional ultrastructure of the vascular endothelium: changes in various pathologies. *In*: S. Moncada and A. Higgs [eds.]. Handbook of Experimental Pharmacology. The Vascular Endothelium Vol. No. 176/I: Springer-Verlag Berlin, Heidelberg, pp. 41–69.

Simionescu, M., A. Gafencu and F. Antohe. 2002. Transcytosis of plasma macromolecules in endothelial cells: a cell biological survey. Microsc Res Tech 57: 269–288.

Simionescu, M. 1988. Receptor-Mediated Transcytosis of Plasma Molecules by Vascular Endothelium. *In:* N. Simionescu and M. Simionescu [eds.]. Endothelial Cell Biology in Health and Disease. Plenum Press, New York (USA) and London (UK), pp. 69–97.

Simionescu, N., M. Simionescu and G.E. Palade. 1981. Differentiated microdomains on the luminal surface of the capillary endothelium. I. Preferential distribution of anionic sites. J Cell Biol 90: 605–613.

Simionescu, N., F. Lupu and M. Simionescu. 1983. Rings of membrane sterols surround the openings of vesicles and fenestrae, in capillary endothelium. J Cell Biol 97: 1592–1600.

Simons, K. and E. Ikonen. 1997. Functional rafts in cell membranes. Nature 387: 569–572.

Smart, E.J., G.A. Graf, M.A. McNiven et al. 1999. Caveolins, liquid-ordered domains, and signal transduction. Mol Cell Biol 19: 7289–7304.

Soga, N., J.O. Connolly, M. Chellaiah et al. 2001. Rac regulates vascular endothelial growth factor stimulated motility. Cell Commun Adhes 8: 1–13.

Song, K.S., S. Li, T. Okamoto et al. 1996. Co-purification and direct interaction of Ras with caveolin, an integral membrane protein of caveolae microdomains: detergent-free purification of caveolae microdomains. J Biol Chem 271: 9690–9697.

Stahlhut, M. and B. van Deurs. 2000. Identification of filamin as a novel ligand for caveolin-1: evidence for the organization of caveolin-1-associated membrane domains by the actin cytoskeleton. Mol Biol Cell 11: 325–337.

Sverdlov, M., V. Shinin, A.T. Place et al. 2009. Filamin a regulates caveolae internalization and trafficking in endothelial cells. Molecular Biology of the Cell 20: 4531–4540.

Tagawa, A., A. Mezzacasa, A. Hayer et al. 2005. Assembly and trafficking of caveolar domains in the cell: caveolae as stable, cargo-triggered, vesicular transporters. J Cell Biol 170: 769–779.

Tzima, E., M.A. del Pozo, S.J. Shattil et al. 2001. Activation of integrins in endothelial cells by fluid shear stress mediates Rho dependent cytoskeletal alignment. EMBO J 20: 4639–4647.

Uehara, K. and A. Uehara. 2010. Vimentin intermediate filaments: the central base in sinus endothelial cells of the rat spleen. Anat Rec (Hoboken) 293: 2034–2043.

Uyy, E., F. Antohe, L. Ivan et al. 2010. Upregulation of caveolin-1 expression is associated with structural modifications of endothelial cells in diabetic lung. Microvasc Res 79: 154–159.

van Buul, J.D., J. van Rijssel, F.P. van Alphen et al. 2010. Inside-out regulation of ICAM-1 dynamics in TNF-alpha-activated endothelium. PLoS One 5: e11336.

Vasanji, A., P.K. Ghosh, L.M. Graham et al. 2004. Polarization of plasma membrane microviscosity during endothelial cell migration. Dev Cell 6: 29–41.

Walton, K.A., A.L. Cole, M. Yeh et al. 2003. Specific phospholipid oxidation products inhibit ligand activation of toll-like receptors 4 and 2. Arterioscler Thromb Vasc Biol 23: 1197–1203.

Waschke, J., F.E. Curry, R.H. Adamson et al. 2005. Regulation of actin dynamics is critical for endothelial barrier functions. Am J Physiol Heart Circ Physiol 288: H1296–305.

Wickström, S.A., K. Alitalo and J. Keski-Oja. 2003. Endostatin associates with lipid rafts and induces reorganization of the actin cytoskeleton via down-regulation of RhoA activity. J Biol Chem 278: 37895–37901.

Wojciak-Stothard, B. and A.J. Ridley. 2003. Shear stress–induced endothelial cell polarization is mediated by Rho and Rac but not Cdc42 or PI 3-kinases. The Journal of Cell Biology 161: 429–439.

Yamada, H. and H. Ando. 2007. Orientation of apical and basal actin stress fibers in isolated and subconfluent endothelial cells as an early response to cyclic stretching. Mol Cell Biomech 4: 1–12.

Yi, F., S. Jin, F. Zhang et al. 2009. Formation of lipid raft redox signaling platforms in glomerular endothelial cells: an early event of homocysteine-induced glomerular injury. Cell Mol Med 13: 3303–3314.

Yoshida, M., W.F. Westlin, N. Wang et al. 1996. Leukocyte adhesion to vascular endothelium induces E-selectin linkage to the actin cytoskeleton. J Cell Biol 133: 445–455.

Yoshida, M., B.E. Szente, J.M. Kiely et al. 1998. Phosphorylation of the cytoplasmic domain of E-selectin is regulated during leukocyte-endothelial adhesion. J Immunol 161: 933–941.

Zeng, H., D. Zhao and D. Mukhopadhyay. 2002. KDR stimulates endothelial cell migration through heterotrimeric G protein Gq/11-mediated activation of a small GTPase RhoA. J Biol Chem 277: 46791–46798.

Zhao, B., W.D. Ehringer, R. Dierichs et al. 1997. Oxidized low density lipoprotein increases endothelial intracellular calcium and alters cytoskeletal f-actin distribution. Eur J Clin Invest 27: 48–54.

8

Cell-Matrix Adhesion Proteins in the Regulation of Endothelial Permeability

Jurjan Aman, Geerten P. van Nieuw Amerongen and
Victor W.M. van Hinsbergh [a],*

Introduction

Endothelial barrier function and regulation

Throughout the body, the vasculature is lined with a thin, continuous sheet of endothelial cells. This monolayer of tightly adhering cells regulates—in addition to a broad spectrum of other processes—the barrier properties of the vascular wall. The endothelial barrier allows extravasation of water, electrolytes, proteins, monocytes, and leukocytes to supply the surrounding tissues. The balance between barrier function and regulated extravasation results from stringently controlled permeability of the endothelial monolayer, which prevents excessive vascular leak and interstitial edema. Endothelial permeability is regulated at various levels, and is determined by environmental conditions, such as inflammation and hypoxia, and the location of the endothelium in the body.

Department of Physiology, Institute for Cardiovascular Research, VU University Medical Center, Amsterdam, The Netherlands.
[a]Email: v.vanhinsbergh@vumc.nl
*Corresponding author

Plasma constituents pass the endothelial barrier via two pathways, the transcellular and the paracellular pathway. Transcellular exchange predominates under basal conditions—hormones and macromolecules are transported from the luminal to the abluminal side of the endothelium via a system of caveola or caveolar channels in most types of continuous endothelia. During paracellular transport, plasma constituents pass through gaps between neighboring endothelial cells. This type of exchange is limited under basal conditions, but after exposure to vasoactive agents (like histamine), leukocyte-derived cytokines, or angiogenic factors (like vascular endothelial growth factor (VEGF)), the interendothelial cell contacts are loosened permitting local or even extensive leakage of fluid and proteins into the interstitium. Disruption of cell-cell contacts and the consequent edema formation characterizes inflammatory conditions like sepsis and acute lung injury (Goldenberg et al. 2011). Transport of macromolecules via the transcellular pathway takes place at a continuous, low rate, while the paracellular route is more dynamic and subject to various and complex regulatory mechanisms. The paracellular pathway should not only allow passage of larger proteins and leukocytes, requiring large plasticity of the intercellular adhesions, but is also subject to continuous remodeling of cell-cell contacts, even in a resting endothelial monolayer (Mehta and Malik 2006).

Despite high plasticity of cell-cell contacts and large differences in oncotic or hydrostatic pressure, the endothelium effectively seals the vessels and prevents leakage of fluid to the interstitium. This endothelial barrier is the resultant of three major forces: actomyosin contractile force, tethering forces at cell-cell contacts (or junctions), and the adhesive force of the endothelial cell to the extracellular matrix (Fig. 1). In a resting monolayer, the tethering forces in the cell-cell contacts balance the relatively low forces resulting from actomyosin contraction. Adhesion of endothelial cells to the extracellular matrix (ECM), in short cell-matrix interactions, is the third factor, which contributes to the spread-out, flat shape of endothelial cells in a resting monolayer.

This chapter specifically focuses on the contribution of cell-matrix interactions to the regulation of endothelial barrier function. Before discussing the modulation of endothelial permeability by cell-matrix interactions in detail, we shall briefly indicate how endothelial cell-cell junctions and the actin cytoskeleton determine the endothelial barrier function. For an excellent overview of endothelial barrier regulation in general, we refer to the review of Mehta and Malik (2006).

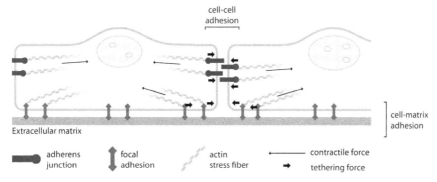

Figure 1. Overview of forces observed in endothelial cells within a resting endothelial monolayer. Actin fibers exert contractile forces by actomyosin interaction. These centripetal forces are balanced by tethering forces in cell-cell adhesions, predominantly the adherens junctions, and in the cell-matrix adhesion sites (focal adhesions). Under basal conditions, these forces are in balance, resulting in maintenance of the spread (flattened) cell shape.

Endothelial junctions

The endothelial monolayer is sealed by junctional structures, which limit the exchange of plasma solutes and circulating cells. Only in pathophysiological conditions, such as upon exposure to vasoactive agents, in tumors, and in chronically inflamed tissues, the junctions (in particular, those in postcapillary venules) widen or dissociate, and thus plasma macromolecules accumulate in the interstitium. In endothelial cells, three types of junctions are found: adherens junctions (AJs), tight junctions, and gap junctions. Similar to those in epithelial cells, the barrier properties of endothelia are governed by adherens and tight junctions, while gap junctions contribute to intercellular communication. However, in most endothelia, AJs are the major determinant of barrier function, and the molecular players in these junctions are specific and specifically regulated.

When a new junction is formed between two adjacent endothelial cells, exploratory lamellipodial protrusions form the initial contact between these cells (Hoelzle and Svitkina 2012, Mattila and Lappalainen 2008). As shown in cultured endothelial cells, these initial contacts contain a branched actin network with Vasodilator-Stimulated Phosphoprotein (VASP) at their leading edge (Hoelzle and Svitkina 2012). Subsequent reorganization causes the retraction of the lamellipodia, accompanied by the formation of interdigitating filopodia-like structures, which accumulate VE-cadherin and maintain the intercellular connection by homophilic VE-cadherin interactions. In this initial stage, VE-cadherin displays a staining pattern that is perpendicular to the cell perimeter, a pattern that is also observed when vasoactive agents have destabilized the junction. Similar to E-cadherin in epithelial cells, endothelial-specific VE-cadherin is the dominant AJ protein in most endothelia. Homophilic VE-cadherin binding triggers interaction with catenins, which in turn facilitates indirect binding to the F-actin cytoskeleton. This is accompanied by a subsequent rearrangement of the cortical F-actin-myosin cytoskeleton along the cell perimeter, eventually resulting in a belt of AJs that seals individual endothelial cells together in a monolayer. Indeed, once stable junctions are formed, VE-cadherin is recognized as a thin lining between adjacent endothelial cells. Tight junctions re-enforce this belt. Only in specialized endothelia, such as in the blood brain barrier, the retina, and the testis, is a sealing belt of tight junctions formed like that in epithelia. In most continuous endothelia of the body, the tight junctions form interrupted mosaic-like structures that can be circumvented by macromolecules (Bundgaard 1984), while the continuously present AJs act as the major determinants of permeability resistance in the junctions of these endothelia (Corada et al. 1999, Taddei et al. 2008). As a consequence, AJs play a dominant role in the tethering forces of most types of endothelial cells (Dudek and Garcia 2001).

Both AJs and tight junctions are coupled via regulatory adaptor proteins to the F-actin cytoskeleton (Fig. 1). In particular, VE-cadherin interacts with p120-catenin and β-catenin, which then binds indirectly to F-actin. The stability of homotypic VE-cadherin interactions is affected by specific phosphorylation events on these proteins (Andriopoulou et al. 1999, Esser et al. 1998, Turowski et al. 2008). α-Catenin is generally thought to act as a connector between β-catenin and the F-actin fibers, although this interaction has been debated (Yamada et al. 2005). Details of the regulation of the AJ complex comprised of VE-cadherin and catenins and its importance in the regulation of endothelial permeability has been described in several recent reviews (Dejana et al. 2008, Mehta and Malik 2006, Vestweber 2008).

The actin cytoskeleton

Actin is the most abundant intracellular protein and accounts together with myosin for approximately 16% of the intracellular protein mass. In undisturbed endothelial cells it is in particular encountered along the cortical rim of the cell. While some of the actin molecules present are soluble inactive G-actin monomers (globular form), these can be readily incorporated into an F-actin form that can bundle into F-actin fibers. The F-actin fibers form thin actin networks, such as the branched F-actin networks in lamellipodia and the more parallel-oriented F-actin fibers in filopodia. Most prominent are the thick, discrete structures generated by assembled F-actin fibers (e.g., stress fibers in migrating and activated endothelial cells) and the cortical F-actin rim at the cell margins of confluent cells. F-actin fibers often form contractile actin-myosin structures with non-muscle myosin. Actin-myosin structures play an important role in cell shape, tensile forces, cell migration, and cell movement in general. Their generation and modification is controlled by many structural and signaling molecules, amongst which the Rho GTPases Rac1, Cdc42 and RhoA play important and individually distinct roles.

The attachment of the actin cytoskeleton to the cell membrane depends on the condition of the endothelial monolayer. In an intact endothelial monolayer, the cortical F-actin network interacts with the cell-cell junctions and condenses into a thin, pericellular F-actin lining at cell borders. The exact dynamics of this interaction remains to be unraveled. At sites where endothelial cells loose their cell-cell contacts, F-actin fibers become attached to cell-matrix adhesion points, the presence of which rapidly increases during loss of cell-cell contact. In the front of a migrating endothelial monolayer (induced after removing several rows of cells by scratching) and in subconfluent endothelial cells, the emerging focal adhesions are the nucleation points for the formation of stress fibers and subsequently provide—in addition to contributing to cell shape and tension—a signaling hub enforcing communication between the cell and underlying matrix. When endothelial cell monolayers become suddenly exposed to shear forces, relatively thick stress fibers also form in the basolateral part of the cell (Boon et al. 2010), most likely a requirement for firm counteraction of new forces on the cell. A new feature directly related to junction (de) stabilization is the formation of so-called radial F-actin stress fibers. Recent data show that endothelial cells, after exposure to a barrier destabilizing stimulus, often acquire this additional type of stress fiber that is connected to discontinuous AJs (Huveneers et al. 2012), and—when it occurs in both cells—connects two adjacent cells (Millán et al. 2010). Using elegant laser ablation techniques, Huveneers et al. showed that these radial F-actin bundles exert pulling forces on AJs (Huveneers et al. 2012). As we shall discuss below, it is plausible that during the destabilization of AJs, vinculin interacts with α-catenin and forms a new interaction site for newly forming stress fibers, which as a safety net may contribute to limiting paracellular permeability and/or restoring junctional organization.

The structure and regulation of F-actin filaments is tightly linked to the other main cytoskeletal component, the microtubules. Microtubules are multimers of the structural protein β-tubulin, and together with microfilaments, are responsible for cytoskeletal maintenance of cell shape (Lee and Gotlieb 2003). Therefore, microtubule disassembly, as observed during endothelial stimulation with vasoactive agents, contributes to disruption of the endothelial barrier (Petrache et al. 2003, Verin et al. 2001).

Endothelial Cell-ECM interactions

Compared to cell-cell junctions and actomyosin contraction, the involvement of cell-matrix interactions in endothelial barrier regulation is less well-defined. Cell-matrix interactions are organized and regulated at four different levels: 1) the composition of the extracellular matrix, 2) the integrins, a large family of dimeric, transmembrane adhesion receptors, 3) the actin cytoskeleton, and 4) the focal adhesion (FA), an intracellular multi-protein signaling complex that connects integrin receptors to the cytoskeleton (Fig. 2). The poor understanding of the role of cell-matrix interactions in endothelial barrier regulation has several explanations, including the complex regulation at the various levels, the multidirectional character of inside-out and outside-in signaling, the methodological difficulty of studying cell-matrix interactions independent of actomyosin contraction, and the often embryonically lethal phenotype of knock-out mice (due to vascular instability) (Wu 2005). Understanding the involvement of cell-matrix interactions in endothelial barrier regulation has direct clinical relevance. It may provide additional targets to intervene in endothelial barrier disruption and contribute to treatment of vascular leak and edema formation. It also yields valuable information related drugs that affect cell-matrix interactions.

In this chapter, we provide an overview of proteins involved in cell-matrix interactions and their role in endothelial barrier regulation. The first part of the chapter discusses endothelial barrier regulation at the different levels of cell matrix interaction. Subsequently, we will discuss the regulation

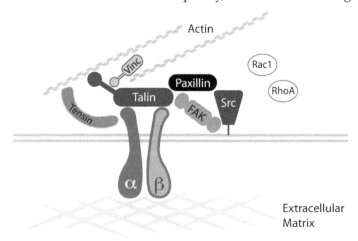

Figure 2. Schematic representation of the focal adhesion. Proteins most important for cell-matrix interaction and its regulation are presented here. α and β represent integrin subunits. The RhoGTPAses RhoA and Rac1 importantly contribute to focal adhesion dynamics, but do not specifically bind to specific FA proteins.

of cell-matrix interactions in the context of other barrier-regulating process, in particular cell-cell junctional dynamics. Integrating existing literature, we will discuss a model in which both cell-cell junctions and FAs exert tethering forces to balance the contractile forces of the actin cytoskeleton. The chapter closes with concluding remarks and clinical implications.

Cell-matrix Interactions and Endothelial Barrier Regulation

The role of the extracellular matrix

Adhesion to extracellular matrix (ECM) is essential for endothelial cells to survive, to proliferate, and to exert their normal function. In blood vessels, this ECM is provided by the basement membrane in mature vessels or by the interstitial or temporary fibrin matrix during sprouting angiogenesis. The ECM consists of a fine meshwork of filamentous proteins, the composition of which is different between a mature basement membrane and the interstitial or wound matrix through which angiogenic vessels meander. In a mature vessel, the basement membrane is mainly composed of laminin, collagen IV, and perlecan (Davis and Senger 2005, Hynes 2007, Kalluri 2003), whereas during tissue repair, the matrix also contains fibronectin, fibrin, and vitronectin. Indeed, vascular stability heavily depends on the presence and composition of the ECM. The phenotype of mice deficient in individual basement membrane proteins has revealed the necessity of these proteins in vascular stability (Costell et al. 1999, Pöschl et al. 2004), as collagen IV- or perlecan knock-out mice suffer from vascular instability and hemorrhage. However, this vascular instability may be attributed to defective vessel formation as well as to aberrant regulation of the endothelial barrier.

The involvement of the ECM in endothelial barrier regulation was stressed in a recent publication, which showed that changing stiffness of the ECM directly alters the barrier of the endothelium. Krishnan et al. (2011) showed that endothelial cells grown on a soft ECM form substantially smaller gaps during stimulation with a barrier disruptive agent than endothelial cells grown on a stiff matrix. They further demonstrated that matrix stiffness is an important contributor to tension development in the cell. Intracellular tension, in turn, has a direct effect on the size and stability of cell-cell junctions, both under resting and stimulated conditions (Liu et al. 2010). Increased ECM stiffness may result from metabolic conditions or vascular aging, and contribute to endothelial permeability as observed in atherosclerosis (Huynh et al. 2011). Details of the influence of extracellular matrix composition and geometry on endothelial barrier integrity were nicely surveyed in earlier reviews (Califano and Reinhart-King 2010, Romer et al. 2006).

These data support a clear, albeit chronic influence of the ECM on endothelial barrier integrity. The ECM is relatively static, and does not importantly change during the time interval of endothelial barrier disruption by vasoactive agents in acute inflammatory conditions (0 hr–18 hr). However, the integrins that connect the cell and matrix can be rapidly activated and subsequently activate regulatory signals that transduce actin-associated changes within the cell. On the basis of this ability, integrins and the associated signaling hub—the FAs—have been considered to contribute to endothelial barrier changes.

The role of integrins

Endothelial cells bind to the ECM via integrins. Integrins form a large family of glycoproteins that serve as cell adhesion receptors. These heterodimeric transmembrane proteins, composed of an alpha and a beta subunit, are the main adhesion receptors mediating adhesion of endothelial cells to the extracellular matrix. The large extracellular domain of integrins directly binds to the filamentous proteins of the ECM, while the small intracellular domain connects via other proteins to the actin skeleton, thus bridging the extracellular matrix with the endothelial cytoskeleton. Although the 18 alpha subunits and eight beta subunits identified in vertebrates yield 24 different integrin heterodimers, only some of these are present in endothelial cells. The combination of alpha and beta subunits determines the affinity of integrins for a specific ECM molecule. In endothelial cells, the following alpha/beta heterodimers have been identified: α1β1, α2β1, α3β1, α4β1, α5β1, α6β1, α6β4, α9β1 (in particular lymphendothelial cells), αvβ3 and αvβ5 (Albelda et al. 1989, Defilippi et al. 1991, Friedlander et al. 1995, Massia and Hubbell 1992, Tarone et al. 1990). Although integrins have been reported to be quite promiscuous with regard to their binding to ECM molecules, α1β1 and α2β1 preferentially bind collagen, α3β1, α6β1 and α9β1 bind laminin, and α5β1 binds to fibronectin, whereas αvβ3 and αvβ5 bind vitronectin and many other proteins that contain the amino acid sequence Agr-Gly-Asp (RGD) (Hynes 2002a, Silva et al. 2008).

Integrin-mediated binding of endothelial cells to the ECM importantly determines cell shape and morphology of the endothelial monolayer. Changes in integrin binding and expression parallel changes in cell shape as observed during formation and disruption of the endothelial barrier. Integrins are observed at the leading edge of spreading endothelial cells, whereas dissociation of integrin binding at the cell edge is required for cell retraction and gap formation. The reported contribution of integrins to endothelial barrier function is quite heterogeneous. Integrin subtype, endothelial cell type, the composition of the extracellular matrix, and the presence of cytokines, growth factors or chemokines all contribute to this heterogeneity. Methods used to elucidate the contribution of integrins to endothelial barrier maintenance and endothelial barrier disruption include RGD-containing peptides (which bind to integrins, thereby preventing integrin binding to the ECM), genetic mutation (functional inhibition) or ablation (either general or endothelial-specific) of integrins, and integrin blocking or activating antibodies.

RGD peptides

Several endothelial integrins bind to the ECM via the specific amino acid sequence Agr-Gly-Asp (RGD). This RGD site is found in various integrin ligands, like fibronectin, vitronectin, laminin, and collagen. RGD-peptides have been constructed that contain this sequence in a specific environment and are shown to compete with integrin binding to (specific) ECM proteins (Qiao et al. 1995). Yet, it should be taken into account that RGD peptides may also affect intracellular signaling independent of integrin/ECM binding (Hynes 2002b). Treating endothelial monolayers with RGD peptides resulted in a loss of barrier function, which was associated with reduced endothelial adhesion strength to the ECM (Qiao et al. 1995). This was also found in an *ex vivo* study of Wu et al., which demonstrated that inhibition of integrin binding to both fibronectin (with the fibronectin-specific RGD peptide, GRGDdSP) or vitronectin (with a vitronectin-specific peptide, GPenGRGDSPCA) increased the permeability of porcine venules. These effects were reversed by co-treatment with fibronectin or vitronectin, respectively (Wu et al. 2001). From this perspective, it is interesting that in humans, the concentration of plasma fibronectin (but not soluble cellular fibronectin) is inversely related to mortality from acute lung injury or acute respiratory distress syndrome (Peters et al. 2011).

Integrin α5β1

The integrins α5β1 and αvβ3 have been studied most extensively in the context of endothelial barrier regulation. Their effect on vascular leakage primarily reflects leakiness because of improper vessel development, but they were also found to play important roles in established endothelial monolayers. Overall genetic ablation of α5 (Yang et al. 1993) or β1 (Fässler and Meyer 1995) subunits, or endothelial-specific ablation of the β1 subunit (Lei et al. 2008), results in early embryonic lethality (at E10–11, E5.5 and E9.5, respectively) due to failure of vascular development. The failed vascular development and early embryonic death of β1 knock-out mice most likely results from impaired α5β1 heterodimer formation, as genetic ablation of α1 (Gardner et al. 1996), α2 (Chen et al. 2002), α3 (Kreidberg et al. 1996), α6

(Georges-Labouesse et al. 1996) or α9 subunits (Huang et al. 2000) did not result in defects in vascular development.

In resting confluent endothelial monolayers, α5β1 is one of the most important integrins for maintenance of barrier function. Several *in vitro* studies have shown that treatment of confluent monolayers with an α5β1-blocking antibody leads to formation of intercellular gaps and enhanced passage of macromolecules in a dose-dependent fashion (Curtis et al. 1995, Lampugnani et al. 1991). The finding that co-incubation with fibronectin (the natural ligand of α5β1) reversed the barrier-disruptive effects of the α5β1-blocking antibody (Curtis et al. 1995) indicates that disruption of this receptor-ligand interaction underlies the loss of barrier dysfunction.

Inhibition of α5β1 may contribute to endothelial barrier disruption by inflammatory cytokines. Exposure of endothelial cells to TNF-α leads to enhanced endothelial permeability, associated with reduced α5β1 integrin localization at FAs (Curtis et al. 1998). TNF-α treated endothelial cells further showed reduced adhesion to the ECM, due to increased internalization and recycling of α5β1 integrins (Gao et al. 2000). Treatment of endothelial cells with α5β1-blocking antibody mimics TNF-α-induced cell detachment, without having additive effects when applied simultaneously, while treatment with an α5β1-activating antibody or fibronectin attenuated the effects of TNF-α (Rotundo et al. 2002). The relative contribution of disturbed cell-matrix interaction to TNF-α-induced endothelial barrier dysfunction remains to be elucidated, however, as effects of TNF-α on F-actin cytoskeleton and junctional integrity may be more prominent in disruption of the barrier (Wójciak-Stothard et al. 1998). Notwithstanding, these studies together show that stable binding of α5β1 to the extracellular matrix is required for endothelial barrier function, and that barrier-disruptive cytokines like TNF-α, by targeting this cell-matrix interaction, reduce endothelial monolayer integrity.

A new view on the involvement of α5β1 in endothelial permeability was recently put forward. Besides binding to ECM molecules, endothelial barrier function may be affected by α5β1 interaction with other proteins. cANGPTL4, a soluble C-terminal fibrinogen-like domain of angiopoietin-like 4, was found to bind the extracellular domain of α5β1, leading to α5β1 activation and internalization (Huang et al. 2011). This α5β1/cANGPTL4 interaction subsequently disturbed VE-cadherin/claudin-5 clustering and disrupted the endothelial barrier. An α5β1-blocking antibody attenuated cANGPTL4-induced endothelial barrier dysfunction and vascular leakage, suggesting the involvement of α5β1 in endothelial barrier disruption. The authors propose a mechanism in which cANGPTL4 binding to α5β1 1) induces Rac1/PAK driven endothelial barrier disruption and 2) clears the way for cANGPTL4 binding to and disruption of VE-cadherin and claudin-5 complexes. The exact role of α5β1 integrin with respect to cell-matrix interactions remains to be elucidated here. Although cANGPTL4 does not contain an RGD-sequence (Source: PubMed Protein) and, according to the authors, cANGPTL4 does not impair α5β1 binding to fibronectin (Goh et al. 2010), cANGPTL4 induces α5β1 internalization, leaving less α5β1 available for cell-matrix interactions. Another non-ECM protein reported to interact with α5β1 is Tie2, the receptor for the angiopoietins Ang-1 and Ang-2. Binding of α5β1 to Tie2 sensitizes Tie2 for Ang-1, a process that is further enhanced by α5β1-binding to fibronectin (Cascone et al. 2005). Ang-1 binding to Tie2 increases endothelial barrier function *in vitro* (Thurston et al. 1999) and protects against edema formation in mouse models of acute lung injury (ALI), whereas plasma Ang-1 is inversely correlated with disease severity in sepsis and ALI (van der Heijden et al. 2008).

In summary, α5β1 and α5β1/ECM-binding are required for maintenance of cell shape under physiologic conditions. Disruption of this α5β1/ECM binding leads to swift dissociation of cell-matrix interactions, cell retraction, and gap formation. This might involve enhanced recycling of α5β1 in inflammatory responses. In contrast, boosting of α5β1/ECM-binding (with additional fibronectin or integrin-activating antibodies) during cytokine stimulation attenuates endothelial barrier disruption, indicating that integrin-mediated cell-matrix interaction counteracts cell-retraction and gap formation. The interaction between α5β1 and cANGPTL4 or Tie2 further demonstrates that binding of α5β1 to cellular proteins can affect endothelial barrier function independent of cell-matrix interactions.

Integrin αvβ3

Unlike α5β1, αvβ3 seems to be dispensable for vascular development and homeostasis. Although genetic ablation of the αv subunit is embryonically lethal at E9.5 due to severe cerebral hemorrhage (Bader et al. 1998), the lethal phenotype of αv knock-out mice most likely results from impaired formation of the αvβ8 heterodimer, which is required for end-feet association of perivascular cells with brain endothelial cells (Silva et al. 2008, Zhu et al. 2002). In line with this, mice with genetic ablation of the β3 subunit (Hodivala-Dilke et al. 1999, Reynolds et al. 2002) or endothelium-specific ablation of the αv subunit (Lacy-Hulbert et al. 2007) show normal vascular development.

Studies on the role of αvβ3 integrin in endothelial barrier regulation have shown heterogeneous results, and have reported both barrier disruptive- and barrier-preserving effects of αvβ3 integrin. Blocking αvβ3 does not perturb the integrity of a resting endothelial monolayer. Mice lacking the β3 show no edema or vascular leakage under non-stimulated conditions (Su et al. 2012) and incubation of cultured endothelial cells with a αvβ3-blocking antibody does not impair endothelial barrier function (Lampugnani et al. 1991, Su et al. 2012). In contrast, β3 knock-out mice or endothelial cells pretreated with an αvβ3-blocking antibody show enhanced vascular leakage and endothelial barrier disruption in response to stimulation with permeability-inducing agents (Li et al. 2009, Su et al. 2012). Similarly, endothelial stimulation with TNF-α and IFN-γ reduced the αvβ3-dependent adhesion of endothelial cells to the extracellular matrix (Rüegg et al. 1998). These data suggest that αvβ3-mediated cell-matrix adhesion exerts a barrier stabilizing effect during endothelial stimulation. Li and Gamble have shown that stimulation of endothelial cells with thrombin directly activates signaling pathways that target and disrupt αvβ3-dependent cell-matrix interactions (Li et al. 2009). In addition, Thomas et al. (2010) demonstrated that upon endothelial stimulation with Ang-2, αvβ3 forms a complex with Tie2 (an Ang-2 receptor) and is consequently internalized and degraded. As Ang-2 is generally known to induce (Parikh et al. 2006, Roviezzo et al. 2005) or prime for (van der Heijden et al. 2011) endothelial barrier disruption, Ang-2 may use αvβ3 to induce endothelial monolayer destabilization and barrier dysfunction (Felcht et al. 2012).

A number of studies have shown that αvβ3 contributes to endothelial barrier disruption. Measuring brain edema formation, Shimamura et al. (2006) reported that αvβ3 inhibition with an αvβ3-specific RGD sequence (cRGDfv) attenuated brain edema following stroke. In line with this finding, the αvβ3 antagonist BS-1417 reduced vascular leakage resulting from choroidal neovascularization (Honda et al. 2009), while αvβ3 blockade (*in vitro*) or genetic αvβ3 ablation (*in vivo*) attenuated endothelial barrier disruption induced by fibrinogen-γ C-terminal fragments (Guo et al. 2009). The interaction of αvβ3 with the VEGFR2, as shown by Soldi et al. (1999), further supports the involvement of αvβ3 in endothelial barrier disruption. αvβ3 binding to its ligand vitronectin is required for phosphorylation and activation of the VEGFR2 upon binding with its ligand VEGF-A165, suggesting that αvβ3/VEGFR2 complex formation is a prerequisite for VEGF-induced vascular leakage.

Multiple explanations may underlie the paradoxical role of αvβ3 in endothelial barrier regulation. As large differences exist between the regulation of blood-brain barrier function and barrier function in the rest of the endothelium, αvβ3 may act differently in brain endothelial cells, in particular in combination with astrocyte end-feet. In addition, extracellular modulation of αvβ3 by, for example, ECM proteins or peptides (outside-in signaling), may differentially affect integrin function and intracellular signaling due to agonist-induced activation (inside-out signaling).

Integrin αvβ5

Considerably less is known about the involvement of other integrins in endothelial barrier regulation. In a series of publications Pittet et al. pointed towards a role of αvβ5 in mediating the pathogenesis of acute lung injury and pulmonary edema. Blockage of αvβ5 and αvβ6 attenuated IL-1β-induced disruption of lung endothelial cell monolayers and reduced pulmonary edema formation during acute lung injury resulting from IL-1β overexpression in the murine lung (Ganter et al. 2008). Similarly, αvβ5 blocking antibodies attenuated endothelial barrier disruption induced by VEGF, thrombin and TNF-α, whereas genetic ablation of the β5 subunit attenuated lung vascular permeability resulting from ventilator-induced lung injury (Su et al. 2007). As αvβ5 was also shown to be involved in pulmonary edema formation related to *Pseudomonas aeruginosa* pneumonia (Ganter et al. 2009), these studies indicate αvβ5 as an important and general mediator of endothelial barrier disruption and pulmonary edema formation during acute lung injury.

Other integrins

Although a direct association of α2β1, α3β1, α4β1 and α6β1 integrins with endothelial barrier disruption or edema formation is lacking, indirect evidence indicates their involvement in endothelial barrier regulation. Cailleteau et al. (2010) have shown that α2β1 activity prevents Rac1 anchorage to the cell membrane, thereby hampering endothelial cell quiescence, a condition required for proper barrier function. Integrin α3β1 localizes to endothelial cell-cell junctions and regulates endothelial cell migration and cell-matrix adhesion, and α3β1 blocking antibodies inhibit endothelial cell migration and increase cell-matrix interaction (Yáñez-Mó et al. 1998). α4β1 blocking antibodies attenuate diabetes-induced upregulation of VEGF, TNF-α and NFκB, leukocyte adhesion, and vascular leakage (Iliaki et al. 2009). However, even though α4β1 is present in endothelial cells, α4β1 is generally considered to be a leukocyte-specific integrin (Hynes 2007). For this reason, the protective effects of α4β1 may also be attributed to α4β1 inhibition of leukocyte binding. Last, α6β1-deficient endothelial cells show increased VEGFR2 expression and VEGF-dependent signaling, whereas genetic ablation of α6β1 in mice leads to enhanced angiogenesis (Germain et al. 2010). As stated before, no direct relation to endothelial barrier disruption is known for these integrins. Together, the presence or activity of these integrins seems to directly modulate pathways that affect endothelial barrier integrity. However, the precise involvement in endothelial barrier regulation remains to be elucidated.

Conclusion integrins

These findings stress the relevance of integrins in endothelial barrier regulation, although the role of integrins is quite heterogeneous. Interaction of α5β1 and αvβ3 with the ECM clearly protects the endothelial barrier, whereas αvβ5 contributes to endothelial barrier disruption. In addition, α5β1 and αvβ3 integrins may interact with proteins other than ECM proteins. Integrin binding results in intracellular signaling and structural changes, which in turn affect endothelial barrier function. The involvement of the remaining endothelial integrins in barrier regulation is less clear. Although knock-out models—either whole-body or endothelial-specific—have yielded considerable insight into the role of integrins in vasculatur development and angiogenesis, the effect on endothelial barrier regulation is difficult to extract from these studies.

Focal adhesions as signaling hubs of matrix-integrin interactions

Inside the cell, integrins bind to the actin cytoskeleton via FAs, the large multiple-protein complex containing scaffold proteins, kinases, phosphatases and a large number of GTPases (Brakebusch and Fässler 2003, Wehrle-Haller 2012). The core of this integrin-cytoskeleton connection is formed by talin and vinculin (Fig. 2). The binding of integrin via talin and vinculin to actin is regulated by allosteric changes in talin and vinculin conformation. The head domain of talin binds to the cell membrane and to the cytoplasmic domain of the integrin in a $PI(4,5)P_2$-dependent manner, whereas the talin tail domain binds to the actin cytoskeleton, inducing tension and unfolding of talin. This stretched allosteric conformation of talin uncovers binding sites for vinculin, which cross links talin and actin and enhances the integrin/talin/actin connection. In addition, vinculin activation (separation of its head and tail regions) enhances FA growth by clustering activated integrins and slowing down FA turnover (Humphries et al. 2007). The FA protein complex is further enriched with structural proteins like tensin, α-actinin, and kindlin, and adaptor proteins like paxillin (Fig. 2). In the FA, paxillin facilitates cell signaling from integrin/ECM to the cytosol and vice versa (Turner 2000). Proteins involved in these signaling pathways are the tyrosine kinases Src and Focal Adhesion Kinase (FAK). FAK binds to paxillin and talin, and is activated by autophosphorylation upon ECM binding or by phosphorylation by Src. Additional signaling is mediated by the RhoGTPases Rac1, Cdc42 and RhoA. Apart from these classical and well-characterized members, the FA protein complex is comprised of many additional proteins, altogether forming a complex from over 50 proteins (Lo 2006).

Contribution of individual FA proteins to endothelial barrier regulation

The involvement of the FA in endothelial barrier regulation can be considered in two ways: a) by looking at the role of the individual FA proteins in endothelial barrier regulation, and b) by looking at the role of the complex as a whole. Although most FA proteins have been characterized in mouse knock-out models (see excellent review by Lo 2006), little is known about their vascular effects. Genetic ablation of proteins like talin, vinculin, and paxillin induces embryonic lethality between E8.5 and E10, strongly suggestive of impaired vascular development, but as far as we know, this has not been confirmed in endothelial-specific knock-out models. In addition, little is known about the effect of genetic depletion of structural proteins like vinculin and talin in cultured endothelial monolayers. Kindlins activate integrins, and as such, may contribute to endothelial barrier regulation (Malinin et al. 2012). Kindlin-2$^{+/-}$ mice show increased basal endothelial permeability but no additive permeability response to VEGF (Pluskota et al. 2011).

The opposite is true for the regulatory proteins Src, FAK and the RhoGTPases. In particular, the role of FAK in endothelial barrier regulation has been extensively addressed in the last decade. The regulation of FAK and its functional consequences during endothelial barrier maintenance and disruption are extremely complex. The complexity of spatial and temporal regulation may account for the fact that studies indicate both barrier-disrupting (Chen et al. 2012) and barrier-protective effects of FAK (Knezevic et al. 2009). The role of FAK in endothelial barrier regulation is addressed in a series of recent reviews on this topic (Belvitch and Dudek 2012, Thennes and Mehta 2012, Zebda et al. 2012). Likewise, the roles of Src and the RhoGTPases have been extensively studied in the role of endothelial barrier regulation (for review see Mehta and Malik 2006). Yet, these proteins —although frequently present in FAs—are found throughout the whole cell. The effect of modulating their action is not limited to changes in FA function, but also affects actomyosin contraction and AJ dynamics.

Studying the role of FAs in endothelial barrier regulation by targeting individual FA proteins is limited by decreased cell viability upon inhibition of essential FA proteins, redundancy of proteins in the FA (loss of function of a specific FA protein is compensated by other proteins), and altered functionality in other parts of the cell in the case where the presence of a protein is not limited to FAs (Wu 2005). Alternatively, the involvement of FAs in endothelial barrier regulation can be studied by following these complexes as a whole (mainly immunofluorescence or live cell imaging) during changes in endothelial barrier function.

Involvement of FAs in endothelial barrier regulation

A hallmark study in the regulation of FAs has been the 1992 paper of Ridley and Hall (Ridley and Hall 1992). They showed that stimulation of fibroblasts with the growth factor PDGF leads to fast formation of actin stress fibers and FAs. Both events could be abrogated by transfection of fibroblasts with a dominant negative RhoA mutant. Later studies have demonstrated that similar processes are displayed during endothelial barrier disruption (Molony and Armstrong 1991, van Nieuw Amerongen et al. 2004). Stimulation of confluent endothelial monolayers with permeability-inducing factors like VEGF, thrombin and TNF-α induces rapid formation of stress fibers and FAs associated with cell contraction. The close spatio-temporal relation between stress fiber formation, FA assembly, actomyosin contractility, cell retraction, and endothelial barrier disruption suggested that these processes form a chain of events in which each event plays a mediating role. This was supported by the finding that during endothelial barrier disruption, stress fibers are capped at both ends by thick and dense FAs (Ridley and Hall 1992, Shikata et al. 2003a), giving rise to the general idea that FAs form anchoring points for stress fibers. Stress fibers, firmly attached to FAs, can exert pulling force on the cell periphery (in particular the AJ), leading to AJ disruption, cell retraction, and gap formation (van Nieuw Amerongen et al. 2004, Yuan 2000).

This idea, however, seems difficult to match with the earlier discussed work that showed that integrins, as the transmembrane component of FAs, are required for maintenance of endothelial barrier function (α5β1) and that disturbed integrin binding can enhance endothelial barrier disruption (αvβ3). Further evidence against the idea that FAs mediate endothelial barrier disruption came from the work of Birukov and Garica in assessing the barrier-enhancing phospholipids (Romer et al. 2006). They demonstrated that endothelial barrier enhancement with S1P and oxidized phosphoplipids is associated with an increase density of FAs, in particular at the periphery of the cell (Shikata et al. 2003a, 2003b). These peripheral FAs stay in close contact with and seem to support the cortical actin band, a dense ring of actin fibers located next to the cell membrane and characteristic of a resting endothelial monolayer (Dudek et al. 2004, Garcia et al. 2001). The phospholipid-induced relocalization of FAs was shown to be Rac1 dependent (Birukov et al. 2004, Shikata et al. 2003b), which agrees with other studies that showed involvement of Rac1 in the formation of new FAs at the cell periphery during cell spreading (Allen et al. 1997, Tan et al. 2008). These studies suggest that the presence of FAs contributes to endothelial barrier function and prevents further disruption of the barrier during endothelial stimulation. In the majority of these studies, FA dynamics were studied by immunofluoresence microscopy and correlated in time to effects on endothelial barrier function, predominantly yielding associative conclusions and leaving open causal questions. These merely associative conclusions, together with the limitations of studying FA dynamics in endothelial barrier regulation discussed above, lead to the comment that *"…the specific contribution of focal adhesions to the regulatory mechanism of microvascular barrier function remains a mystery…"* (Wu 2005).

However, the difference in FA location during endothelial junction disassembly (FAs located predominantly at ends of actin stress fibers) and endothelial barrier enhancement (FAs located predominantly at the cell periphery) suggests an important role of FA spatial distribution in endothelial barrier regulation.

Subcellular localization of FAs associated with endothelial barrier function

The relation between the spatial distribution of FAs and endothelial barrier function can be predicted by the tensegrity model as described by Ingber (2000). According to this model, cells can only maintain their shape when the internal tension is balanced by distension through compression-resistant elements or through extracellular adhesions. When translating this principle to endothelial cells in a monolayer, one can say that the cell shape as required for monolayer integrity (spread cell morphology) is maintained by a low internal tensile force (low actomyosin contraction) counterbalanced by FAs. Presence of FAs only at the center of the cell is not sufficient to maintain the spread morphology, but induces cell retraction and rounding up. However, the presence of FAs at the periphery is sufficient to maintain the spread shape of the cell required for monolayer integrity (Chen et al. 1997, Ingber 2002). In addition, peripheral redistribution of FAs not only maintains cell shape, but also enhances cell adhesion strength to the ECM, as demonstrated in the elegant study of Elineni and Gallant (2011). Seeding fibroblasts on fibronectin islands with a solid circular versus an annular shape, they demonstrated that cells grown on an annular fibronectin island have greater adhesive strength than cells grown on solid circular islands, even though the adhesive area was constant. Although these studies need to be confirmed with endothelial cells, they indicate that peripheral redistribution of FAs during endothelial barrier restoration or fortification contributes to improved endothelial barrier function. Otherwise, preservation of peripheral FAs during endothelial stimulation may attenuate gap formation and barrier dysfunction.

Two issues should be taken into account when considering spatial distribution of FAs, one being theoretical and the other methodological. The theoretical issue is the bidirectional signaling of FAs. FAs convert

ECM-dependent integrin conformation into intracellular signaling (outside-in signaling), but also convert intracellular signals to changes in integrin conformation (inside-out signaling). Outside-in signaling is required for sensing of the extracellular environment and the consequent intracellular response (Chen et al. 1997), while inside-out signaling mediates growth factor-induced changes in cell adhesion. Both pathways make use of the same FA proteins, but may differentially affect endothelial barrier function. The methodological issue is the identification of FAs. In the past, several proteins have been used to identify FAs, as these proteins were proposed to be FA-specific (e.g., FAK and vinculin), and to study cell-matrix interactions, in particular by immunofluoresence staining. This approach carries two problems: a) it is not clear how the spatial distribution of these single proteins relates to true cell-matrix receptors (integrins) and to true cell-matrix adhesive forces, and b) ongoing research has put into question the FA-specificity of these proteins. Vinculin, for example, was recently shown to be also present in AJs (Peng et al. 2011, Huveneers et al. 2012). These problems warrant care in the interpretation of experiments in which proteins like FAK and vinculin are used to identify FAs or cell-matrix interactions.

Conclusion

From the foregoing two mechanisms, comes an explanation of the role of integrins and FAs in endothelial barrier function. First, reinforcing cell-matrix adhesions, in particular at the cell periphery, compensates the loss of tethering forces due to dissociation of cell-cell junction. Secondly, integrin-activation causes intracellular signaling (FAK, Src, Rho GTPases and other) and altered F-actin dynamics, which may directly or indirectly affect the intracellular tension dynamics and cell junction proteins. Various proteins that are classically associated with FAs cluster in the region of the AJ, in particular immediately after destabilization of these junctions by vasoactive and inflammatory stimuli.

AJ Proteins Reinforce Cell-matrix Interactions during Endothelial Barrier Disruption

The Adherens Junction structure

Endothelial AJs resemble FAs in that they form a connection between the extracellular environment (in this case neighboring cells) and the cytoskeleton (Fig. 3). As indicated above, the core of the AJ is the transmembrane protein VE-cadherin, which binds in a homophilic, calcium-dependent fashion to

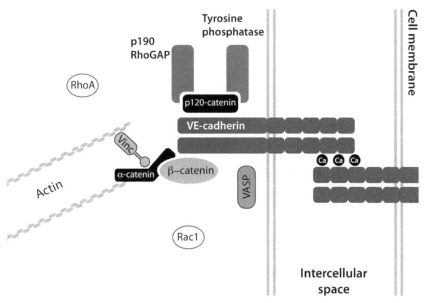

Figure 3. Schematic presentation of an adherens junction in the endothelium. Calcium-dependent homophilic VE-cadherin binding between neighbouring cells forms the core of this complex. VE-cadherin is a transmembrane protein that binds inside the cell to the actin cytoskeleton via a number of linker proteins. In addition, VE-cadherin binds to scaffolding proteins like p120-catenin that link regulatory proteins like phosphatases and p190RhoGAP.

VE-cadherin in a neighboring cell. Inside the cell, VE-cadherin binds to three members of the catenin family. α-Catenin and β-catenin directly or indirectly bind to VE-cadherin and both mediate the attachment of actin to the AJ. Attachment to the actin cytoskeleton is required for AJ stabilization as absence or dysfunction of these catenins hampers vascular development. The third is p120-catenin which is predominantly involved in binding phosphatases to the AJ. Phosphatases keep VE-cadherin and β-catenin in a dephosphorylated state, and since tyrosine phosphorylation of VE-cadherin and beta-catenin is an initial event in AJ dissociation (Esser et al. 1998), p120-catenin further stabilizes the AJ.

A number of proteins present in mature or destabilized AJs are also found in FAs. Among these proteins are vinculin, p190RhoGAP, VASP and FAK. Other proteins, which are neither AJ- nor FA-specific, show a remarkable translocation from one site to the other during endothelial barrier disruption and recovery. Examples of these proteins are the regulatory RhoGTPases Rac1 and RhoA. The next paragraphs will discuss the role of vinculin, p190RhoGAP, and other proteins in AJ and FA regulation during endothelial barrier disruption. During this process, these proteins relocate from the dissociating AJ to the FA. Stimulated endothelial cells show a large increase in the number of FAs, and FA formation and reinforcement of cell-matrix interactions may therefore compensate the loss of tethering forces due to AJ dissociation.

Vinculin

Vinculin, a classical FA protein, was demonstrated to be present in cell-cell junctions decades ago (Geiger et al. 1980, Lampugnani et al. 1991), although its function in AJs remained poorly understood. Recent studies in both epithelial and endothelial monolayers have made fast progress in this area and indicate there may be a clear role for vinculin in stabilization of AJs (Chervin-Pétinot et al. 2012, Peng et al. 2011). In a resting monolayer, AJs are subject to low actomyosin tension required for AJ stability. This tension induces a slight allosteric change in α-catenin configuration, which uncovers binding sites for vinculin (Gomez et al. 2011, Yonemura et al. 2010), similar to the tension-induced allosteric changes in talin which facilitate vinculin incorporation in the FA. In addition to its AJ-stabilizing effect under non-stimulated conditions, the presence of vinculin in AJs was recently shown to attenuate endothelial barrier disruption by thrombin, VEGF and TNF-α. Vinculin makes AJs more resistant to tensile forces exerted on the junction during actomyosin contraction. Mutation of the vinculin binding site on α-catenin (which prevents vinculin recruitment to the AJ) enhances gap formation upon agonist-induced actomyosin contraction (Huveneers et al. 2012).

In a resting monolayer and during the early phase of barrier disruption, vinculin localizes predominantly at the cell perimeter (suggestive of its presence in the AJ). The vinculin present in the cytosol has an inactive configuration. During prolonged stimulation with barrier disruptive agents, vinculin disappears from the AJs and accumulates in FAs. Sakurai et al. (2006) showed that acute disruption of the VE-cadherin bond using the calcium-chelator EGTA leads to an immediate loss of vinculin from the cell perimeter and consequent vinculin accumulation in FAs. Reversing this condition by calcium supplementation resulted in fast (within 20 min) disappearance of vinculin from FAs and restoration of vinculin localization at the lateral membrane. As changes in expression are precluded by the short interval of stimulation, this study is strongly suggestive for a common pool of vinculin serving both FAs and AJs. Even more, the relocalization of vinculin from the dissociating AJ to the FA readily explains the reinforcement of cell-matrix interactions upon loss of cell-cell adhesion.

p190RhoGAP

Like vinculin, p190RhoGAP is a protein present in both AJs and FAs. p190RhoGAP is a GTPase Activating Protein (GAP) that inactivates RhoA by converting it from a GTP- to a GDP-bound state. p190RhoGAP activity is regulated by tyrosine phosphorylation and nitration. In endothelial cells, p190RhoGAP is observed in AJs where it locally inactivates RhoA, thereby attenuating RhoA-dependent actomyosin contraction in the junctions. The presence of p190RhoGAP in the AJ is required for maintenance of AJ integrity (Holinstat et al. 2006), and redistribution of p190RhoGAP to the membrane mediates the effects of barrier-enhancing agents like oxidized phospholipids (Birukova et al. 2011). In contrast, during endothelial barrier disruption with thrombin, p190RhoGAP is nitrated and inactivated, leading to increased RhoA activity in the AJ and disruption of the AJ (Siddique et al. 2011).

In addition to the AJ, p190RhoGAP is active in the cytosol, in particular at sites of cell-matrix adhesion. p190RhoGAP co-localizes with paxillin at peripheral FAs, and knock-down of p190RhoGAP prevents the formation of peripheral FAs (Tomar et al. 2009). As RhoA inhibits cell spreading and initial FA formation (Bass et al. 2008), local inhibition of RhoA by p190RhoGAP is required for formation of new FAs and cell spreading. Impaired 190RhoGAP localization to FAs reduces FA turnover, leading to formation of large FAs, probably due to enhanced and prolonged local RhoA activity (Schober et al. 2007). The spread phenotype, due to peripheral localization of FAs and stable AJs, is characteristic of Rac1 activity. The opposite phenotype—a rounded-up morphology with disrupted AJs and contractile stress fibers connected to large FAs—characterizes cells with high RhoA activity. In between, p190RhoGAP acts a central regulator of the antagonism between Rac1 and RhoA as described for epithelial cells (Wildenberg et al. 2006). Membrane-bound p190RhoGAP may thus enhance AJ stability in the resting endothelium, whereas cytosolic, FA-bound p190RhoGAP enhances FA turnover and cell spreading in recovering endothelium. Although these observations strongly suggest a mechanistic link between AJ and FA regulation, evidence for such a link is less established than for vinculin. Live cell imaging of p190RhoGAP during endothelial barrier disruption and recovery is required to confirm this circumstantial evidence. As Rac1 (Birukova et al. 2011, Wildenberg et al. 2006) and Focal Adhesion Kinase (FAK) (Holinstat et al. 2006, Lim et al. 2010, Playford et al. 2008, Tomar et al. 2009) both regulate p190RhoGAP activity, and both are present in FAs and AJs, dynamic imaging of these proteins may further clarify its role in cell-matrix reinforcement upon AJ disruption (Sun et al. 2009).

VASP

A third protein involved in the regulation of both the AJs and the FAs is VASP. In confluent cells, VASP localizes to stabilizing junctions (Hoelzle and Svitkina 2012), as discussed in section 1.2., and stable AJs contain VASP. When endothelial cells grow to confluence, more and more VASP is recruited to AJs. Here, VASP contributes to AJ stability and endothelial barrier function by enhancing the formation of the cortical actin band (Benz et al. 2008). VASP is required for stability of the endothelial barrier, as demonstrated by reduced endothelial barrier function in VASP-deficient mice (Furman et al. 2007) and VASP-depleted endothelial cells (Schlegel et al. 2008). In sparsely seeded cells (Benz et al. 2008, Draijer et al. 1995) or cells stimulated with vasoactive agents (Han et al. 2009), however, VASP is found in FAs. The latter study demonstrated that the thrombin-induced translocation of VASP from AJs to FAs is mediated by zyxin. Depletion of zyxin enhanced the thrombin-induced endothelial barrier disruption and retarded recovery after thrombin stimulation (Han et al. 2009). These studies indicate that re-localization of VASP from the AJ to the FA during thrombin stimulation protects against endothelial barrier dysfunction. This may be explained by two mechanisms. First VASP may attenuate cell retraction and gap formation by actin polymerization and cytoskeletal reinforcement: at the level of the AJ (by forming the cortical actin band and the consequent AJ stabilization) in resting, confluent cells or at the level of the FA (by stabilizing the contact of FAs with the cytoskeleton) in sparse or stimulated cells. Second, as shown by Schlegel and Waschke, VASP is required for integrin-mediated adhesion (Schlegel and Waschke 2009). This study suggests that upon AJ dissociation, VASP may reinforce integrin-mediated adhesion.

Conclusion

Altogether, these studies demonstrate a close relationship between loss of cell-cell contact and reinforcement of cell-matrix interactions. FAs and AJs share proteins that have a similar function in both protein complexes (e.g., in both the AJ and FA, vinculin links the transmembrane receptor to the actin cytoskeleton in a manner which is dependent on tension-induced allosterical changes in another linker protein). Assuming that a common pool serves both complexes, loss of protein from the one complex (in this case AJ) may lead to accumulation of that protein in the other complex (in this case FA). The increased tethering force in the FA (due to FA reinforcement) thus compensates the loss of tethering force in the AJ (Fig. 4).

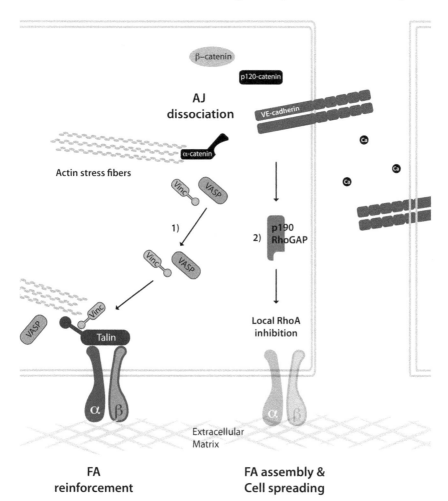

Figure 4. Possible mechanisms for focal adhesion reinforcement upon adherens junction dissociation. As described in the text, two mechanisms may mediate reinforcement of binding of the cell to the extracellular matrix. Upon adherens junction dissociation: 1) vinculin and VASP are released in the cytosol and relocate to focal adhesions where they stabilize focal adhesions and increase binding strength to the extracellular matrix, and 2) p190RhoGAP is released in the cytosol and may contribute to formation of novel focal adhesions and cell spreading by local RhoA inhibition.

As we have seen before, cell shape and endothelial barrier function depend on tethering forces in cell-cell contacts, actomyosin contractile force, and adhesive forces in association with the ECM. Under basal conditions (in a resting endothelial monolayer) the sum of these forces is zero, resulting in the spreading out of the endothelial cell and an intact endothelial barrier. Upon stimulation with permeability factors (Goldenberg et al. 2011) there is an acute increase in actomyosin contraction and dissociation of cell-cell junctions. The resulting dysfunction of the endothelial barrier is determined by the size of the intercellular gaps, which in turn is determined by the extent of cell retraction. In the absence of functional cell-cell junctions, the extent of cell retraction mainly depends on the adhesion of cell edges to the ECM. The influence of cell-matrix interactions (reflected by the number of FAs) on cell retraction, and thus on endothelial barrier disruption, is illustrated in Fig. 5. A low number of FAs at the cell periphery facilitates swift cell retraction and formation of large intercellular gaps (Fig. 5A), whereas a high number of peripheral FAs prevents cell retraction and reduces the size of the intercellular gaps (Fig. 5B). Reinforcement of cell-matrix interactions either by preservation of existing FAs or formation of new FAs may therefore serve as a 'hand brake' that prevents rigorous retraction of the cell. An increase in tethering force in the FAs thus compensates for the loss of tethering forces at cell-cell junctions during endothelial barrier disruption.

A clear illustration of such a protective compensation between tethering forces in cell-cell contacts and cell-matrix interactions was recently shown for epithelial cells. Lehembre et al. showed that loss of E-cadherin resulted in upregulation of NCAM. NCAM and E-cadherin were reciprocally expressed, and the increased expression of NCAM induced assembly of β1-integrin-dependent FAs (Frame and Inman 2008, Lehembre et al. 2008).

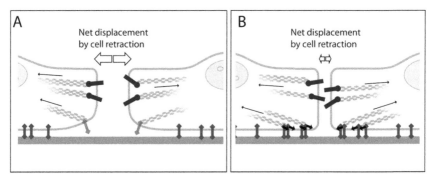

Figure 5. Overview of forces observed in endothelial cells within an endothelial monolayer stimulated with barrier-disruptive agents. Stimulation of endothelial cells with barrier disruptive agents leads to dissociation of adherens junctions and increased actomyosin contraction. Consequently, intercellular gaps are formed, the size of which depends on tethering forces in cell-matrix adhesions. A) Low tethering forces in cell-matrix adhesions facilitates cell retraction, resulting in larger net cell displacement and intercellular gaps. B) High tethering forces in cell-matrix adhesions hampers swift cell retraction and limits the net cell displacement.

Conclusions and Future Perspectives

Summary of the proposed model

In summary, this chapter illustrates the significance of proteins involved in cell-matrix interactions in endothelial barrier regulation. Binding of cells to the ECM is required for maintenance of the spread cell shape that characterizes a functional monolayer. As a tethering force counteracting actomyosin contraction, cell-matrix interactions protect against endothelial barrier disruption. During endothelial barrier disruption, cells may reinforce their adhesion to the ECM as a defensive mechanism to protect against rigorous cell retraction. Attenuating cell retraction thus diminishes gap formation and prevents cell rounding up (and thus apoptosis). This reinforcement of cell-matrix interaction is closely related in time to AJ disruption. Reviewing dynamics of vinculin, p190RhoGAP, and VASP, we propose that AJ disruption and FA reinforcement are mechanistically linked. The observations that 1) enhanced cell-matrix interaction attenuates endothelial barrier disruption and 2) AJ disruption is mechanically linked to FA reinforcement support the idea that reinforcement of cell-matrix interactions are a defensive mechanism by which endothelial cells protect the monolayer against barrier dysfunction (Fig. 5).

Remaining questions

Yet, several important issues remain unresolved, the first of which is integrin signaling during endothelial barrier regulation. Although a lot is known about the functional relevance of integrins in endothelial barrier regulation (as elucidated by knock-out models, RGD peptides, and integrin function-modulating antibodies), little is known about the regulation of integrin function during agonist-induced endothelial barrier disruption. To the best of our knowledge, there are very few studies that address the effect of permeability on the spatiotemporal distribution of integrins, the changes in integrin affinity and avidity for ECM ligands, or on integrin internalization and recycling. Integrin activation may play a central role in these processes. As known from circulating blood cells like platelets and leukocytes, inflammatory cytokines can activate or deactivate integrins by so-called inside-out signaling. Cytokine receptor activation induces intracellular signaling, leading to changes in integrin conformation, which in turn changes affinity and avidity of integrins for their ECM ligands, and thus affects the binding of endothelial cells to the ECM. Evaluation of integrin activity (with antibodies specific for epitopes uncovered in activated integrins) upon endothelial stimulation with permeability factors will contribute to the understanding of integrin regulation during vascular leakage and edema formation.

A second point, closely related to this matter, is the composition of FAs during endothelial barrier maintenance and disruption. Most studies have considered FAs as rather uniform in composition, but functional analyses of cell-matrix interactions and barrier function raise doubts about this approach. FA heterogeneity is particularly illustrated by two observations discussed in this chapter: 1) integrins play a heterogeneous role in endothelial barrier regulation—α5β1 and αvβ3 usually protect against endothelial barrier disruption, whereas αvβ5 evidently contributes to endothelial barrier disruption, and 2) during endothelial barrier disruption some FAs bind to stress fibers, whereas other FA-like complexes are observed independent of stress fibers, at the periphery of the cell. Together with spatial distribution, the idea of heterogeneous FA composition explains and combines both observations: central, stress fiber-bound FAs containing 'deleterious' integrins like αvβ5 contribute to endothelial barrier disruption, whereas peripheral FAs containing 'beneficial' integrins like α5β1 maintain basal barrier function, attenuate barrier disruption, and mediate barrier restoration.

Therapeutic implications

Given these data, it is striking that a drug like volociximab (an FDA-approved α5β1-blocking antibody designed to inhibit integrin-mediated tumor angiogenesis (Ramakrishnan et al. 2006)) gained fast access to clinical trials. The barrier-disrupting properties of α5β1-blockage observed in *in vitro* studies may give rise to edema in the clinical setting, in particular in patients prone to edema because of increased systemic inflammation (like sepsis and acute lung injury). Indeed, clinical trials on volociximab reported (peripheral) edema as a side-effect in 10%-35% of treated patients (Ricart et al. 2008, http://www.evaluatepharma.com/Universal/View.aspx?type=Story&id=86561).

From a clinical point of view, the data reviewed in this chapter indicate that proteins involved in cell-matrix interactions may form a target to attenuate endothelial barrier disruption and vascular leakage. In the search for specific treatments, improvement of integrin binding to the ECM may be a suitable option, as interfering with, e.g., vinculin or RhoGTPase dynamics will interfere with cell-matrix independent processes. Enhancing cell-matrix binding may seal the endothelial barrier to prevent inflammatory charges, or reinforce the defensive response to treat barrier disruption.

References

Albelda, S.M., M. Daise, E.M. Levine et al. 1989. Identification and characterization of cell-substratum adhesion receptors on cultured human endothelial cells. J Clin Invest 83: 1992–2002.

Allen, W.E., G.E. Jones, J.W. Pollard et al. 1997. Rho, Rac and Cdc42 regulate actin organization and cell adhesion in macrophages. J Cell Sci 110: 707–20.

Andriopoulou, P., P. Navarro, A. Zanetti et al. 1999. Histamine induces tyrosine phosphorylation of endothelial cell-to-cell adherens junctions. Arterioscler Thromb Vasc Biol 19: 2286–97.

Bader, B.L., H. Rayburn, D. Crowley et al. 1998. Extensive vasculogenesis, angiogenesis, and organogenesis precede lethality in mice lacking all alpha v integrins. Cell 13: 507–19.

Bass, M.D., M.R. Morgan, K.A. Roach et al. 2008. p190RhoGAP is the convergence point of adhesion signals from alpha 5 beta 1 integrin and syndecan-4. J Cell Biol 16: 1013–26.

Belvitch, P. and S.M. Dudek. 2012. Role of FAK in S1P-regulated endothelial permeability. Microvasc Res 83: 22–30.

Benz, P.M., C. Blume, J. Moebius et al. 2008. Cytoskeleton assembly at endothelial cell-cell contacts is regulated by alphaII-spectrin-VASP complexes. J Cell Biol 180: 205–19.

Birukov, K.G., V.N. Bochkov, A.A. Birukova et al. 2004. Epoxycyclopentenone-containing oxidized phospholipids restore endothelial barrier function via Cdc42 and Rac. Circ Res 29: 892–901.

Birukova, A.A., N. Zebda, I. Cokic et al. 2011. p190RhoGAP mediates protective effects of oxidized phospholipids in the models of ventilator-induced lung injury. Exp Cell Res 317: 859–72.

Boon, R.A., T.A. Leyen, F.D. Fontijn et al. 2010. KLF2-induced actin shear fibers control both alignment to flow and JNK signaling in vascular endothelium. Blood 115: 2533–42.

Brakebusch, C. and R. Fässler. 2003. The integrin-actin connection, an eternal love affair. EMBO J 22: 2324–33.

Bundgaard, M. 1984. The three-dimensional organization of tight junctions in a capillary endothelium revealed by serial-section electron microscopy. J Ultrastruct Res 88: 1–17.

Cailleteau, L., S. Estrach, R. Thyss et al. 2010. alpha2beta1 integrin controls association of Rac with the membrane and triggers quiescence of endothelial cells. J Cell Sci 123: 2491–501.

Califano, J.P. and C.A. Reinhart-King. 2010. Exogenous and endogenous force regulation of endothelial cell behavior. J Biomech 43: 79–86.

Cascone, I., L. Napione, F. Maniero et al. 2005. Stable interaction between alpha5beta1 integrin and Tie2 tyrosine kinase receptor regulates endothelial cell response to Ang-1. J Cell Biol 170: 993–1004.

Chen, C.S., M. Mrksich, S. Huang et al. 1997. Geometric control of cell life and death. Science 276: 1425–8.

Chen, J., T.G. Diacovo, D.G. Grenache et al. 2002. The alpha(2) integrin subunit-deficient mouse: a multifaceted phenotype including defects of branching morphogenesis and hemostasis. Am J Pathol 161: 337–44.

Chen, X.L., J.O. Nam, C. Jean et al. 2012. VEGF-induced vascular permeability is mediated by FAK. Dev Cell 22: 146–57.

Chervin-Pétinot, A., M. Courçon, S. Almagro et al. 2012. Epithelial protein lost in neoplasm (EPLIN) interacts with α-catenin and actin filaments in endothelial cells and stabilizes vascular capillary network *in vitro*. J Biol Chem 287: 7556–72.

Corada, M., M. Mariotti, G. Thurston et al. 1999. Vascular endothelial-cadherin is an important determinant of microvascular integrity *in vivo*. Proc Natl Acad Sci USA 96: 9815–20.

Costell, M., E. Gustafsson, A. Aszódi et al. 1999. Perlecan maintains the integrity of cartilage and some basement membranes. J Cell Biol 147: 1109–22.

Curtis, T.M., P.J. McKeown-Longo, P.A. Vincent et al. 1995. Fibronectin attenuates increased endothelial monolayer permeability after RGD peptide, anti-alpha 5 beta 1, or TNF-alpha exposure. Am J Physiol 269: L248–60.

Curtis, T.M., R.F. Rotundo, P.A. Vincent et al. 1998. TNF-alpha-induced matrix Fn disruption and decreased endothelial integrity are independent of Fn proteolysis. Am J Physiol 275: L126–38.

Davis, G.E. and D.R. Senger. 2005. Endothelial extracellular matrix: biosynthesis, remodeling, and functions during vascular morphogenesis and neovessel stabilization. Circ Res 97: 1093–107.

Defilippi, P., V. van Hinsbergh, A. Bertolotto et al. 1991. Differential distribution and modulation of expression of alpha 1/beta 1 integrin on human endothelial cells. J Cell Biol 114: 855–63.

Dejana, E., F. Orsenigo and M.G. Lampugnani. 2008. The role of adherens junctions and VE-cadherin in the control of vascular permeability. J Cell Sci 121: 2115–22.

Draijer, R., A.B. Vaandrager, C. Nolte, H.R. de Jonge, U. Walter and V.W. van Hinsbergh. HYPERLINK "/pubmed/7554143" Expression of cGMP-dependent protein kinase I and phosphorylation of its substrate, vasodilator-stimulated phosphoprotein, in human endothelial cells of different origin. Circ Res 1995 Nov; 77(5): 897–905.

Dudek, S.M. and J.G. Garcia. 2001. Cytoskeletal regulation of pulmonary vascular permeability. J Appl Physiol 91: 1487–500.

Dudek, S.M., J.R. Jacobson, E.T. Chiang et al. 2004. Pulmonary endothelial cell barrier enhancement by sphingosine 1-phosphate: roles for cortactin and myosin light chain kinase. J Biol Chem 279: 24692–700.

Elineni, K.K. and N.D. Gallant. 2011. Regulation of cell adhesion strength by peripheral focal adhesion distribution. Biophys J 101: 2903–11.

Esser, S., M.G. Lampugnani, M. Corada et al. 1998. Vascular endothelial growth factor induces VE-cadherin tyrosine phosphorylation in endothelial cells. J Cell Sci 111: 1853–65.

Fässler, R. and M. Meyer. 1995. Consequences of lack of beta 1 integrin gene expression in mice. Genes Dev 9: 1896–908.

Felcht, M., R. Luck, A. Schering et al. 2012. Angiopoietin-2 differentially regulates angiogenesis through TIE2 and integrin signaling. J Clin Invest May 15. Epub ahead of print.

Frame, M.C. and G.J. Inman. 2008. NCAM is at the heart of reciprocal regulation of E-cadherin- and integrin-mediated adhesions via signaling modulation. Dev Cell 15: 494–6.

Friedlander, M., P.C. Brooks, R.W. Shaffer et al. 1995. Definition of two angiogenic pathways by distinct alpha v integrins. Science 270: 1500–2.

Furman, C., A.L. Sieminski, A.V. Kwiatkowski et al. 2007. Ena/VASP is required for endothelial barrier function *in vivo*. J Cell Biol 179: 761–75.

Ganter, M.T., J. Roux, B. Miyazawa et al. 2008. Interleukin-1beta causes acute lung injury via alphavbeta5 and alphavbeta6 integrin-dependent mechanisms. Circ Res 102: 804–12.

Ganter, M.T., J. Roux, G. Su et al. 2009. Role of small GTPases and alphavbeta5 integrin in Pseudomonas aeruginosa-induced increase in lung endothelial permeability. Am J Respir Cell Mol Biol 40: 108–18.

Gao, B., T.M. Curtis, F.A. Blumenstock et al. 2000. Increased recycling of (alpha)5(beta)1 integrins by lung endothelial cells in response to tumor necrosis factor. J Cell Sci 113: 247–57.

Garcia, J.G., F. Liu, A.D. Verin et al. 2001. Sphingosine 1-phosphate promotes endothelial cell barrier integrity by Edg-dependent cytoskeletal rearrangement. J Clin Invest 108: 689–701.

Gardner, H., J. Kreidberg, V. Koteliansky et al. 1996. Deletion of integrin alpha 1 by homologous recombination permits normal murine development but gives rise to a specific deficit in cell adhesion. Dev Biol 175: 301–13.

Geiger, B., K.T. Tokuyasu, A.H. Dutton et al. 1980. Vinculin, an intracellular protein localized at specialized sites where microfilament bundles terminate at cell membranes. Proc Natl Acad Sci USA 77: 4127–31.

Georges-Labouesse, E., N. Messaddeq, G. Yehia et al. 1996. Absence of integrin alpha 6 leads to epidermolysis bullosa and neonatal death in mice. Nat Genet 13: 370–3.

Germain, M., A. de Arcangelis, S.D. Robinson et al. 2010. Genetic ablation of the alpha 6-integrin subunit in Tie1Cre mice enhances tumour angiogenesis. J Pathol 220: 370–81.

Goh, Y.Y., M. Pal, H.C. Chong et al. 2010. Angiopoietin-like 4 interacts with integrins beta1 and beta5 to modulate keratinocyte migration. Am J Pathol 177: 2791–803.

Goldenberg, N.M., B.E. Steinberg, A.S. Slutsky et al. 2011. Broken barriers: a new take on sepsis pathogenesis. Sci Transl Med 3: 88ps25.

Gomez, G.A., R.W. McLachlan and A.S. Yap. 2011. Productive tension: force-sensing and homeostasis of cell-cell junctions. Trends Cell Biol 21: 499–505.

Guo, M., D. Daines, J. Tang et al. 2009. Fibrinogen-gamma C-terminal fragments induce endothelial barrier dysfunction and microvascular leak via integrin-mediated and RhoA-dependent mechanism. Arterioscler Thromb Vasc Biol 29: 394–400.

Han, J., G. Liu, J. Profirovic et al. 2009. Zyxin is involved in thrombin signaling via interaction with PAR-1 receptor. FASEB J 23: 4193–206.

Hodivala-Dilke, K.M., K.P. McHugh, D.A. Tsakiris et al. 1999. Beta3-integrin-deficient mice are a model for Glanzmann thrombasthenia showing placental defects and reduced survival. J Clin Invest 103: 229–38.

Hoelzle, M.K. and T. Svitkina. 2012. The cytoskeletal mechanisms of cell-cell junction formation in endothelial cells. Mol Biol Cell 23: 310–23.

Holinstat, M., N. Knezevic, M. Broman et al. 2006. Suppression of RhoA activity by focal adhesion kinase-induced activation of p190RhoGAP: role in regulation of endothelial permeability. J Biol Chem 281: 2296–305.

Honda, S., T. Nagai and A. Negi. 2009. Anti-angiogenic effects of non-peptide integrin alphavbeta3 specific antagonist on laser-induced choroidal neovascularization in mice. Graefes Arch Clin Exp Ophthalmol 247: 515–22.

Huang, R.L., Z. Teo, H.C. Chong et al. 2011. ANGPTL4 modulates vascular junction integrity by integrin signaling and disruption of intercellular VE-cadherin and claudin-5 clusters. Blood 118: 3990–4002.

Huang, X.Z., J.F. Wu, R. Ferrando et al. 2000. Fatal bilateral chylothorax in mice lacking the integrin alpha9beta1. Mol Cell Biol 20: 5208–15.

Humphries, J.D., P. Wang, C. Streuli et al. 2007. Vinculin controls focal adhesion formation by direct interactions with talin and actin. J Cell Biol 179: 1043–57.

Huveneers, S., J. Oldenburg, E. Spanjaard et al. 2012. Vinculin associates with endothelial VE-cadherin junctions to control force-dependent remodeling. J Cell Biol 196: 641–52.

Huynh, J., N. Nishimura, K. Rana et al. 2011. Age-related intimal stiffening enhances endothelial permeability and leukocyte transmigration. Sci Transl Med 3:112ra122.

Hynes, R.O. 2002a. Integrins: bidirectional, allosteric signaling machines. Cell 110: 673–87.

Hynes, R.O. 2002b. A reevaluation of integrins as regulators of angiogenesis. Nat Med 8: 918–21.

Hynes, R.O. 2007. Cell-matrix adhesion in vascular development. J Thromb Haemost 5 Suppl 1: 32–40.

Iliaki, E., V. Poulaki, N. Mitsiades et al. 2009. Role of alpha 4 integrin (CD49d) in the pathogenesis of diabetic retinopathy. Invest Ophthalmol Vis Sci 50: 4898–904.

Ingber, D.E. 2000. Opposing views on tensegrity as a structural framework for understanding cell mechanics. J Appl Physiol 89: 1663–70.

Ingber, D.E. 2002. Mechanical signaling and the cellular response to extracellular matrix in angiogenesis and cardiovascular physiology. Circ Res 91: 877–87.

Kalluri, R. 2003. Basement membranes: structure, assembly and role in tumour angiogenesis. Nat Rev Cancer 3: 422–33.

Knezevic, N., M. Tauseef, T. Thennes et al. 2009. The G protein betagamma subunit mediates reannealing of adherens junctions to reverse endothelial permeability increase by thrombin. J Exp Med 206: 2761–77.

Kreidberg, J.A., M.J. Donovan, S.L. Goldstein et al. 1996. Alpha 3 beta 1 integrin has a crucial role in kidney and lung organogenesis. Development 122: 3537–47.

Krishnan, R., D.D. Klumpers, C.Y. Park et al. 2011. Substrate stiffening promotes endothelial monolayer disruption through enhanced physical forces. Am J Physiol Cell Physiol 300: C146–54.

Lacy-Hulbert, A., A.M. Smith, H. Tissire et al. 2007. Ulcerative colitis and autoimmunity induced by loss of myeloid alphav integrins. Proc Natl Acad Sci USA 104: 15823–8.

Lampugnani, M.G., M. Resnati, E. Dejana et al. 1991. The role of integrins in the maintenance of endothelial monolayer integrity. J Cell Biol 112: 479–90.

Lee, T.Y. and A.I. Gotlieb. 2003. Microfilaments and microtubules maintain endothelial integrity. Microsc Res Tech 60: 115–27.

Lehembre, F., M. Yilmaz, A. Wicki et al. 2008. NCAM-induced focal adhesion assembly: a functional switch upon loss of E-cadherin. EMBO J 27: 2603–15.

Lei, L., D. Liu, Y. Huang et al. 2008. Endothelial expression of beta1 integrin is required for embryonic vascular patterning and postnatal vascular remodeling. Mol Cell Biol 28: 794–802.

Li, X., M. Stankovic, B.P. Lee et al. 2009. JAM-C induces endothelial cell permeability through its association and regulation of {beta}3 integrins. Arterioscler Thromb Vasc Biol 29: 1200–6.

Lim, S.T., X.L. Chen, A. Tomar et al. 2010. Knock-in mutation reveals an essential role for focal adhesion kinase activity in blood vessel morphogenesis and cell motility-polarity but not cell proliferation. J Biol Chem 285: 21526–36.

Liu, Z., J.L. Tan, D.M. Cohen et al. 2010. Mechanical tugging force regulates the size of cell-cell junctions. Proc Natl Acad Sci USA 107: 9944–9.

Lo, S.H. 2006. Focal adhesions: what's new inside. Dev Biol 294: 280–91.

Malinin, N.L., E. Pluskota and T.V. Byzova. 2012. Integrin signaling in vascular function. Curr Opin Hematol 19: 206–11.

Massia, S.P. and J.A. Hubbell. 1992. Vascular endothelial cell adhesion and spreading promoted by the peptide REDV of the IIICS region of plasma fibronectin is mediated by integrin alpha 4 beta 1. J Biol Chem 267:14019–26.

Mattila, P.K. and P. Lappalainen. 2008. Filopodia: molecular architecture and cellular functions. Nat Rev Mol Cell Biol 9: 446–54.

Mehta, D. and A.B. Malik. 2006. Signaling mechanisms regulating endothelial permeability. Physiol Rev 86: 279–367.

Millán, J., R.J. Cain, N. Reglero-Real et al. 2010. Adherens junctions connect stress fibres between adjacent endothelial cells. BMC Biol 8: 11.

Molony, L. and L. Armstrong. 1991. Cytoskeletal reorganizations in human umbilical vein endothelial cells as a result of cytokine exposure. Exp Cell Res 196: 40–8.

Parikh, S.M., T. Mammoto, A. Schultz et al. 2006. Excess circulating angiopoietin-2 may contribute to pulmonary vascular leak in sepsis in humans. PLoS Med 3: e46.

Peng, X., E.S. Nelson, J.L. Maiers et al. 2011. New insights into vinculin function and regulation. Int Rev Cell Mol Biol 287: 191–231.

Peters, J.H., M.N. Grote, N.E. Lane et al. 2011. Changes in plasma fibronectin isoform levels predict distinct clinical outcomes in critically ill patients. Biomark Insights 6: 59–68.

Petrache, I., A. Birukova, S.I. Ramirez et al. 2003. The role of the microtubules in tumor necrosis factor-alpha-induced endothelial cell permeability. Am J Respir Cell Mol Biol 28: 574–81.

Playford, M.P., K. Vadali, X. Cai et al. 2008. Focal adhesion kinase regulates cell-cell contact formation in epithelial cells via modulation of Rho. Exp Cell Res 314: 3187–97.

Pluskota, E., J.J. Dowling, N. Gordon et al. 2011. The integrin coactivator kindlin-2 plays a critical role in angiogenesis in mice and zebrafish. Blood 117: 4978–87.

Pöschl, E., U. Schlötzer-Schrehardt, B. Brachvogel et al. 2004. Collagen IV is essential for basement membrane stability but dispensable for initiation of its assembly during early development. Development 131: 1619–28.

Qiao, R.L., W. Yan, H. Lum et al. 1995. Arg-Gly-Asp peptide increases endothelial hydraulic conductivity: comparison with thrombin response. Am J Physiol 269: C110–7.

Ramakrishnan, V., V. Bhaskar, D.A. Law et al. 2006. Preclinical evaluation of an anti-alpha5beta1 integrin antibody as a novel anti-angiogenic agent. J Exp Ther Oncol 5: 273–86.

Reynolds, L.E., L. Wyder, J.C. Lively et al. 2002. Enhanced pathological angiogenesis in mice lacking beta3 integrin or beta3 and beta5 integrins. Nat Med 8: 27–34.

Ricart, A.D., A.W. Tolcher, G. Liu et al. 2008. Volociximab, a chimeric monoclonal antibody that specifically binds alpha5beta1 integrin: a phase I, pharmacokinetic, and biological correlative study. Clin Cancer Res 14: 7924–9.

Ridley, A.J. and A. Hall. 1992. The small GTP-binding protein rho regulates the assembly of focal adhesions and actin stress fibers in response to growth factors. Cell 70: 389–99.

Romer, L.H., K.G. Birukov and J.G. Garcia. 2006. Focal adhesions: paradigm for a signaling nexus. Circ Res 98: 606–16.

Rotundo, R.F., T.M. Curtis, M.D. Shah et al. 2002. TNF-alpha disruption of lung endothelial integrity: reduced integrin mediated adhesion to fibronectin. Am J Physiol Lung Cell Mol Physiol 282: L316–29.

Roviezzo, F., S. Tsigkos, A. Kotanidou et al. 2005. Angiopoietin-2 causes inflammation *in vivo* by promoting vascular leakage. J Pharmacol Exp Ther 314: 738–44.

Rüegg, C., A. Yilmaz, G. Bieler et al. 1998. Evidence for the involvement of endothelial cell integrin alphaVbeta3 in the disruption of the tumor vasculature induced by TNF and IFN-gamma. Nat Med 4: 408–14.

Sakurai, A., S. Fukuhara, A. Yamagishi et al. 2006. MAGI-1 is required for Rap1 activation upon cell-cell contact and for enhancement of vascular endothelial cadherin-mediated cell adhesion. Mol Biol Cell 17: 966–76.

Schlegel, N. and J. Waschke. 2009. Impaired integrin-mediated adhesion contributes to reduced barrier properties in VASP-deficient microvascular endothelium. J Cell Physiol 220: 357–66.

Schlegel, N., S. Burger, N. Golenhofen et al. 2008. The role of VASP in regulation of cAMP- and Rac 1-mediated endothelial barrier stabilization. Am J Physiol Cell Physiol 294: C178–88.

Schober, M., S. Raghavan, M. Nikolova et al. 2007. Focal adhesion kinase modulates tension signaling to control actin and focal adhesion dynamics. J Cell Biol 176: 667–80.

Shikata, Y., K.G. Birukov, A.A. Birukova et al. 2003a. Involvement of site-specific FAK phosphorylation in sphingosine-1 phosphate- and thrombin-induced focal adhesion remodeling: role of Src and GIT. FASEB J 17: 2240–9.

Shikata, Y., K.G. Birukov and J.G. Garcia. 2003b. S1P induces FA remodeling in human pulmonary endothelial cells: role of Rac, GIT1, FAK, and paxillin. J Appl Physiol 94: 1193–203.

Shimamura, N., G. Matchett, I. Solaroglu et al. 2006. Inhibition of integrin alphavbeta3 reduces blood-brain barrier breakdown in focal ischemia in rats. J Neurosci Res 84: 1837–47.

Siddiqui, M.R., Y.A. Komarova, S.M. Vogel et al. 2011. Caveolin-1-eNOS signaling promotes p190RhoGAP-A nitration and endothelial permeability. J Cell Biol 193: 841–50.

Silva, R., G. D'Amico, K.M. Hodivala-Dilke et al. 2008. Integrins: the keys to unlocking angiogenesis. Arterioscler Thromb Vasc Biol 28: 1703–13.

Soldi, R., S. Mitola, M. Strasly et al. 1999. Role of alphavbeta3 integrin in the activation of vascular endothelial growth factor receptor-2. EMBO J 18: 882–92.

Su, G., M. Hodnett, N. Wu et al. 2007. Integrin alphavbeta5 regulates lung vascular permeability and pulmonary endothelial barrier function. Am J Respir Cell Mol Biol 36: 377–86.

Su, G., A. Atakilit, J.T. Li et al. 2012. Absence of integrin $\alpha v\beta 3$ enhances vascular leak in mice by inhibiting endothelial cortical actin formation. Am J Respir Crit Care Med 185: 58–66.

Sun, X., Y. Shikata, L. Wang et al. 2009. Enhanced interaction between focal adhesion and adherens junction proteins: involvement in sphingosine 1-phosphate-induced endothelial barrier enhancement. Microvasc Res 77: 304–13.

Taddei, A., C. Giampietro, A. Conti et al. 2008. Endothelial adherens junctions control tight junctions by VE-cadherin-mediated upregulation of claudin-5. Nat Cell Biol 10: 923–34.

Tan, W., T.R. Palmby, J. Gavard et al. 2008. An essential role for Rac1 in endothelial cell function and vascular development. FASEB J 22: 1829–38.

Tarone, G., G. Stefanuto, P. Mascarello et al. 1990. Expression of receptors for extracellular matrix proteins in human endothelial cells. J Lipid Mediat 2 Suppl: S45–53.

Thennes, T. and D. Mehta. 2012. Heterotrimeric G proteins, focal adhesion kinase, and endothelial barrier function. Microvasc Res 83: 31–44.

Thomas, M., M. Felcht, K. Kruse et al. 2010. Angiopoietin-2 stimulation of endothelial cells induces alphavbeta3 integrin internalization and degradation. J Biol Chem 285: 23842–9.

Thurston, G., C. Suri, K. Smith et al. 1999. Leakage-resistant blood vessels in mice transgenically overexpressing angiopoietin-1. Science 286: 2511–4.

Tomar, A., S.T. Lim, Y. Lim et al. 2009. A FAK-p120RasGAP-p190RhoGAP complex regulates polarity in migrating cells. J Cell Sci 122: 1852–62.

Turner, C.E. 2000. Paxillin and focal adhesion signalling. Nat Cell Biol 2: E231–6.

Turowski, P., R. Martinelli, R. Crawford et al. 2008. Phosphorylation of vascular endothelial cadherin controls lymphocyte emigration. J Cell Sci Jan 1; 121(Pt 1): 29–37.

van der Heijden, M., G.P. van Nieuw Amerongen, P. Koolwijk et al. 2008. Angiopoietin-2, permeability oedema, occurrence and severity of ALI/ARDS in septic and non-septic critically ill patients. Thorax 63: 903–9.

van der Heijden, M., G.P. van Nieuw Amerongen, J. van Bezu et al. 2011. Opposing effects of the angiopoietins on the thrombin-induced permeability of human pulmonary microvascular endothelial cells. PLoS One 6: e23448.

van Nieuw Amerongen, G.P., K. Natarajan, G. Yin et al. 2004. GIT1 mediates thrombin signaling in endothelial cells: role in turnover of RhoA-type focal adhesions. Circ Res 94 1041–9.

Verin, A.D., A. Birukova, P. Wang et al. 2001. Microtubule disassembly increases endothelial cell barrier dysfunction: role of MLC phosphorylation. Am J Physiol Lung Cell Mol Physiol 281: L565–74.

Vestweber, D. 2008. VE-cadherin: the major endothelial adhesion molecule controlling cellular junctions and blood vessel formation. Arterioscler Thromb Vasc Biol 28: 223–32.

Wehrle-Haller, B. 2012. Structure and function of focal adhesions. Curr Opin Cell Biol 24: 116–24.

Wildenberg, G.A., M.R. Dohn, R.H. Carnahan et al. 2006. p120-catenin and p190RhoGAP regulate cell-cell adhesion by coordinating antagonism between Rac and Rho. Cell 127: 1027–39.

Wójciak-Stothard, B., A. Entwistle, R. Garg et al. 1998. Regulation of TNF-alpha-induced reorganization of the actin cytoskeleton and cell-cell junctions by Rho, Rac, and Cdc42 in human endothelial cells. J Cell Physiol 176: 150–65.

Wu, M.H. 2005. Endothelial focal adhesions and barrier function. J Physiol 569: 359–66.

Wu, M.H., E. Ustinova and H.J. Granger. 2001. Integrin binding to fibronectin and vitronectin maintains the barrier function of isolated porcine coronary venules. J Physiol 532: 785–91.

Yamada, S., S. Pokutta, F. Drees et al. 2005. Deconstructing the cadherin-catenin-actin complex. Cell 123: 889–901.

Yáñez-Mó, M., A. Alfranca, C. Cabañas et al. 1998. Regulation of endothelial cell motility by complexes of tetraspan molecules CD81/TAPA-1 and CD151/PETA-3 with alpha3 beta1 integrin localized at endothelial lateral junctions. J Cell Biol 141: 791–804.

Yang, J.T., H. Rayburn and R.O. Hynes. 1993. Embryonic mesodermal defects in alpha 5 integrin-deficient mice. Development 119: 1093–105.

Yonemura, S., Y. Wada, T. Watanabe et al. 2010. alpha-Catenin as a tension transducer that induces adherens junction development. Nat Cell Biol 12: 533–42.

Yuan, S.Y. 2000. Signal transduction pathways in enhanced microvascular permeability. Microcirculation 7: 395–403.

Zebda, N., O. Dubrovskyi and K.G. Birukov. 2012. Focal adhesion kinase regulation of mechanotransduction and its impact on endothelial cell functions. Microvasc Res 83: 71–81.

Zhu, J., K. Motejlek, D. Wang et al. 2002. beta8 integrins are required for vascular morphogenesis in mouse embryos. Development 129: 2891–903.

9

Endothelial Cells and the Regulation of Platelet Function

P.C. Redondo, N. Dionisio, E. Lopez, A. Berna-Erro and *J.A. Rosado*[a], *

Introduction

The endothelial cells (ECs) are subjected to several mechanical forces and chemical insults that might result in the loss of their integrity, evoking a quick degradation of the subjacent layers, which often causes blood extravasation and complications in blood circulation. The study of molecules and mechanisms underlying ECs and blood cell signaling has been crucial for designing new drugs, novel surgery procedures, or therapies that enhance the life-span and health of the patients, who have suffered several insults on the integrity of the ECs, hence compromising the integrity and subsequent circulation in their vessels. In this sense, certain diseases like atherosclerosis, heart attack, coronary thrombosis, deep venous thrombosis, idiopathic thrombocytopenic purpura, diabetic feet, etc. share common elements during their development.

ECs layer disruption exposes the subjacent layer of tissue containing collagen and Von Willebrand factor (vWF), among other molecules that stimulate deep changes in platelet physiology. All these messengers released

Cellular Physiology Research Group, Department of Physiology, Veterinary Faculty, University of Extremadura, Cáceres, (10003), Spain.
[a]Email: jarosado@unex.es
*Corresponding author

by ECs evoke the activation of several intracellular mechanisms, like changes in cytosolic calcium concentration ($[Ca^{2+}]_c$), protein phosphorylation, secretion of granules containing autocrine and paracrine factors, and cytoskeleton reorganization. These phenomena contribute to enhance the adherence of platelets to the ECs, and subsequently, promote shape change, platelet-platelet (and other blood cells) adherence, fibrin deposition and finally, clot generation. Simultaneously, during platelet activation, several growth factors are produced and secreted, including the platelet-derived growth factor (PDGF), which enhances cell proliferation and division, all of them crucial to regenerate the ECs integrity.

In the present chapter we have compiled recent discoveries regarding the complex relationship between ECs and platelets; furthermore, we describe the current knowledge regarding the role of the endothelial and platelet cytoskeleton in the interaction between both cell types.

Platelet Structure

Platelets are defined as non-nucleated elements; nevertheless, this statement is not a general feature, since in less evolved animals like birds and reptiles, platelets have a prominent nucleus and are designated as thrombocytes. This non-nucleated condition of human platelets resulted in their going unnoticed for centuries; hence, studies of these particular blood cells began later than erythrocytes or leukocytes. The first observation of platelets is attributed to Sir William Hewson. However, the first scientists to describe the presence of corpuscles without color in the blood preparations were the coetaneous Alfred Donné and George Gulliver. Later on, Friederich Arnold did the first representation of a platelet and Gustav Zimmermann (1846) and Max Schultze (1862) supported his initial observations. In 1873, Dr. Edme Felix Alfred Vulpian identified an interesting instance of platelet behavior when exposed to glass, thereby discovering their ability to aggregate. A year later, William Osler reported that platelets are independent elements in the blood stream that aggregate under several circumstances. This observation was reinforced by George Hayem, who, in 1883, described that platelets or "plaquette", upon activation, were able to change their shape and bind fibrin fibers. Understanding of platelet physiology and morphology further improved with the studies done by James Homer Wright (1906), who described that platelets were derived from megakaryocytes by cytosolic fragmentation. This finding has remained undisputed. Now, the mechanisms by which platelets are derived are under intense investigation, as a result of the effort invested in identifying the causes underlying certain diseases that lead to severe thrombocytopenia among other symptoms (Izaguirre Avila 1997).

Under the microscope, platelets appear as discoid elements with granulated cytoplasm and lacking nuclei, but contrary to first scientific hypothesis, the absence of nuclei is not necessary linked to the absence of new protein synthesis (Weyrich et al. 2009). In 1967, Dr. Andrew L. Warshaw's group reported the first evidence supporting the production of proteins by platelets (Warshaw et al. 1967); later on, two independent groups demonstrated that platelets contain ribosomes and the necessary machinery required for protein transduction. In 1971, Dr. Chung-hsin Ts'ao published evidence showing ribosomes and the rough endoplasmic reticulum in platelets (Ts'ao 1971). Nowadays, experimental evidence supports the fact that platelets are able to induce protein translation, and that, among others, proteins like COX-1, Bcl3, IL-β, PAI-1, or proteins that are commonly secreted from alpha-granules like fibrinogen, thrombospondin, albumin and von Willebrand factor are synthesized de novo by platelets (Kieffer et al. 1987). These interesting new findings regarding *de novo* synthesis of proteins by platelets, led scientists to believe that platelets have acquired unknown extranuclear mechanisms to process and efficiently translate mRNAs into proteins (Belloc et al.1982, Schwertz et al. 2012, Weyrich et al. 2004). Moreover, the possibility that platelets synthesize new proteins by using mRNA inherited from progenitor megakaryocytes remains accepted till today (Weyrich et al. 2009).

More controversial is the fact that non-nucleated cells, like platelets, would be able to generate progeny. Dr. Andrew S. Weyrich and his group have discovered that the platelet number increases during platelet storage, which requires synthesis of new proteins and other cellular elements. According to their results, newborn platelets are almost indistinguishable from their progenitors and they are able to adhere, secrete CD62 or to expose annexin V on their cell surface. Evidence of platelet capacity to produce functional progeny was provided by isolating aged platelets placed in culture for six hours *in vitro*. Under this experimental condition, platelet numbers increased, demonstrating this capacity. Nevertheless, these authors were not able to establish whether this mechanism also takes place *in vivo*, which would reinforce the idea of the existence of circulating proplatelets (Schwertz et al. 2010). Furthermore, the hypothesis, that proplatelets generate from megakaryocytes and later degrade into smaller entities—platelets—while still circulating in the blood stream, continues to be widely supported.

A membrane system is observed in platelets, which has been identified as dense tubular system (DTS), similar to the rough endoplasmic reticulum or the surface-connected open canalicular system (OCS), which consists of a shape changing structure that is protruded upon platelets stimulation (Choi et al. 2010). Despite all their properties not having been completely defined yet, it is supposed that both membrane systems play an important role in the regulation of platelet reactivity and the subsequent platelet function during clot generation. Thus, DTS is the main Ca^{2+} store in platelets, containing all the signaling machinery involved in the regulation of Ca^{2+} homeostasis present in other cells, like Ca^{2+} channels sensitive to second messengers like inositol 1,4,5-trisphophate (IP_3R), Ca^{2+} pumps that reuptake Ca^{2+} to the lumen of the stores, the sarconendoplasmic reticulum Ca^{2+} ATPases, SERCAs, or intraluminal Ca^{2+} sensors like calreticulin and stromal interaction molecule-1, and STIM1 (Arber et al. 1992, Elton et al. 2002, Jardin et al. 2008, Lopez et al. 2006, 2008b). Regarding OCS, as previously commented, it is widely accepted that the membranes that constitute the OCS system are protruded upon platelet activation. Several groups also suggest that secretory granules may fuse to OCS and release their content into it. This secretion to an extracellular compartment would be then accelerating the autocrine response, thus enhancing the speed of the first steps towards aggregation (Choi et al. 2010). This last fact was evidenced when comparing responses of stimulated platelets between mammalian species lacking OCS, such as horse, cow and camel, and those containing OCS like human, mouse and dog (Choi et al. 2010, Stenberg et al. 1984).

Several authors consider OCS as an extension of the platelet plasma membrane; nevertheless the distribution of proteins in both pools and structural properties lead us to define both membrane pools as separate entities. It has been reported that during platelet activation OCS could be extruded due to a mechanism that requires the physical support provided by the cytoskeleton, although previous to cytoskeleton reorganization, Ca^{2+} pumps and other proteins are recruited to the proximity of the OCS in order to produce a Ca^{2+} gradient that favors cytoskeleton reorganization. Additionally, OCS has been described to be involved in platelet plasma membrane regeneration, since dense granules, as well as alpha granules, under certain circumstances, are secreted into the OCS.

On the other hand, endocytic organelles have also been associated with OCS, as determined by using electron microscopy, this being the mechanism the model proposed, by which tissue factor and other molecules would be uptaken by platelets (Escolar et al. 2008, White 1999).

In addition to mitochondria that are present in scarce number in platelets, other internal membranes define intracellular organelles, which are easily identified when observed under the microscope. Thus small, dark granules are known as dense-granules (δ-granules; DG) and contain calcium, ATP, ADP and serotonin, which are very important during platelet activation. However, this classic definition has been questioned in recent times by James G. White. This author has recently published a guideline for detecting dense granules. According to his observations there are several platelet structures like dense rings, glycosomes, "fuzzy" balls, chains, clusters and other dense elements, including some alpha granules (AG, α-granules), that might be wrongly identified as δ-granules (DG), hence some platelet storage pool deficiency disorders would remain undiagnosed (White 2008). One very early research performed in 1988, using electron microscopy determined that δ-granule size was reduced and at the same time OCS size was increased during platelet activation, which may indicate that release of δ-granules might take place in the OCS, contributing to add membrane portions for the subsequent pseudopodia sprouting that occurs during platelet activation (Riboni et al. 1988).

On the contrary, α-granules are easily identified under the microscope. α-granules contain several important molecules and messengers, either produced by platelets or stored by reuptake from the extracellular medium. As much as 300 proteins are released by human platelets following thrombin activation according to the proteomic study performed by Dr. Maguire's group in 2004 (Coppinger et al. 2004). Rab4, GMP33, rap1, GPIIbIIIa, GPIB-IX, GPIV, p24, PECAM, GLUT-3 vitronectin receptor, osteonectin, β-thromboglobulin, platelet factor 4, serglycin, HRGP, PBP, CTAP-II, NAP-2, fibronectin, vitronecting, vWF, thrombospondin, fibrinogen, coagulation cascade factors V, VII, XI, XIII, kininogen, protein S, plaminogen, cellular mitogens PDGE, TGF, ECGF, EGF, VEGF/VSF, IGE, interleukine b, α2-antiplasmin, PAI1, TFPI, PN2/APP, C1 inhibitor, Ig G, Ig A, Ig M, albumin and GPI/multimerin are examples of proteins accumulated by platelets within the α-granules (Coppinger et al. 2004, Suzuki et al. 1990).

A recent publication has shown the presence of different secretory granule sub-populations. They were identified according to the protein located in their membranes. In this sense, α-granules containing VAMP3 and VAMP8 were shown to dock and to secrete their content into the central granulomere portion which is surrounded by microtubules, while other molecules like serotonin are localized by immunofluorescence during platelet spreading. Contrary to this, α-granules containing VAMP7 as a molecular marker in their membranes were localized in the spreading areas, indicating that these granules may have a role in platelet spreading. This hypothesis was then confirmed in patients lacking α-granules whose platelets do not spread as normal. These patients suffer a syndrome known as gray platelet syndrome (Peters et al. 2012). Hence, fusion of these secretory granules (α- and δ-granules) contribute to regenerate the plasma membrane, or at least provide the extra membrane portions required to fully allow platelet spreading over the damaged ECs (Peters et al. 2012).

Additionally, other important organelles found in platelets are glycogen granules, which are crucial during platelet activation. Platelet metabolism is very low under resting conditions, but it quickly increases during activation leading to the mobilization of the stored glucose. Most of this energy demand is used to enhance the complex signaling machinery that promotes granule secretion. Thus, a very interesting research performed in 1983 by

Holmsen's group reported that an incremental energy consumption of 2.5, 4.2 and 6.7 µmol of ATP equivalent $\times (10^{11}$ platelets$)^{-1}$ is required to induce the secretion of dense-, alpha- and acid hydrolase-granules (Akkerman et al. 1983). Many of this required ATP is produced by the scarce number of mitochondria, due to degradation of mobilized glucose from glycogen granules. Despite the low number of mitochondria found in platelets, they behave similarly to those found in other cell types. In fact, it has been evidenced that mitochondria are able to accumulate Ca^{2+}, and even more surprisingly, they are able to conduct activation of intrinsic apoptotic pathways like those that occur in other cell types, by releasing cytochrome c and activating the proapoptotic caspases, bax, bid, bad (Amor et al. 2006, Bertino et al. 2003, Lopez et al. 2007, 2008a, Perrotta et al. 2003, Rosado et al. 2006, Wolf et al. 1999). Apoptosis in platelets or non-nucleated cells is a very recent finding that is still under intense debate; nevertheless it has been demonstrated by different research groups that platelets undergo an *"apoptotic-like mechanism"* upon thrombin stimulation (Wolf et al. 1999). Similarly, apoptotic techniques and mitochondria fluorescence dyes used to evaluate mitochondrial membrane potential can also be used in platelets (Lopez et al. 2007, 2008).

Lysosomes and peroxisomes have also been described in platelets, which behave similarly to those found in other cell types. Hence, lysosomes act by degrading uptake molecules, and peroxisomes participate in lipid metabolism and in synthesis of platelet activating factors (Farstad and Sander 1971, Lopez et al. 2007, White and Clawson 1980).

All these organelles previously described, that can be easily identified when platelets are observed under the microscope (Fig. 1), are crucial for platelet function. There are a group of hemorrhagic disorders characterized by the alteration of either granules secretion, or by the lack of some of these platelet organelles. For instance, gray-platelet syndrome is characterized by the lack of alpha granules, and is the origin of two pathologies, Quebec and Jacobsen or Paris-Trousseau syndromes. On the other hand, syndromes associated to altered dense granules are Hermansky-Pudlack, in which eight different variants have been described (HP1-HP8), while others like Chediak-hygashi, Wiskott-Aldrich and Tar syndromes constitute examples of dense granules abnormalities (Cattaneo 2003).

Finally, the cytoskeleton plays an important role in the physiology of platelets. Its role begins even prior to platelet generation from megakaryocytes. Nowadays, beside to who for the very first time described megakaryocytes maturation and platelets generation, a wide number of studies have supported the idea of platelet generation from polyploid megakaryocytes. During polyploidation, several mitotic spindles organized by centrosomes can be identified in the cytosol of megakaryocytes, and growing spindle generate the forces involved in membrane demarcation that

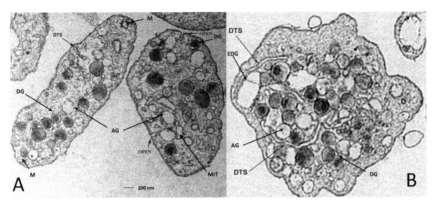

Figure 1. Structure of platelets. Platelets electron mycrography showing the most relevant internal organelles and structures. The structure of platelets was examined by electron microscopy, under resting conditions (A) and stimulated with Thr (0.1 U/ml (5 min) (B). Upon stimulation samples were further processed as described by Alexandru et al.[166] Platelets samples were taken by using TEM. Abbreviations: DTS, dense tubular system; M, microtubule; DG, dense granules; EDG: Empty dense granules; AG, α-granules; OPEN, OCS, open canalicular system; MIT, mitochondria.

subsequently ends in generation of platelets. In this mechanism, β1-tubulin seems to be crucial to support platelets generation. Despite this, however, the hypothesis of proplatelets generation described above is supported by many research studies. The latest evidences using life fluorescence microscopy have reported controversial results and thus the precise model by which platelets are generated remains under debate (Kosaki and Kambayashi 2011, Thon et al. 2010). Upon platelet fragmentation from megakaryocytes, the cytoskeleton of tubulin organizes just below the platelet membrane, but it is particularly evident in a structure known as tubulin ring, which supports the discoid shape of resting platelets. It has been recently described that microtubules control the size of platelets together with actin-myosin-spectrin cortex forces (Thon et al. 2012). Once platelets become active, the tubulin ring quickly reorganizes appearing several gaps or interruptions in the ring structure, undergoing several ramifications that later will adopt a diffuse distribution through the platelet cytoplasm. Simultaneously, an alternative intense tubulin staining can be observed in a more central disposition, which it is believed constrain granules in the centre of platelets. Despite this evident tubulin structure, tubulin acts like in other cell types, promoting secretory granules mobilization and promoting Ca^{2+} influx as recently demonstrated (Patel-Hett et al. 2008, Redondo et al. 2007).

On the other hand, actin cytoskeleton modification by using drugs like jasplakinolide or cytokalasins have revealed that actin cytoskeleton is subjected to an intense process of remodeling in platelets; subsequently, it supports several important mechanisms like fusion of secretory granules, shape change, platelets adhesion, and reinforcing platelet aggregation (Ge et al. 2012). It has been shown that the function of the cytoskeleton lasts further after aggregation process, since it favors clot stabilization by a mechanism called clot retraction that entirely relies on cytoskeleton reorganization. These well known characteristics of tubulin and actin cytoskeleton will be revisited in the following sections.

Finally, intermediate filaments constitute the third structural element that can be observed in platelets. These filaments are constituted of small fibers, which contain type 2 intermediate filaments consisting of vimentin, desmin and others. Intermediate filaments seem to localize just underneath the platelet membrane in order to support pseudopodia emission during activation. However, no clear evidences of such a role have been described yet.

Platelet Physiology

Platelets are the main blood cells involved in hemostasis. They do not usually interact with the intact vessel wall (May et al. 2008), but after injury of this wall and exposure of the disrupted ECs once the vascular injury is produced, the thrombus formation take place in four steps: activation of the ECs, tethering and rolling of platelets, activation of platelets, and firm adhesion. Hence, platelets produced in the bone marrow circulate in the blood stream for at least 8–10 days in a dormant state, as reflected by the low basal metabolism recorded which ranged from 0.5 to 3.5 μmol of ATP equivalents/min. Only upon activation using a strong agonist such as thrombin does energy consumption reach values of 2 to 16 μmol of ATP equivalents/min (Akkerman et al. 1983). This consumed energy is used to support all the structural changes required for completing its functions as well as for activating several intracellular kinases, Ca^{2+} channels and Ca^{2+} pumps that maintains the complex intracellular signaling cascades required for platelet clot and subsequently, blood hemostasis.

In vitro experiments using flow chamber and other alternative techniques to circulate platelets over a surface for facilitating contact with a simulated disrupted epithelium, such as coverslips coated with collagen or fibrinogen, have been used by scientists to monitor the immediate cell changes and other events that occur just when platelets become active (van Kruchten et al. 2012, Yuan et al. 1999). This approach led to the hypothesis of *platelet rolling activation mechanisms,* which is supposed to occur as an adaptive mechanism to avoid the stretch forces that exists under flow forces (Malik et al. 1983, McEver and Zhu 2010). Alternatively, *in vivo* techniques performed to corroborate platelet adhesion to vessel wall in response to mechanical and chemical insults, have been well documented. Two examples of these techniques are the Doppler flow probe to monitor vessel occlusion, and direct blood vessel visualization by intravital microscopy performed on rodents and other small animal models (Furie and Furie 2005). This phenomenon is widely supported by the tight relation that exists between ECs and platelets, but is not exclusive of platelets, since it has also been observed during leukocyte interaction with ECs as initially described by Dr. Alan Warnerg in 1983, and later confirmed by several authors (Diacovo et al. 1996, Langer and Chavakis 2009, Malik 1993).

More surprising is the fact that platelets might favor the adhesion of metastatic cells through a mechanism that would also be enhanced by interacting molecules expressed in the platelet surface (the contribution of platelets to cancer metastasis is described in depth later in this chapter). It is not clear whether pseudopodia emission by platelets is required for platelet-ECs interaction, or if it occurs prior to the interaction, which would favor contact between ECs and platelets even before adhesion takes place (Riboni et al. 1988). A very recent study suggests that PAR4 (proteinase activated receptor 4, operated by thrombin) is the key molecule involved in platelet pseudopodia emission and subsequently, it would be responsible for platelet spreading. Hence, the remaining molecules involved in this process, like collagen or fibrinogen receptor, conditioned to PAR4 preactivation (Lee et al. 2012a). In this process, actin cytoskeleton remodeling is a key element, as will be described in following sections. Furthermore, recent evidences reinforce the platelet rolling adherence model, as the first events of platelet clot generation; events like Ca^{2+} channels opening, operated by stretching of the membrane where they are inserted, together with the fact that this Ca^{2+} entry is needed prior to platelets spreading, help to consolidate this model (Berrout et al. 2012, Matsumura et al. 2011). According to this model, either platelet-platelet or ECs-platelet contacts would evoke platelet plasma membrane stretching as a consequence of blood-flow forces, in which interaction points generated by adhesive molecules would be able to support up to 16 pN. Platelets are capable of supporting these forces, due to contact between the von Willebrand factor and its receptor, GPIbalpha and other molecule-receptor contacts described in the following section (Reininger et al. 2006). Additionally, stretching forces also affect ECs, since these cells express in their membranes PECAM-1 (Platelet endothelial cell adhesion molecule 1), which has been shown to act as a mechanochemical converter transducing mechanical forces into biochemical signals in endothelial cells, through activating effector proteins like Fyn tyrosine kinase (Chiu et al. 2008). Fyn also act as linking proteins between some platelet glycoproteins and intracellular kinases in platelets; hence it is suspected that these confluent mechanisms would act in parallel to facilitate the production and secretion of further stimulatory molecules, which would support the contact between both cell types.

Platelet Adhesive Molecules and other ECs-platelet Linking Molecules

The plasma membrane of platelets is a complex structure, which further extends platelet contour; in fact, the extracellular region close to the platelet membrane was designated very early as *atmosphere plasmatique* by White, an eminence in the field of platelet structure and physiology, who termed this platelet extracellular region as a peripheral zone that can be divided into three parts: exterior coat, extracellular membrane, and sub-membrane zone (White 1923). This peripheral zone is very important for platelet activation and adhesion, since it contains many proteins and molecules that help the platelet-ECs interaction. Hence, these molecules contribute to the reactions that occur during coagulation cascade, such as conversion of factor IX, and X to factor IX active and X active, which take place in the peripheral zone of platelets upon complexing with structural protein included in the exterior coat (Luchtman-Jones and Broze 1995, Muszbek et al. 2011, Zhu 2007). Additionally, factor XIII is recruited to the peripheral zone of platelets to stabilize fibrin-platelet contact and helps to consolidate the blood clot (Muszbek et al. 2011). The external coat of platelets, also called glycocalyx, is a complex structure characterized by the inclusion of several glycoproteins and molecules that are crucial for platelet recognition and interaction. Furthermore glycocalyx is present in most cell types allowing contact between molecules and subsequently, promoting signaling between cells, such as occurs between platelet-ECs and platelets leukocytes. We list below, the main molecules integrated in this peripheral zone that participate in platelet adhesion. Among them are included integrins, glycoproteins and proteins like P-selectin.

Integrins are formed by α and β heterodimers, with non-covalent association between them. Platelets mainly express the following integrins:

-$\alpha_{IIb}\beta_3$ or GPIIb/IIIa: It acts as a vWF, fibrinogen, fibrin, fibronectin, thrombospondin, collagen and vitronectin receptor. It can form complexes in the presence of the ion Ca^{2+} (Fujimura and Phillips 1983, Santoro 1988) and, after association with its ligands, $\alpha_{IIb}\beta_3$ becomes active, causing granule secretion and hence regulating platelet adhesion, aggregation and thrombus formation (Shattil and Newman 2004).

-$\alpha_V\beta_3$ or αV/GPIIIa: It represents a versatile receptor able to bind fibronectin, osteopontin and vitronectin. Although this integrin is not expressed as much as other types in platelets, it seems to be crucial during aggregation. *In vitro* experiments have concluded that this integrin would require Ca^{2+} to bind vitronectin, but not other divalent cations such as magnesium or manganese (Bennett et al.1997, Coller et al. 1991, Ekmekci and Ekmekci 2006).

-$\alpha_2\beta_1$ or GPIa/IIa: It is formally the collagen receptor, but it also binds vWF. This integrin has a structural particularity which domain is termed as MIDAS (metal ion-dependent adhesion site), through which it is able to bind ions like calcium, manganese, magnesium and nickel (Coller et al. 1989, Humphries 2000, Pulcinelli et al. 1998).

-$\alpha_5\beta_1$ or GPIc/IIa: Formal fibronectin receptor. This integrin acts as an extra supporter of the adhesion of platelets to the vessel (Beumer et al. 1994, Piotrowicz et al. 1988).

-$\alpha_6\beta_1$ or GPIc*/IIa: Capable of binding laminin, a structural protein belonging to vessel subendothelium matrix (Li et al. 2010, Nieswandt et al. 2011), this integrin also plays a crucial role in megakaryocyte differentiation.

On the other hand, *Glycoproteins* also participate in platelet adhesion and among them, the most important ones are GPIb/IX (vWF and collagen receptor), GPIV or CD36 (a receptor for thrombospondin and collagen) and GPVI (collagen receptor, which acts as a GPIa/IIa receptor activator (Lowenberg et al. 2010)).

vWF deserves special consideration during contact platelet-ECs. vWF is synthesized by endothelial cells and megakaryocytes, and can be secreted or targeted to ECs storage organelles (Weibel-Palade bodies), or α-granules in megakaryocytes or platelets (Ruggeri 1999). In its mature form, vWF is formed by disulphide-linked multimers (Wagner 1990). The inactive form of vWF circulates in the plasma, whereas the active form is found in storage organelles. Interchange between the inactive and active form occurs after previous adhesion to collagen of the ECM, upon recognition by the vWF-GPIb-V-IX receptor complex, which locates in the surface of platelets in their inactive form, therefore acting like an initiator of platelet activation and transient adhesion to the vessel wall (Li et al. 2010, Nuyttens et al. 2011, Phillips 1980, Ware and Heistad 1993). Once platelets bind to vWF, they suffer a conformational change, which causes them to produce some of the substances that are stored in their organelles, like ADP or fibrinogen. This mechanism enhances platelet activation (Nurden and Nurden 2003). vWF is also a ligand for GPIIb/IIIa ($\alpha_{IIb}\beta_3$), an integrin complex which participates in platelet activation, forming a bridge to fibrinogen, fibronectin or vWF once the conformational change is produced and thus mediating stable adhesion of platelets, aggregation and thrombus formation (Offermanns 2006). It is usually in a low-affinity or resting state in platelets, but it turns into an activated form after platelet activation (Fig. 2).

During endothelial cell activation a number of substances participate, such as E-selectin, intercellular adhesion molecule-1 (ICAM-1) and vascular cell adhesion molecule-1 (VCAM-1), which act as pro-thrombotic and pro-

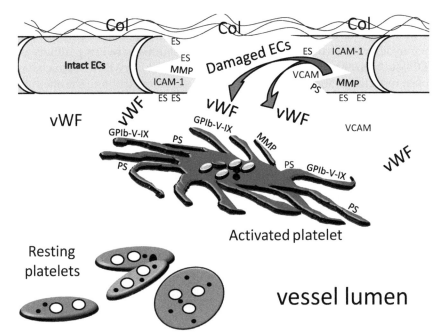

Figure 2. Endothelium-platelets interplay. Main proteins that favor interaction and regulation of platelet-ECs contact are highlighted (Black-letters). Additional inhibitory molecules are also represented (Gray- letters). Abbreviations: vWF, Von Willebrand Factor; GPIb-V-IX; glycoprotein Ib, glycoprotein V and IX complex: receptor of vWF; ES: E-selectin; PS- P-selectins; MMP, matrix metalloproteinase; Col: collagen.

inflammatory factors (Breitenstein et al. 2011, Gawaz et al. 1998, Kaplan and Jackson 2011). A ligand of this ICAM-1 is leukosialin, also known as sialiphirin or CD43. Upon platelet and ECs activation, P-selectin (expressed on platelets and endothelial cells) that remains under resting conditions stored within Weibel-Palade bodies is exposed in the cell surface. E-selectin (CD62E) is expressed on endothelial cells (Lowenberg et al. 2010), thus both molecules constitute main platelets-ECs selectins (Frenette et al. 1995, 1998). Alternatively, P-selectin glycoprotein ligand-1 is expressed on the surface of platelets, which mediates leukocyte rolling via its interaction with P- and E-selectin in the ECs. Therefore, P-selectin acts as an important mechanism for platelet tethering and rolling, but it does not play an important role in firm adhesion (Xia et al. 2002, Yang et al. 1999). Some proteoglycans belonging to the endothelial membrane are involved in thrombosis, such as perlecan, which participates in angiogenesis and wound healing; further, it is involved in platelet adhesion. Its abundance has been reported to increase upon injury (Kinsella et al. 2003).

Once the initial platelet-ECs contact has been consolidated, it is followed by the firm adhesion phenomenon that is also regulated by integrins, like

$\alpha_{IIb}\beta_3$, which mediates platelet adhesion to endothelial cells *in vitro* (Bombeli et al. 1998). This interaction is enhanced by fibrinogen, fibronectin and vWF. Some other molecules are involved in interactions between platelets and ECs, such as the metalloproteinases MMP-1, MMP-2 or MMP-9, which have been reported to stimulate mechanisms of adhesion, through the alteration of cell surface proteins, ECs transmembrane proteins, and receptors.

Tetraspanins expressed in platelets, such as CD9 (co-localized with GPIIb/IIIa in the inner surface of α-granules and in pseudopodia during activation) (Brisson et al. 1997), CD63 (also called LAMP-3 or granulophysin (that are located in membranes of lysosomes and dense granules), act as a marker of platelet activation. Additionally, other activation markers detected in the platelet membrane upon activation are PETA-3 or CD151 (platelet-endothelial cell tretra-Span Antigen, involved in the signaling pathway in platelets) (Nishibori et al. 1993, Sincock et al. 1997). Alternatively, there are also some molecules, like thrombomodulin, mainly synthesized by endothelial cells and formally expressed in the vascular ECs (with the exception of the brain ECs) (van Hinsbergh 2012), which act as a thrombin receptor on the surface of the ECs, reducing blood coagulation and also participating in the activation of tissue-associated fibrinolysis inhibitor (Bajzar et al. 1996, van Hinsbergh 2012).

Nevertheless, not all molecules produced by either ECs or platelets act in order to promote contact between both cells. Comprehensive balance between promoting molecules and negative regulators would result in a healthy state where the lumen of the vessel is clear of obstacles. Regarding negative regulators of platelet function produced by ECs we should mention molecules like nitric oxide (NO), prostacyclin (PGI_2, PGE_2), TFPI in combination with TFPI-2, and finally, ecto-ADPase, also known as CD39. These molecules regulate platelet function via different extracellular or intracellular signaling pathways, either by up-regulating AMPc and GMPc, altering activation of factor X of coagulation, or by reducing ADP circulating level which acts as a platelet agonist, respectively (Gayle et al. 1998, Loscalzo and Welch 1995). It has also been described that these agents impair platelet function by a mechanism which requires that newly synthesized thrombin interact with thrombomodulin, located in the membrane of ECs. Subsequent reaction consists of the release of protein C that is able to degrade coagulation factors (Factor V and VIII), but also interacts with protein S evoking its activation, and ends with the degradation of factors V and X (Griffin et al. 2012).

Role of Cytoskeleton in Platelets-ECs Interaction

As previously described, three main elements are present in platelets: microtubule, actin cytoskeleton, and intermediate filaments. During platelet function, these elements must be remodeled in order to support the structural shape of resting platelets, and also changes in the internal organelles localization.

Microtubules, the most evident of these three structural elements, have been reported to be distributed in the periphery of platelets under resting conditions. This marginal distribution was regularly observed by White under electronic microscopy, as early as 1965 (White and Krivit 1965). This microtubule structure has been designated as tubular ring. According to the studies performed by Italiano et al. in mature megakaryocytes, microtubules reorganized in three different structures (Italiano et al. 1999). The most abundant microtubule structure is responsible for the actual round shape of megakaryocytes, but their maturation process evokes the appearance of cytoplasmic protrusions supported by microtubules reorganization. These microtubule elongations present at the very end of small bundles that upon fragmentation result in platelet tubular rings. In contrast, during platelet stimulation, the tubular ring drastically reorganized, first suffering a depolymerization process, and then, suddenly repolymerized in order to allow different platelet mechanisms to take place, like platelet secretion and filopodia emission (Patel-Hett et al. 2008). Another crucial mechanism during platelet activation that relies on microtubule reorganization is Ca^{2+} influx through plasma membrane channels operated by intracellular Ca^{2+} store depletion, also known as store operated Ca^{2+} entry (SOCE). In human platelets, alteration in the normal pattern of microtubule reorganization, by incubating human platelets with either chemical microtubule stabilizers or disassemblers such as paclitaxel or colchicine altered both SOCE and platelet aggregation (Bouaziz et al. 2007). This might be a feedback mechanism, since microtubule reorganization also relies on changes in the cytosolic Ca^{2+} concentration, as demonstrated by Dr. Vlado Buljian et al. in experiments performed *in vitro*. These authors showed that in the presence of calcium, microtubules reorganize in larger and more branched filaments, showing higher complex structures; meanwhile in the presence of Ca^{2+} chelators (EGTA), microtubule filaments were shorter and less branched (Buljan et al. 2009). Nowadays, it is known that microtubules associated proteins (MAPs), like dynein and STOP, are substrates of ser/thr kinases, like PKC and Ca^{2+}/ CaM kinase (CaMK); thus, high cytosolic Ca^{2+} concentration evokes CaMK type II activation and subsequently, phosphorylation of dynein and others of these microtubule disassembly proteins. This general mechanism has been demonstrated in a wide range of species; in fact there was evidence in the fungus *Ustilago maydis* that presents altered expression of endoplasmic-reticulum ATPase and elevated cytosolic Ca^{2+} concentration. In this cell model, addition of CaMK antagonist or silencing of dynein protein restores the normal pattern of microtubule reorganization (Adamikova et al. 2004, Baratier et al. 2006).

Italiano's group, by using fluorescent dyes, has been able to visualize microtubule reorganization during murine platelet activation. These observations confirm that a microtubule ring is present under resting conditions, and upon platelet stimulation microtubules reorganize to support filopodia structure generation (Patel-Hett et al. 2008). In 2001, a case report was presented of a patient with a long history of easy skin bruising. In this case, circulating platelets were reported lacking microtubule rings under resting conditions and hence appearing as spherically shaped platelets, thus highlighting the role of microtubules in human platelet functionality (White and de Alarcon 2002). In this context, microtubule dependent surface exposition of certain ECs, cancer cell-plasma membrane molecules, and their role in platelets adhesion will be discussed in following sections.

On the other hand, the actin cytoskeleton seems to be as relevant as microtubules for platelet function. Italiano's group, in their studies, often reported data regarding actin filaments organization in megakaryocytes and platelets under both conditions, resting and upon stimulation. It is well known that during megakaryocyte maturation and fragmentation into platelets, membrane demarcation appears in the cytosol. In order to allow this fragmentation, actin cytoskeleton pulls the lobulated nuclei to a final marginal region, thus avoiding possible interference of nuclei during megakaryocyte fragmentation process (Italiano et al. 1999). Platelets newly generated also present F-actin ring underneath the plasma-membrane known as cortical actin or actin sub-membrane barrier.

The presence of actin filaments in platelets was confirmed by electron microscopy in 1971 (Behnke et al. 1971); two years later, it was stated that actin interacts with other cytoskeletal elements (Booyse et al. 1973), but Dr. Isaac Cohen, and later on, Dr. Saul Puszkin, who introduced the use of the cytochalasin D as actin filament disassembler, were the very first to describe the role of actin filaments in platelet function (Cohen et al. 1973). Together with the improvement of the confocal fluorescence microscopy techniques and the synthesis of better F-actin dyes, researchers have been able to understand the actin cytoskeleton function under resting conditions and upon stimulation. Thus, this solid structure that was firstly described as forming a cortical barrier just underneath the plasma membrane has been demonstrated to suffer an intense process of reorganization during platelets activation with physiological agonist. Actin filament reorganization is supported by the activation of calcium-dependent and -independent pathways. In 2000, Dr. Juan A. Rosado and Stewart O. Sage showed that actin remodeling occurred upon activation of small GTPases belonging to the Ras superfamily, like Rac, Ras, Rho, etc.; also, that cortical F-actin barriers self-reorganize in order to allow the contact between elements that participate in Ca^{2+} entry (Marcu et al. 1996, Rosado and Sage 2000a, 2000b, 2000c). These initial observations have since then been confirmed, and today we know that upon GPVI activation, Rac requires other intermediate molecules like Scar/WAVE to act over Arp2/3 complex and finally to reorganize actin filaments (Calaminus et al. 2007). Furthermore, previous publications described the presence of severing Ca^{2+}-dependent proteins, like scinderin A or gelsolin, whose function is to degrade actin filament in the presence of Ca^{2+} (Rodriguez Del Castillo et al. 1992). By using rapid kinetic techniques like quench-flow and stopped-flow methodology, Rosado's group determined that during thrombin stimulation, actin filament partially depolymerized in just 1.5 second and subsequently, along with the raise in cytosolic Ca^{2+} concentration, there was a hyperpolymerization of actin filaments (Redondo et al. 2006). In other cell types, PKC activation has been involved in remodeling of actin cortical barrier (Vitale et al. 1992). This model of PKC-dependent actin remodeling has also been demonstrated in platelets, although other Ser/Thr kinases also seem to regulate this mechanism, like PI_3K, ERK, etc. In recent years, this field has been subjected to intense study (Harper and Poole 2010, Konopatskaya et al. 2009, Pandey et al. 2006). Consistent with the above observations, tyrosine phosphorylation inhibition, using non-specific inhibitors, has been seen to reduce actin filament reorganization. It has been demonstrated by using LFM-A13, which is a specific inhibitor of Bruton tyrosine kinase (Btk), that depolymerization of actin filaments rely on the activation of this protein. Btk is a tyrosine kinase activated downstream of several platelets receptors like GPIb or PARs (Jackson et al. 2007, Redondo et al. 2005), and also requires activation of Phospholipase $C\gamma_2$. Furthermore, BTK requires generating complex with p72 syk, Fyn and Lyn, to become fully active and to phosphorylate proteins, like p60src that require actin recruitment to be fully active, and subsequently to participate in Ca^{2+} entry through SOCE among other mechanisms (Gibbins 2004, Redondo et al. 2005). On the other hand, platelets contain other mechanisms that allow actin filament to hyperpolarize due to the phosphorylation of ser/thr kinases, like LIMK-1 that is downstream of Ca^{2+}/CaMK, and phosphorylates cofilin, avoiding its severing role on actin filaments in platelets and other cell types (Okamoto et al. 2009, Pandey et al. 2006, Redondo et al. 2006).

Major roles for the actin depolymerization and repolymerization processes are to promote cell shape change and facilitate contact between inner membranes and the plasma membranes, crucial for mechanisms like platelet Ca^{2+} influx through SOCE and platelet secretion.

Two types of membranes protrusions can be defined in platelets—filopodia, whose spike-like shape is supported by parallel bundles of actin filaments, and lamellipodia that are sheet-like protrusions inside which a branched network of actin has been observed (Lee et al. 2012b, Mattila and Lappalainen 2008). With regard to actin filament-dependent shape change of platelets, there is a shared function with microtubules for promoting filopodia generation. According to Steve P. Watson's model, a clear ring shape structure appears in a particular and delimited place underneath the plasma membrane, just before filopodia generation becomes evident (Calaminus et al. 2008). Hence, it has been recently reported that platelets spreading over a substrate require actin reorganization; in this process Rac seems to be very important due to its role in cytoskeleton reorganization, and has been proposed to be downstream of Ser/Thr proteins like mTOR. (Aslan and McCarty 2012). In other cell types, mTOR regulates cell cycle (Appenzeller-Herzog and Hall 2012), division and apoptosis, and has even been involved in cancer (Alvarado et al. 2011, Mahalingam et al. 2009, Vemulapalli et al. 2011). In human platelets, mTOR has been involved in platelets production due to its participation in megakaryocyte maturation, as well as in platelet function (Guerriero et al. 2006, Liu et al. 2011).

Cortical barrier reorganization allows proteins that belong to the "fusion complex core" to interact and further facilitate the membrane fusion required for granules secretion. Scinderin function and other severing proteins have a crucial role in this mechanism, since it has been demonstrated that scinderin A inhibitions, by adding peptides that interfere with its actin binding region, abolished platelet secretion. Similarly, introduction of recombinant scinderin enhanced serotonin secretion from δ-granules. The role of actin in platelets secretion, and in particular in δ-granule secretion is exclusive of this structural element, since microtubules do not seem to be involved in this process, according to a very recent publication (Ge et al. 2012).

Platelet-ECs Interaction in Heath and Disease

As described above, a correct interaction between ECs and platelets results in a healthy state and avoids blood extravasation. However, abnormalities in this interaction might evoke serious pathologies, such as those found in inflammatory lung diseases where the existence of circulating platelet-leukocyte aggregates aggravates the illness. These aggregates can be also found in patients suffering cystic fibrosis, and in experimental lung injury models (Tabuchi and Kuebler 2008). In this sense, acute lung injury is associated with a high mortality rate, and despite the origin of such malignancy, platelets have been described to firstly interact with damaged ECs and to promote subsequent leukocyte adherence. Upon leukocyte interaction, leukocytes extravasate and/or promote degradation of the ECs continuity due to release of pro-inflammatory molecules after the stimulation of receptors located on their surfaces (Zarbock and Ley 2009). A far more surprising result is the ability of platelets to move along a surface, probably in response to chemotactic molecules. In very early studies, performed in *in vivo* models where bronchial tissue of experimental animals were stimulated with allergenous or with platelet activating factor (PFA), platelets were

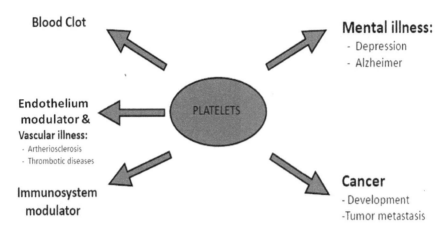

Figure 3. Platelet role in diseases. Schematic representation of the illnesses where platelets have been presented to be involved in the past (left hand side). New illness where the participation of platelets have been suggested or they have been considered as markers of illness progression (Right hand side).

found extravasated and in contact with smooth muscle cells (Lellouch-Tubiana et al. 1985). These observations were then confirmed in human patients suffering asthma and other lung illnesses like bronchospasm, hence showing the ability of platelets to pass across the ECs in lung tissue. Later on, it has been demonstrated that platelets can extravasate almost in any tissue, particularly those undergoing severe inflammatory or tumorogenic processes, like the liver. Thus, these previous observations, together with the fact that platelets can also release chemokines, molecules that attract all type of leucocytes to their proximity (like serotonin, PDGF, TNFβ, PF-4, RANTES, MIP-1αβ, TG Ag, IL-8, IL-1, histamine, lysosomal enzymes, PAF, PGE2, TXA2, PECAM-1, CD40L), allows a new hypothesis to be raised regarding inflammation and vessel wall disruption, both symptoms that might be particularly relevant in diseases such as atherosclerosis and atheroma plaque generation and breakdown. In fact, a very interesting work by Dr. Junichi Masuda and Dr. Robert Ross in 1990 showed platelets inside atheroma plaques of primates that contain low levels of cholesterol (Masuda and Ross 1990a, 1990b). More recent publications have shown that platelets control the progression of the atherosclerotic plaques, which opens a new field of investigation where platelets might be markers for diagnosis of these diseases (Gawaz et al. 2008).

Additionally, platelets have been demonstrated to be involved in brain diseases. For instance, serotonin is a very important neurotransmitter whose imbalance has been proposed to lead to depression. Surprisingly, different studies have reported that as much as 90% of serotonin found in our bodies is in the digestive tract and in blood platelets. Platelets seem to be the reservoir for serotonin, uptaken upon a peak of serotonin release by the neurons in the brain (Paraskevaidis et al. 2012). Upon platelets stimulation, serotonin is released into the blood stream and it has been reported to enhance the time of recovery of patients suffering this illness. Another illness related to serotonin like migraine also shows altered platelet levels of serotonin and platelet function (Ayalp et al. 2012). On the other hand, since serotonin is a platelet agonist it has been demonstrated as posing a higher risk for vascular illness linked to thrombotic disorder in patients suffering alteration in the concentration of circulating serotonin (Vashadze 2011a, 2011b).

Another mental illness in which alteration of platelets has been reported is Alzheimer's disease. Current studies are now looking for novel and early markers of this disease in the blood corpuscles and elements (Casoli et al. 2010). In fact, APP and several enzymes that participate in the progression of Alzheimer's disease have been found expressed in platelets; more important, in platelets of patients diagnosed for Alzheimer's disease, the enzymes and protein markers for this disease, like α- and β-secretases, and APP have been found to be altered (Catricala et al. 2012, Laske et al. 2008, Zellner et al. 2012). This evidence has opened up an entirely new and exciting field of investigation that is now under intense debate.

Role of Platelets in Cancer Development and Metastasis

The function of platelets in coagulation, homeostasis, and inflammation is very well known. However, nowadays a number of studies are considering the possibility that platelets play an important role in the progress of tumors, according to the body of experimental evidence presented below.

The first evidence that linked platelets with neoplasia was based on the fact that patients affected by neoplasia also reported hypercoagulability problems associated with an increased platelet count (Varki 2007). Thrombocytosis has been described in the majority of all common cancers; furthermore, it is considered as an adverse prognostic factor and inversely correlated with survival (Buergy et al. 2012). Thrombopoiesis takes place due to the interaction between tumor cells and megakaryocytes within the bone marrow. Tumor cells secrete, among other substances, cytokines such as interleukins: (IL)-1, IL-3, IL-6, and IL-11, which enhance megakaryocyte growth and platelet production (Ciurea and Hoffman 2007). Among them, the most important is IL-6, which can be expressed directly by the tumor cells; an elevated concentration in the bloodstream has been associated with a malignant progression (Nakano et al. 1998). Thrombopoietin is the main stimuli during thrombopoiesis, but its expression by human tumor cells is scarce, and can be detected only in hepatoblastoma, hepatocellular carcinoma cell line, and ovarian carcinoma. Other factors, such as vascular endothelial growth factor (VEGF), leukemia inhibitory factor (LIF), granulocytemacrophage colony-stimulating factor (GM-CSF), FMS like tyrosinkinase 3 (FLT3) ligand, fibroblast growth factor (FGF), KitL, oncostatin M (OM) and erythropoietin (EPO), can contribute to thrombopoiesis (Buergy et al. 2012).

Cancer metastasis is a critical point that might compromise the survival of the patient; in this process platelets have been shown to contribute to tumor spreading and extravasation of tumor cells. In order to participate in this mechanism, platelets may undergo aggregation induced by the secretion by tumor cells of pro-aggregation factors like ADP, thrombin, cathepsin B, and thromboxane A2 among others (Bambace and Holmes 2011).

Tumor cells reach the lumen of a blood vessel by a process called intravasation, where they will be in contact with all blood factors. The activated coagulation cascade generates thrombin and fibrin that support clot formation. This process causes the expression of adhesion molecules on the surface of the platelet membrane; for instance, there is a body of evidence regarding contact between platelets and tumor cells through P-selectin, as well as other selectins which, as described above, facilitate interaction with other platelets in order to establish a thrombus and, together with fibrin, favor stabilization and lead to clot generation, where tumor cells are incorporated (Borsig et al. 2001). Very recently, it has been described that tumor cells express a high amount of CD44, as well as HCELL (hematopoietic cell E-/L-selectin ligand), both enhancing their interaction with ECs, leukocytes, and platelets (Jacobs and Sackstein 2011). Early studies reported that tumor cells, like Walker 256 carcinosarcoma cells and platelets may interact due to the surface expression of a molecule similar to the platelet glycoprotein IIb/IIIa. In these cells, by disrupting actin and intermediate filaments, but not microtubules, GPIIb/IIIa-like protein expression in the surface is impaired, and platelets recognition of tumor cells decrease, together with the ability of these tumor cells to evoke platelet aggregation (Chopra et al. 1988). Thus, platelets coat and shield tumor cells, allowing them to evade the immune system effectors known as natural killers (NK).

NK cells are responsible for eliminating tumor cells from the bloodstream. However, metastasis occurs because they cannot gain access to them in the clot due to the physical barrier presented by represent fibrin, platelets and other blood cells (Palumbo and Degen 2007). NK cells are inactivated by other mechanisms, such as secretion of immunoregulatory molecules from platelets like PDGF, which has been found to reduce NK cell reactivity *in vitro*. TGF-β contained within the platelets releasate also diminished NK cells by altering their cytotoxicity and IFN-γ production via downregulation of the NKG2DL receptor on NK cells. Molecules on the platelets surface, like a GTR or histocompatibility complex (MHC) class I, can inhibit receptors on NK cells (Buergy et al. 2012).

The next step of the tumor metastasis is called extravasation phase, and it requires the attachment of tumor cells on endothelial cells of the vessel wall. Binding occurs through the interaction between platelets, tumor cells and sub-endothelial extracellular matrix, due to the expression of proteins on the outer surface of cellular membranes. The glycoprotein IIb-IIIa (GPIIb/IIIa) is highly expressed on the surface of activated platelets and participates in tumor cell-platelet interaction (Nierodzik et al. 1995) and platelet-ECs binding (Menter et al. 1987). P-selectin also appears in activated platelets and endothelial cells in order to facilitate the interaction between these cells and the vessel wall; in a similar manner, they can mediate the association between tumor cells and the ECs (Bendas and Borsig 2012). Inhibition of both types of interaction decreases the metastasis by reducing the interaction with the vessels (Dardik et al. 1998).

On the other hand, platelets can also modulate the permeability of the vasculature through the content of the secretory granules that allow tumor cells extravasation and colonization of the target organs, hence favoring the cancer metastasis. Serotonin and thrombospondin are two potent bioactive factors that are both released by activated platelets, inducing vasoconstriction and vasodilatation that promote the survival of tumor cells. There are other molecules also secreted by activated platelets that participate in this process by increasing the vascular permeability such a sphingosine 1–phosphate (S1P), lysophosphatidic acid (LPA), C-C motif chemokine 5 (CCL5) and histamine (Gay and Felding-Habermann 2011).

Extravasated tumor cells need a vascular system to consolidate the metastasis. The growth of solid tumors depends on angiogenesis in order to allow the nutrition and development of tumor cells. Platelets secrete many pro-angiogenic factors such as platelet-derived growth factor (PDGF), fibroblast growth factor (FGF), epidermal growth factor (EGF), hepatocyte growth factor (HGF), insulin-like growth factor (IGF), angiopoietin, LPA, S1P, CD40 ligand, matrix metalloproteinase-1 (MMP–1), MMP–2, MMP–9, gelatinase A, and heparanase (Jain et al. 2010); however, the most studied and most potent is VEGF. This factor plays an important role in metastasis since it participates in proliferation of tumor cells, and also supports their adhesion to the ECs, the subsequent vascular permeabilization and final cellular extravasation. It has also been involved in megakaryocyte maturation and angiogenesis. Circulating concentrations of this factor are useful for detecting metastasis since VEGF is secreted by platelets upon stimulation of PAR-1 by thrombin (Nash et al. 2002), and is able to promote endothelial cell proliferation (Ma et al. 2005).

Pro-angiogenic factors promote signaling events by interacting with integrins that are involved in endothelial cell shape regulation and cell–cell interactions, hence controlling the spatial three-dimensional disposition of tubes. This signaling mechanism that requires cytoskeleton reorganization is under control of Rho GTPases, which connect integrins with the cytoskeleton. There are three actin regulatory proteins that are upregulated during endothelial cell morphogenesis: gelsolin, vasoactive-stimulated phosphoprotein (VASP), and profiling (Davis et al. 2002). Contrary to this, platelets contains angiogenic inhibitors such as platelets factor-4 (PF-4), TSP-1 and endostatin, which may be in continued balance with the promoting agents mentioned above (Bambace and Holmes 2011). Hence, a rational use of platelets inhibitors and anticoagulants in therapies against cancer metastasis might be useful for anti-tumoral therapies (Nash et al. 2002).

Conclusion

Platelets are key elements of blood cells since they control blood hemostasis. Regulation of platelet-ECs interactions is crucial to avoid bleeding disorders. In this process, interactive molecules located within the surface of both cells like GPIα and many others are crucial for allowing interaction and stabilization of platelets to the injured ECs. Both microtubule and actin cytoskeleton participate in the shape changes that allow platelets to exert their function. It has been well documented that the cytoskeleton undergoes continuous structural changes, these changes being both calcium-dependent and independent mechanisms. Among other functions, cytoskeleton reorganization allows secretion of platelet granules and cell shape. Finally, we have summarized the current knowledge regarding the role of platelets in the development of diseases that are linked to ECs disruption, like atherosclerosis.

Acknowledgments

Our thanks to Dr. Nicoleta Alexandru and her colleagues for providing us the image of the electron microscopy included in Fig. 1.

Our thanks to MINECO (BFU-2010-21043-C02-01), Junta de Extremadura-FEDER (GR10010 & PRIBS10020) for their support. Pedro C. Redondo is supported by MINECO (RYC2007-00349). Natalia Dionisio and Esther López, are supported by pre-doctoral fellowships from Junta de Extremadura (PRE09020) and MEC-Carlos III Health Institute (FI10/00573).

Berna-Erro A. is supported by a post-doctoral research contract from the University of Extremadura (D-01).

Abbreviation

$[Ca^{2+}]c$	cytosolic calcium concentration
DTS	dense tubular system
M	microtubule
DG	dense granules
EDG	Empty dense granules
AG	α-granules
OPEN or OCS	open canalicular system
MIT	mitochondria
PAR4	Proteinase activated receptor 4
PECAM-1	Platelet endothelial cell adhesion molecule 1
VCAM-1	vascular cell adhesion molecule-1
ICAM-1	intercellular adhesion molecule-1
CD62E	E-selectin
PSGL-1 or CD162	P-selectin glycoprotein ligand-1
PETA-3 or CD151	platelet-endothelial cell tretra-Span Antigen
MAP	microtubules associated proteins
BTK	Bruton's tyrosine kinase
VEGF	endothelial growth factor
LIF	leukemia inhibitory factor
FLT3	FMS like tyrosinkinase
FGF	fibroblast growth factor
OM	oncostatin M
EPO	erythropoietin
NK	natural killer
MHC	histocompatibility complex
S1P	sphingosine 1–phosphate
LPA	lysophosphatidic acid
CCL5	C-C motif chemokine 5
PDGF	platelet-derived growth factor
FGF	fibroblast growth factor
EGF	epidermal growth factor
HGF	hepatocyte growth factor
IGF	insulin-like growth factor
MMP-1	matrix metalloproteinase-1
VASP	vasoactive-stimulated phosphoprotein
PF-4	platelets factor-4

References

Adamikova, L., A. Straube, I. Schulz et al. 2004. Calcium signaling is involved in dynein-dependent microtubule organization. Mol Biol Cell 15: 1969–1980.

Akkerman, J.W., G. Gorter, L. Schrama et al. 1983. A novel technique for rapid determination of energy consumption in platelets. Demonstration of different energy consumption associated with three secretory responses. Biochem J 210: 145–155.

Alvarado, Y., M.M. Mita, S. Vemulapalli et al. 2011. Clinical activity of mammalian target of rapamycin inhibitors in solid tumors. Target Oncol 6: 69–94.

Amor, N.B., J.A. Pariente, G.M. Salido et al. 2006. Thrombin-induced caspases 3 and 9 translocation to the cytoskeleton is independent of changes in cytosolic calcium in human platelets. Blood Cell Mol Dis 36: 392–401.

Appenzeller-Herzog, C. and M.N. Hall. 2012. Bidirectional crosstalk between endoplasmic reticulum stress and mTOR signaling. Trends Cell Biol 22: 274–282.

Arber, S., K.H. Krause and P. Caroni. 1992. s-cyclophilin is retained intracellularly via a unique COOH-terminal sequence and colocalizes with the calcium storage protein calreticulin. J Cell Biol 116: 113–125.

Aslan, J.E. and O.J. McCarty. 2012. Regulation of the mTOR-Rac1 axis in platelet function. Small Gtpases 3: 67–70.

Ayalp, S., S. Sahin, F. Benli Aksungar et al. 2012. Evaluation of platelet serotonin levels in migraine without aura. Agri: Agri 24: 117–122.

Bajzar, L., J. Morser and M. Nesheim. 1996. TAFI, or plasma procarboxypeptidase B, couples the coagulation and fibrinolytic cascades through the thrombin-thrombomodulin complex. J Biol Chem 271: 16603–16608.

Bambace, N.M. and C.E. Holmes. 2011. The platelet contribution to cancer progression. J Thromb Haemost 9: 237–249.

Baratier, J., L. Peris, J. Brocard et al. 2006. Phosphorylation of microtubule-associated protein STOP by calmodulin kinase II. J Biol Chem 281: 19561–19569.

Behnke, O., B.I. Kristensen and L.E. Nielsen. 1971. Electron microscopical observations on actinoid and myosinoid filaments in blood platelets. J Ultrastruct Res 37: 351–369.

Belloc, F., P. Hourdille, M.R. Boisseau et al. 1982. Protein synthesis in human platelets correlation with platelet size. Nouvelle revue francaise d'hematologie 24: 369–373.

Bendas, G. and L. Borsig. 2012. Cancer cell adhesion and metastasis: selectins, integrins, and the inhibitory potential of heparins. Int J Cell Biol 2012: 676731.

Bennett, J.S., C. Chan, G. Vilaire et al. 1997. Agonist-activated alphavbeta3 on platelets and lymphocytes binds to the matrix protein osteopontin. J Biol Chem 272: 8137–8140.

Berrout, J., M. Jin and R.G. O'Neil. 2012. Critical role of TRPP2 and TRPC1 channels in stretch-induced injury of blood-brain barrier endothelial cells. Brain Res 1436: 1–12.

Bertino, A.M., X.Q. Qi, J. Li et al. 2003. Apoptotic markers are increased in platelets stored at 37 degrees C. Transfusion 43: 857–866.

Beumer, S., I. J. MJ, P.G. de Groot et al. 1994. Platelet adhesion to fibronectin in flow: dependence on surface concentration and shear rate, role of platelet membrane glycoproteins GP IIb/IIIa and VLA-5, and inhibition by heparin. Blood 84: 3724–3733.

Bombeli, T., B.R. Schwartz and J.M. Harlan. 1998. Adhesion of activated platelets to endothelial cells: evidence for a GPIIbIIIa-dependent bridging mechanism and novel roles for endothelial intercellular adhesion molecule 1 (ICAM-1), alphavbeta3 integrin, and GPIbalpha. J Exp Med 187: 329–339.

Booyse, F.M., T.P. Hoveke and M. E. Rafelson, Jr. 1973. Human platelet actin. Isolation and properties. J Biol Chem 248: 4083–4091.

Borsig, L., R. Wong, J. Feramisco et al. 2001. Heparin and cancer revisited: mechanistic connections involving platelets, P-selectin, carcinoma mucins, and tumor metastasis. Proc Natl Acad Sci USA 98: 3352–3357.

Bouaziz, A., N.B. Amor, G.E. Woodard et al. 2007. Tyrosine phosphorylation/dephosphorylation balance is involved in thrombin-evoked microtubular reorganisation in human platelets. Thromb & Haemost 98: 375–384.

Breitenstein, A., S. Stein, E.W. Holy et al. 2011. Sirt1 inhibition promotes *in vivo* arterial thrombosis and tissue factor expression in stimulated cells. Cardiovasc Res 89: 464–472.

Brisson, C., D.O. Azorsa, L.K. Jennings et al. 1997. Co-localization of CD9 and GPIIb-IIIa (alpha IIb beta 3 integrin) on activated platelet pseudopods and alpha-granule membranes. Histochem J 29: 153–165.

Buergy, D., F. Wenz, C. Groden et al. 2012. Tumor-platelet interaction in solid tumors. Int J Cancer 130: 2747–2760.

Buljan, V., E.P. Ivanova and K.M. Cullen. 2009. How calcium controls microtubule anisotropic phase formation in the presence of microtubule-associated proteins *in vitro*. Biol Bioph Res Commun 381: 224–228.

Calaminus, S.D., O.J. McCarty, J.M. Auger et al. 2007. A major role for Scar/WAVE-1 downstream of GPVI in platelets. J Thromb Haemost 5: 535–541.

Calaminus, S.D., S. Thomas, O.J. McCarty et al. 2008. Identification of a novel, actin-rich structure, the actin nodule, in the early stages of platelet spreading. J Thromb Haemost 6: 1944–1952.

Casoli, T., G. Di Stefano, M. Balietti et al. 2010. Peripheral inflammatory biomarkers of Alzheimer's disease: the role of platelets. Biogerontology 11: 627–633.

Catricala, S., M. Torti and G. Ricevuti. 2012. Alzheimer disease and platelets: how's that relevant. Immunity & ageing: I & A 9: 20.

Cattaneo, M. 2003. Inherited platelet-based bleeding disorders. J Thromb Haemost: JTH 1: 1628–1636.

Chiu, Y.J., E. McBeath and K. Fujiwara. 2008. Mechanotransduction in an extracted cell model: Fyn drives stretch- and flow-elicited PECAM-1 phosphorylation. J Cell Biol 182: 753–763.

Choi, W., Z.A. Karim and S.W. Whiteheart. 2010. Protein expression in platelets from six species that differ in their open canalicular system. Platelets 21: 167–175.

Chopra, H., J.S. Hatfield, Y.S. Chang et al. 1988. Role of tumor cytoskeleton and membrane glycoprotein IRGpIIb/IIIa in platelet adhesion to tumor cell membrane and tumor cell-induced platelet aggregation. Cancer Res 48: 3787–3800.

Ciurea, S.O. and R. Hoffman. 2007. Cytokines for the treatment of thrombocytopenia. Semin Hematol 44: 166–182.

Cohen, I., E. Kaminski and A. De Vries. 1973. Actin-linked regulation of the human platelet contractile system. FEBS letters 34: 315–317.

Coller, B.S., J.H. Beer, L.E. Scudder et al. 1989. Collagen-platelet interactions: evidence for a direct interaction of collagen with platelet GPIa/IIa and an indirect interaction with platelet GPIIb/IIIa mediated by adhesive proteins. Blood 74: 182–192.

Coller, B.S., D.A. Cheresh, E. Asch et al. 1991. Platelet vitronectin receptor expression differentiates Iraqi-Jewish from Arab patients with Glanzmann thrombasthenia in Israel. Blood 77: 75–83.

Coppinger, J.A., G. Cagney, S. Toomey et al. 2004. Characterization of the proteins released from activated platelets leads to localization of novel platelet proteins in human atherosclerotic lesions. Blood 103: 2096–2104.

Dardik, R., N. Savion, Y. Kaufmann et al. 1998. Thrombin promotes platelet-mediated melanoma cell adhesion to endothelial cells under flow conditions: role of platelet glycoproteins P-selectin and GPIIb-IIIA. Br J Cancer 77: 2069–2075.

Davis, G.E., K.J. Bayless and A. Mavila. 2002. Molecular basis of endothelial cell morphogenesis in three-dimensional extracellular matrices. Anat Rec 268: 252–275.

Diacovo, T.G., S.J. Roth, J.M. Buccola et al. 1996. Neutrophil rolling, arrest, and transmigration across activated, surface-adherent platelets via sequential action of P-selectin and the beta 2-integrin CD11b/CD18. Blood 88: 146–157.

Ekmekci, O.B. and H. Ekmekci. 2006. Vitronectin in atherosclerotic disease. Clin Chim Acta 368: 77–83.

Elton, C.M., P.A. Smethurst, P. Eggleton et al. 2002. Physical and functional interaction between cell-surface calreticulin and the collagen receptors integrin alpha2beta1 and glycoprotein VI in human platelets. Thromb & Haemost 88: 648–654.

Escolar, G., I. Lopez-Vilchez, M. Diaz-Ricart et al. 2008. Internalization of tissue factor by platelets. Thromb Res 122 Suppl 1: S37–41.

Farstad, M. and J. Sander. 1971. The existence of a long-chain acyl-CoA synthetase in homogenates of human blood platelets. Scv J Clinic and Lab Invest 28: 261–265.

Frenette, P.S., R.C. Johnson, R.O. Hynes et al. 1995. Platelets roll on stimulated endothelium *in vivo*: an interaction mediated by endothelial P-selectin. Proc Natl Acad Sci USA 92: 7450–7454.

Frenette, P.S., C. Moyna, D.W. Hartwell et al. 1998. Platelet-endothelial interactions in inflamed mesenteric venules. Blood 91: 1318–1324.

Fujimura, K. and D.R. Phillips. 1983. Calcium cation regulation of glycoprotein IIb-IIIa complex formation in platelet plasma membranes. J Biol Chem 258: 10247–10252.

Furie, B. and B.C. Furie. 2005. Thrombus formation *in vivo*. J Clin Invest 115: 3355–3362.

Gawaz, M., F.J. Neumann, T. Dickfeld et al. 1998. Activated platelets induce monocyte chemotactic protein-1 secretion and surface expression of intercellular adhesion molecule-1 on endothelial cells. Circulation 98: 1164–1171.

Gawaz, M., K. Stellos and H.F. Langer. 2008. Platelets modulate atherogenesis and progression of atherosclerotic plaques via interaction with progenitor and dendritic cells. J Thromb and Haemost: JTH 6: 235–242.

Gay, L.J. and B. Felding-Habermann. 2011. Contribution of platelets to tumour metastasis. Nat Rev Cancer 11: 123–134.

Gayle, R.B., 3rd, C.R. Maliszewski, S.D. Gimpel et al. 1998. Inhibition of platelet function by recombinant soluble ecto-ADPase/CD39. J Clinic Invest 101: 1851–1859.

Ge, S., J.G.White and C.L. Haynes. 2012. Cytoskeletal F-actin, not the circumferential coil of microtubules, regulates platelet dense-body granule secretion. Platelets 23: 259–263.

Gibbins, J.M. 2004. Platelet adhesion signalling and the regulation of thrombus formation. J Cell Sci 117: 3415–3425.

Griffin, J.H., B.V. Zlokovic and L.O. Mosnier. 2012. Protein C anticoagulant and cytoprotective pathways. Int J Hematol 95: 333–345.

Guerriero, R., I. Parolini, U. Testa et al. 2006. Inhibition of TPO-induced MEK or mTOR activity induces opposite effects on the ploidy of human differentiating megakaryocytes. J Cell Sci 119: 744–752.

Harper, M.T. and A.W. Poole. 2010. Diverse functions of protein kinase C isoforms in platelet activation and thrombus formation. J Thromb Haemost: JTH 8: 454–462.

Humphries, M.J. 2000. Integrin structure. Biochem Soc Trans 28: 311–339.

Italiano, J.E., Jr., P. Lecine, R.A. Shivdasani et al. 1999. Blood platelets are assembled principally at the ends of proplatelet processes produced by differentiated megakaryocytes. J Cell Biol 147: 1299–1312.

Izaguirre Avila, R. 1997. Evolution of the knowledge about platelets. Archivos del Instituto de Cardiologia de Mexico 67: 511–520.

J, R. 1923. Contribution a l'etude de la physiologie normale et pathologique du globulin. Arch Inst Card Mexican 20: 241–257.

Jackson, S.P., S. Cranmer, P. Mangin et al. 2007. Are Erk, Btk, and PECAM-1 major players in GPIb signaling? The challenge of unraveling signaling events downstream of platelet GPIb. Blood 109: 846–847; discussion 847–848.

Jacobs, P.P. and R. Sackstein. 2011. CD44 and HCELL: preventing hematogenous metastasis at step 1. FEBS letters 585: 3148–3158.

Jain, S., J. Harris and J. Ware. 2010. Platelets: linking hemostasis and cancer. Arterioscler Thromb Vasc Biol 30: 2362–2367.

Jardin, I., J.J. Lopez, G.M. Salido et al. 2008. Orai1 mediates the interaction between STIM1 and hTRPC1 and regulates the mode of activation of hTRPC1-forming Ca^{2+} channels. J Biol Chem 283: 25296–25304.

Kaplan, Z.S. and S.P. Jackson. 2011. The role of platelets in atherothrombosis. Hematology Am Soc Hematol Educ Program 2011: 51–61.

Kieffer, N., J. Guichard, J.P. Farcet et al. 1987. Biosynthesis of major platelet proteins in human blood platelets. Eur J Biochem/FEBS 164: 189–195.

Kinsella, M.G., P.K.Tran, M.C. Weiser-Evans et al. 2003. Changes in perlecan expression during vascular injury: role in the inhibition of smooth muscle cell proliferation in the late lesion. Arterioscler Thromb Vasc Biol 23: 608–614.

Konopatskaya, O., K. Gilio, M.T. Harper et al. 2009. PKCalpha regulates platelet granule secretion and thrombus formation in mice. J Clinic Invest 119: 399–407.

Kosaki, G. and J. Kambayashi. 2011. Thrombocytogenesis by megakaryocyte; Interpretation by protoplatelet hypothesis. Proceedings of the Japan Academy. Series B, Physic and Biol Sci 87: 254–273.

Langer, H.F. and T. Chavakis. 2009. Leukocyte-endothelial interactions in inflammation. J Cell Mol Med 13: 1211–1220.

Laske, C., K. Stellos, E. Stransky et al. 2008. Decreased plasma and cerebrospinal fluid levels of stem cell factor in patients with early Alzheimer's disease. Journal of Alzheimer's disease: JAD 15: 451–460.

Lee, D., K.P. Fong, M.R. King et al. 2012a. Differential dynamics of platelet contact and spreading. Biophys J 102: 472–482.

Lee, D., K.P. Fong, M.R. King et al. 2012b. Differential dynamics of platelet contact and spreading. Biophys J 102: 472–482.

Lellouch-Tubiana, A., J. Lefort, E. Pirotzky et al. 1985. Ultrastructural evidence for extravascular platelet recruitment in the lung upon intravenous injection of platelet-activating factor (PAF-acether) to guinea-pigs. Br J Exp Pathology 66: 345–355.

Li, Z., M.K. Delaney, K.A. O'Brien et al. 2010. Signaling during platelet adhesion and activation. Arterioscler Thromb Vasc Biol 30: 2341–2349.

Liu, Z.J., J. Italiano, Jr., F. Ferrer-Marin et al. 2011. Developmental differences in megakaryocytopoiesis are associated with up-regulated TPO signaling through mTOR and elevated GATA-1 levels in neonatal megakaryocytes. Blood 117: 4106–4117.

Lopez, J.J., G.M. Salido, J.A. Pariente et al. 2006. Interaction of STIM1 with endogenously expressed human canonical TRP1 upon depletion of intracellular Ca2+ stores. J Biol Chem 281: 28254–28264.

Lopez, J.J., G.M. Salido, E. Gomez-Arteta et al. 2007. Thrombin induces apoptotic events through the generation of reactive oxygen species in human platelets. J Thromb and Haemost: JTH 5: 1283–1291.

Lopez, J.J., G.M. Salido, J.A. Pariente et al. 2008a. Thrombin induces activation and translocation of Bid, Bax and Bak to the mitochondria in human platelets. J Thromb and Haemost: JTH 6: 1780–1788.

Lopez, J.J., I. Jardin, R. Bobe et al. 2008b. STIM1 regulates acidic Ca^{2+} store refilling by interaction with SERCA3 in human platelets. Biochem Pharmacol 75: 2157–2164.

Loscalzo, J. and G. Welch. 1995. Nitric oxide and its role in the cardiovascular system. Progr Card Diseases 38: 87–104.

Lowenberg, E.C., J.C. Meijers and M. Levi. 2010. Platelet-vessel wall interaction in health and disease. Neth J Med 68: 242–251.

Luchtman-Jones, L. and G.J. Broze, Jr. 1995. The current status of coagulation. Ann Med 27: 47–52.

Ma, L., R. Perini, W. McKnight et al. 2005. Proteinase-activated receptors 1 and 4 counter-regulate endostatin and VEGF release from human platelets. Proc Natl Acad Sci USA 102: 216–220.

Mahalingam, D., K. Sankhala, A. Mita et al. 2009. Targeting the mTOR pathway using deforolimus in cancer therapy. Future Oncol 5: 291–303.

Malik, A.B. 1993. Endothelial cell interactions and integrins. New Horiz 1: 37–51.

Malik, A.B., A. Johnson and F.A. Blumenstock. 1983. Interaction between formed elements and the pulmonary endothelium. Gen Pharmacol 14: 197–200.

Marcu, M.G., L. Zhang, K. Nau-Staudt et al. 1996. Recombinant scinderin, an F-actin severing protein, increases calcium-induced release of serotonin from permeabilized platelets, an effect blocked by two scinderin-derived actin-binding peptides and phosphatidylinositol 4,5-bisphosphate. Blood 87: 20–24.

Masuda, J. and R. Ross. 1990a. Atherogenesis during low level hypercholesterolemia in the nonhuman primate. I. Fatty streak formation. Arteriosclerosis 10: 164–177.

Masuda, J. and R. Ross. 1990b. Atherogenesis during low level hypercholesterolemia in the nonhuman primate. II. Fatty streak conversion to fibrous plaque. Arteriosclerosis 10: 178–187.

Matsumura, C.Y., A.P. Taniguti, A. Pertille et al. 2011. Stretch-activated calcium channel protein TRPC1 is correlated with the different degrees of the dystrophic phenotype in mdx mice. American journal of physiology. Cell physiology 301: C1344–1350.

Mattila, P.K. and P. Lappalainen. 2008. Filopodia: molecular architecture and cellular functions. Nat Rev Mol Cell Biol 9: 446–454.

May, A.E., P. Seizer and M. Gawaz. 2008. Platelets: inflammatory firebugs of vascular walls. Arterioscler Thromb Vasc Biol 28: s5–10.

McEver, R.P. and C. Zhu. 2010. Rolling cell adhesion. Ann rev Cell & Develop Biol 26: 363–396.

Menter, D.G., B.W. Steinert, B.F. Sloane et al. 1987. Role of platelet membrane in enhancement of tumor cell adhesion to endothelial cell extracellular matrix. Cancer Res 47: 6751–6762.

Muszbek, L., Z. Bereczky, Z. Bagoly et al. 2011. Factor XIII: a coagulation factor with multiple plasmatic and cellular functions. Physiol Rev 91: 931–972.

Nakano, T., A.P. Chahinian, M. Shinjo et al. 1998. Interleukin 6 and its relationship to clinical parameters in patients with malignant pleural mesothelioma. Br J Cancer 77: 907–912.

Nash, G.F., L.F. Turner, M.F. Scully et al. 2002. Platelets and cancer. Lancet Oncol 3: 425–430.

Nierodzik, M.L., A. Klepfish and S. Karpatkin. 1995. Role of platelets, thrombin, integrin IIb-IIIa, fibronectin and von Willebrand factor on tumor adhesion *in vitro* and metastasis *in vivo*. Thromb Haemost 74: 282–290.

Nieswandt, B., I. Pleines and M. Bender. 2011. Platelet adhesion and activation mechanisms in arterial thrombosis and ischaemic stroke. J Thromb Haemost 9 Suppl 1: 92–104.

Nishibori, M., B. Cham, A. McNicol et al. 1993. The protein CD63 is in platelet dense granules, is deficient in a patient with Hermansky-Pudlak syndrome, and appears identical to granulophysin. J Clin Invest 91: 1775–1782.

Nurden, A.T. and P. Nurden. 2003. Advantages of fast-acting ADP receptor blockade in ischemic heart disease. Arterioscler Thromb Vasc Biol 23: 158–159.

Nuyttens, B.P., T. Thijs, H. Deckmyn et al. 2011. Platelet adhesion to collagen. Thromb Res 127 Suppl 2: S26–29.

Offermanns, S. 2006. Activation of platelet function through G protein-coupled receptors. Circ Res 99: 1293–1304.

Okamoto, K., M. Bosch and Y. Hayashi. 2009. The roles of CaMKII and F-actin in the structural plasticity of dendritic spines: a potential molecular identity of a synaptic tag? Physiology 24: 357–366.

Palumbo, J.S. and J.L. Degen. 2007. Mechanisms linking tumor cell-associated procoagulant function to tumor metastasis. Thromb Res 120 Suppl 2: S22–28.

Pandey, D., P. Goyal, J.R. Bamburg et al. 2006. Regulation of LIM-kinase 1 and cofilin in thrombin-stimulated platelets. Blood 107: 575–583.

Paraskevaidis, I., J. Palios, J. Parissis et al. 2012. Treating depression in coronary artery disease and chronic heart failure: what's new in using selective serotonin re-uptake inhibitors? Card & Haemat agents in Med Chem 10: 109–115.

Patel-Hett, S., J.L. Richardson, H. Schulze et al. 2008. Visualization of microtubule growth in living platelets reveals a dynamic marginal band with multiple microtubules. Blood 111: 4605–4616.

Perrotta, P.L., C.L. Perrotta and E.L. Snyder. 2003. Apoptotic activity in stored human platelets. Transfusion 43: 526–535.

Peters, C.G., A.D. Michelson and R. Flaumenhaft. 2012. Granule exocytosis is required for platelet spreading: differential sorting of alpha-granules expressing VAMP-7. Blood 120: 199–206.

Phillips, D.R. 1980. An evaluation of membrane glycoproteins in platelet adhesion and aggregation. Prog Hemost Thromb 5: 81–109.

Piotrowicz, R.S., R.P. Orchekowski, D.J. Nugent et al. 1988. Glycoprotein Ic-IIa functions as an activation-independent fibronectin receptor on human platelets. J Cell Biol 106: 1359–1364.

Pulcinelli, F.M., S. Sebastiani, M. Pesciotti et al. 1998. Nickel enhances collagen-induced platelet activation acting by increasing the organization of the cytoskeleton. Thromb Haemost 79: 395–399.

Redondo, P.C., N. Ben-Amor, G.M. Salido et al. 2005. Ca^{2+}-independent activation of Bruton's tyrosine kinase is required for store-mediated Ca^{2+} entry in human platelets. Cellular signalling 17: 1011–1021.

Redondo, P.C., M.T. Harper, J.A. Rosado et al. 2006. A role for cofilin in the activation of store-operated calcium entry by de novo conformational coupling in human platelets. Blood 107: 973–979.

Redondo, P.C., A.G. Harper, S.O. Sage et al. 2007. Dual role of tubulin-cytoskeleton in store-operated calcium entry in human platelets. Cell Signal 19: 2147–2154.

Reininger, A.J., H.F. Heijnen, H. Schumann et al. 2006. Mechanism of platelet adhesion to von Willebrand factor and microparticle formation under high shear stress. Blood 107: 3537–3545.

Riboni, L., E. Ubaldo and H. Nunez-Duran. 1988. Morphometric and three-dimensional study of platelets during activation in the rat. Acta anatomica 132: 28–34.

Rodriguez Del Castillo, A., M.L. Vitale, L. Tchakarov et al. 1992. Human platelets contain scinderin, a Ca^{2+}-dependent actin filament-severing protein. Thromb and Haemost 67: 248–251.

Rosado, J.A. and S.O. Sage. 2000a. The actin cytoskeleton in store-mediated calcium entry. J Physiol 526(Pt 2): 221–229.

Rosado, J.A. and S.O. Sage. 2000b. Farnesylcysteine analogues inhibit store-regulated Ca^{2+} entry in human platelets: evidence for involvement of small GTP-binding proteins and actin cytoskeleton. Biochem J 347(Pt 1): 183–192.

Rosado, J.A. and S.O. Sage. 2000c. A role for the actin cytoskeleton in the initiation and maintenance of store-mediated calcium entry in human platelets. Trends Card Med 10: 327–332.

Rosado, J.A., D. Graves and S.O. Sage. 2000. Tyrosine kinases activate store-mediated Ca^{2+} entry in human platelets through the reorganization of the actin cytoskeleton. Biochem J 351(Pt 2): 429–437.

Rosado, J.A., S. Jenner and S.O. Sage. 2000. A role for the actin cytoskeleton in the initiation and maintenance of store-mediated calcium entry in human platelets. Evidence for conformational coupling. J Biol Chem 275: 7527–7533.

Rosado, J.A., J.J. Lopez, E. Gomez-Arteta et al. 2006. Early caspase-3 activation independent of apoptosis is required for cellular function. J Cell Physiol 209: 142–152.

Ruggeri, Z.M. 1999. Structure and function of von Willebrand factor. Thromb Haemost 82: 576–584.

Santoro, S.A. 1988. Molecular basis of platelet adhesion to collagen. Prog Clin Biol Res 283: 291–314.

Schwertz, H., S. Koster, W.H. Kahr et al. 2010. Anucleate platelets generate progeny. Blood 115: 3801–3809.

Schwertz, H., J.W. Rowley, N.D. Tolley et al. 2012. Assessing protein synthesis by platelets. Methds in Mol Biol 788: 141–153.

Shattil, S.J. and P.J. Newman. 2004. Integrins: dynamic scaffolds for adhesion and signaling in platelets. Blood 104: 1606–1615.

Sincock, P.M., G. Mayrhofer and L.K. Ashman. 1997. Localization of the transmembrane 4 superfamily (TM4SF) member PETA-3 (CD151) in normal human tissues: comparison with CD9, CD63, and alpha5beta1 integrin. J Histochem Cytochem 45: 515–525.

Stenberg, P.E., M.A. Shuman, S.P. Levine et al. 1984. Redistribution of alpha-granules and their contents in thrombin-stimulated platelets. J Cell Biol 98: 748–760.

Suzuki, H., Y. Katagiri, S. Tsukita et al. 1990. Localization of adhesive proteins in two newly subdivided zones in electron-lucent matrix of human platelet alpha-granules. Histochemistry 94: 337–344.

Tabuchi, A. and W.M. Kuebler. 2008. Endothelium-platelet interactions in inflammatory lung disease. Vasc Pharmacol 49: 141–150.

Thon, J.N., A. Montalvo, S. Patel-Hett et al. 2010. Cytoskeletal mechanics of proplatelet maturation and platelet release. J Cell Biol 191: 861–874.

Thon, J.N., H. Macleod, A.J. Begonja et al. 2012. Microtubule and cortical forces determine platelet size during vascular platelet production. Nature Comm 3: 852.

Ts'ao, C.H. 1971. Rough endoplasmic reticulum and ribosomes in blood platelets. Scv. J. Haematol. 8: 134–140.

van Hinsbergh, V.W. 2012. Endothelium—role in regulation of coagulation and inflammation. Semin Immunopathol 34: 93–106.

van Kruchten, R., J.M. Cosemans and J.W. Heemskerk. 2012. Measurement of whole blood thrombus formation using parallel-plate flow chambers—a practical guide. Platelets 23: 229–242.

Varki, A. 2007. Trousseau's syndrome: multiple definitions and multiple mechanisms. Blood 110: 1723–1729.

Vashadze, S. 2011a. Anxiety, depression, panic disorder and bronchial asthma. Georgian medical news 11: 63–67.

Vashadze, S. 2011b. Depression as a risk factor for arterial hypertension. Georgian Med News 201: 45–47.

Vemulapalli, S., A. Mita, Y. Alvarado et al. 2011. The emerging role of mammalian target of rapamycin inhibitors in the treatment of sarcomas. Target Oncol 6: 29–39.

Vitale, M.L., A. Rodriguez Del Castillo and J.M. Trifaro. 1992. Protein kinase C activation by phorbol esters induces chromaffin cell cortical filamentous actin disassembly and increases the initial rate of exocytosis in response to nicotinic receptor stimulation. Neuroscience 51: 463–474.

Wagner, D.D. 1990. Cell biology of von Willebrand factor. Annu Rev Cell Biol 6: 217–246.

Ware, J.A. and D.D. Heistad. 1993. Seminars in medicine of the Beth Israel Hospital, Boston. Platelet-endothelium interactions. N Engl J Med 328: 628–635.

Warshaw, A.L., L. Laster and N.R. Shulman. 1967. Protein synthesis by human platelets. The J Biol Chem 242: 2094–2097.

Weyrich, A.S., S. Lindemann, N.D. Tolley et al. 2004. Change in protein phenotype without a nucleus: translational control in platelets. Semmin. Thromb. & Haemost. 30: 491–498.

Weyrich, A.S., H. Schwertz, L.W. Kraiss et al. 2009. Protein synthesis by platelets: historical and new perspectives. J Thromb and Haemost: JTH 7: 241–246.

White, J.G. 1999. Platelet secretory process. Blood 93: 2422–2425.

White, J.G. 2008. Electron opaque structures in human platelets: which are or are not dense bodies? Platelets 19: 455–466.

White, J.G. and W. Krivit. 1965. Fine structural localization of adenosine triphosphatase in human platelets and other blood cells. Blood 26: 554–568.

White, J.G. and C.C. Clawson. 1980. Overview article: biostructure of blood platelets. Ultrastructural pathology 1: 533–558.

White, J.G. and P.A. de Alarcon. 2002. Platelet spherocytosis: a new bleeding disorder. Am J Hematol 70: 158–166.

Wolf, B.B., J.C. Goldstein, H.R. Stennicke et al. 1999. Calpain functions in a caspase-independent manner to promote apoptosis-like events during platelet activation. Blood 94: 1683–1692.

Xia, L., M. Sperandio, T. Yago et al. 2002. P-selectin glycoprotein ligand-1-deficient mice have impaired leukocyte tethering to E-selectin under flow. J Clin Invest 109: 939–950.

Yang, J., T. Hirata, K. Croce et al. 1999. Targeted gene disruption demonstrates that P-selectin glycoprotein ligand 1 (PSGL-1) is required for P-selectin-mediated but not E-selectin-mediated neutrophil rolling and migration. J Exp Med 190: 1769–1782.

Yuan, Y., S. Kulkarni, P. Ulsemer et al. 1999. The von Willebrand factor-glycoprotein Ib/V/IX interaction induces actin polymerization and cytoskeletal reorganization in rolling platelets and glycoprotein Ib/V/IX-transfected cells. J Biol Chem 274: 36241–36251.

Zarbock, A. and K. Ley. 2009. The role of platelets in acute lung injury (ALI). Front. Biosci.: a journal and virtual library 14: 150–158.

Zellner, M., M. Baureder, E. Rappold et al. 2012. Comparative platelet proteome analysis reveals an increase of monoamine oxidase-B protein expression in Alzheimer's disease but not in non-demented Parkinson's disease patients. J Proteomics 75: 2080–2092.

Zhu, D. 2007. Mathematical modeling of blood coagulation cascade: kinetics of intrinsic and extrinsic pathways in normal and deficient conditions. Blood Coagul Fibrinolysis: an international journal in haemostasis and thrombosis 18: 637–646.

Index

A

Actin 53–73
actin microfilaments 1, 15
actomyosin 1–26
acute lung injury (ALI) 11–13, 20
acute respiratory distress syndrome (ARDS) 12, 20
adherens complexes 7, 15
adherent junction (AJ) 170, 181, 182, 184–188, 191
advanced glycation end-products 153
ammation 1, 7, 12
angiogenesis 1, 3, 7, 13, 53–73, 150, 151, 158
apoptosis 1, 11
Arp2/3 complex 2, 11
asthma 11, 12, 20
atherosclerosis 156–158

B

basement membrane 13

C

cadherin 77–84
calcium 77
calcium signaling 223
calmodulin (CaM) 3, 7, 9
cancer 208, 214, 216, 218–221
catenin 77–82, 84
caveolae 147, 152, 154, 156, 158–161
Caveolin-1 151, 153, 157–161
Cell polarity 147, 148, 150, 152, 160
cortactin 3, 11, 12, 14–19

D

differentiated microdomains 154

E

EC-platelet linking molecules 209
edema 28, 32, 47

Endocytosis 151, 158–161
Endothelial cells 146–166
Endothelial-junctions 76

F

FAK 54, 60, 64, 65
FcRn receptor 151
Focal adhesion (FA) 6, 7, 14, 15, 172, 180–188, 191, 192
force transmission 91–115

G

gap junction 169
G-proteins 15

H

hemodynamic forces 148
HGF 14

I

IgG 149, 151, 152
IL-1β 12
intermediate filaments 1, 2, 15, 149, 151
interstitial 6
intracellular vesicular transport 161
ischemia reperfusion (I/R)-induced injury 12

L

lamellipodial 14, 15
LDL receptors 151
lipid rafts 147, 154–159, 161

M

macromolecule trafficking 146
mathematical modeling 93, 96, 97, 99, 109, 111
mechanical transducers 160

mechanotransduction 1, 11, 93, 94, 96, 97,
     108, 111
membrane protein-cytoskeleton 152
microtubule associated protein 6 (MAP6)
     30, 36, 43–47
Microtubule/microfilament 116–145
microtubules 1, 7, 15
MLC phosphatase 7, 12
Myosin 2–4, 6, 7, 9, 19
myosin light chain (MLC) 6, 7, 9, 12, 13, 18,
     19, 119–121, 124, 125, 131
myosin light-chain kinase (MLCK) 4, 7,
     9–12, 16, 18, 19

N

network models 97
non-muscle isoform of myosin light chain
     kinase (nmMLCK) 3, 9, 11, 12, 14, 15,
     18–20
notch-signalling 58

P

p38 MAPK 118, 121–124, 131–133
p60src 9, 11, 12, 14, 17
paracellular gaps 1, 6, 13
permeability 27, 28, 30, 32
phenotypic heterogeneity of endothelial
     cells 152
platelet cytoskeleton 201

R

RacGTPase 14, 15, 17
raft-cytoskeleton interactions 154, 155
Rho GTPase 118, 124, 153, 156

S

S1P 14, 15, 17–20
sepsis 11, 12, 20
SH2- and SH3-binding domains 3, 9, 11, 12,
     14, 15, 18–20
specialized membrane structures 147, 154,
     160
Stable tubule-only polypeptides (STOP)
     36, 43
stress fibers 2, 6, 7, 8, 11, 13, 14

T

Tau 30, 34–43, 46, 47
TER 17, 18
TGFβ 9
thrombin 4, 9, 12, 18
tight junction 6, 169, 170
TNF-α 12, 13
transcellular transport 146, 152, 161
transcytosis 151, 159, 160
tyrosine kinases 18

V

VE-cadherin 54, 56–58, 60–63
VEGF 9, 13
ventilator-induced lung injury (VILI) 12, 20
viscoelastic 96, 98–100, 103, 106, 108, 110

Color Plate Section

Chapter 1

Vascular Barrier Enhancement

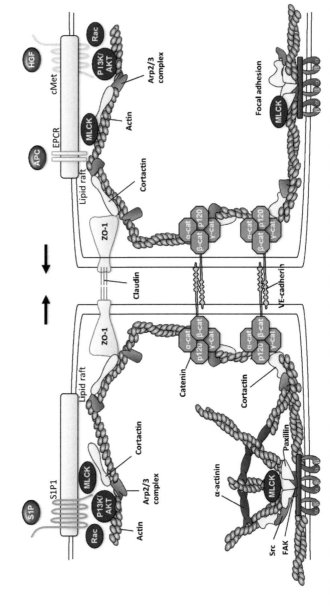

Figure 1. **Schematic representation of junction complexes involved in cell-cell and cell-matrix adhesions.** Barrier enhancing agonists activate intracellular signaling pathways that lead to peripheral cortical actin formation, resulting in barrier enhancement.

Paracellular Gap Formation

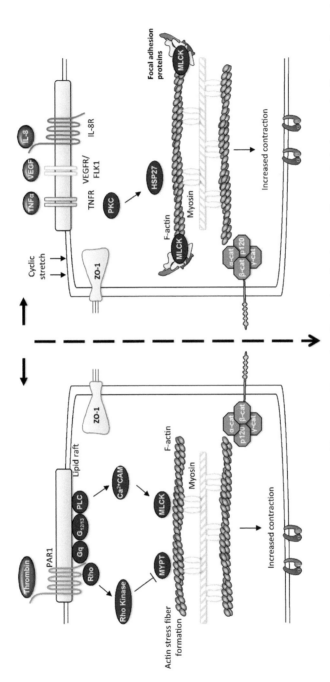

Figure 2. Intracellular signaling pathway activated by contractile agonists. Formation of actin stress fiber results in increased cell contraction and formation of paracellular gap.

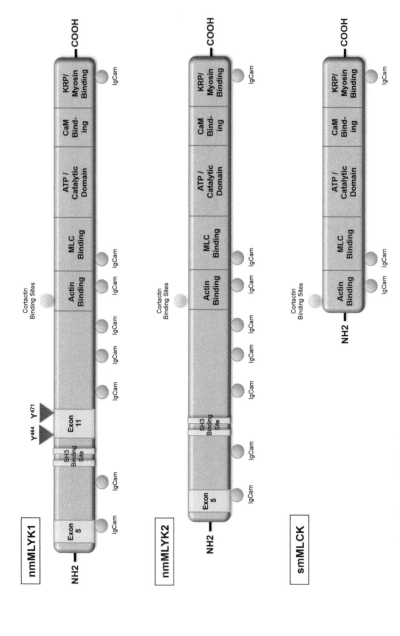

Figure 3. Representation of MLCK isoform/splice-variants.

Vascular Barrier Recovery

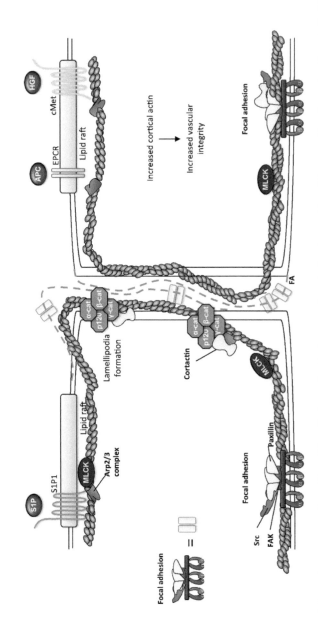

Figure 4. Schematic representation of lamellipodia formation by increased cortical actin at cell periphery and barrier restoration. Actin-binding proteins (i.e., MLCK and cortactin) translocate to this spatially defined region.

Chapter 2

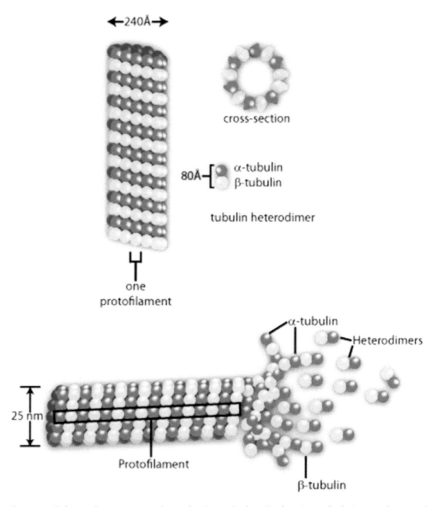

Figure 1. Schematic representation of microtubules. Each microtubule is a polymer of alternating α and β-tubulin heterodimers forming a diameter of about 25 nm.

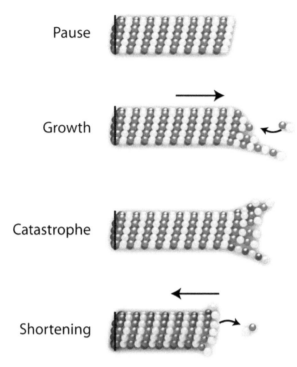

Pause

Growth

Catastrophe

Shortening

Figure 3. Schematic representation of the dynamic instability model. The alternating phases of shortening and growth—known as catastrophes and rescues—respectively are responsible for the dynamic instability behavior of microtubules. Importantly, MAPs contribute to this behavior.

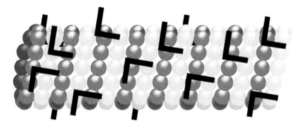

Structural MAPs bind to and stabilize MTs

β-tubulin
α-tubulin   MAP

Figure 4. Structural MAPs bind to and stabilize microtubules.

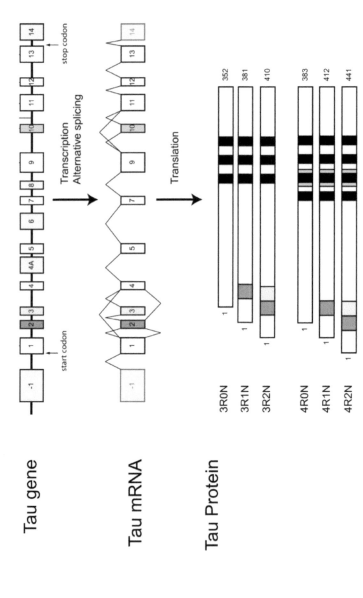

Figure 5. Organization of the Tau gene, Tau mRNA, and Tau proteins. The complex series of intronic regions are shown between exons (numbered). Splice sites are shown in the Tau mRNA that give rise to six Tau proteins. Tau proteins are defined by the number of their microtubule binding domains (R; black) and the number of N-terminal repeat domains (N; orange and yellow).

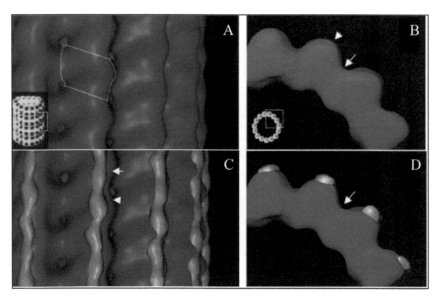

Figure 6. Tau binding to microtubules. [A] 3D reconstruction of cryo-electron microscopy of a microtubule stabilized by taxol. A tubulin monomer is outlined in white. Insert: microtubule. [B] Transversal view of four protofilaments seen from the minus side. Arrowheads: protofilaments. Arrows: ridge. [C] En face view of Tau (orange) decorating the microtubule. [D] View of C from the minus end showing Tau bound to microtubules. Modified from figures originally published in (Al-Bassam et al. 2002).

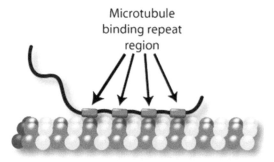

Figure 7. Tau functions as molecular glue. When Tau binds to microtubule fibers, it raises the free energy state of the microtubule that decreases the rate of microtubule breakdown, reducing the frequency of catastrophic events, and increasing the probability for rescues.

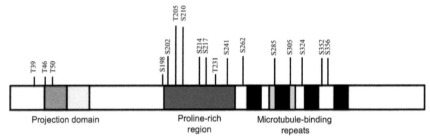

Figure 8. 4R2N Tau phosphorylation sites. The majority of phosphorylation sites are grouped within the proline-rich region (blue) and in the microtubule-binding repeats (black squares). T = Threonine. S = Serine.

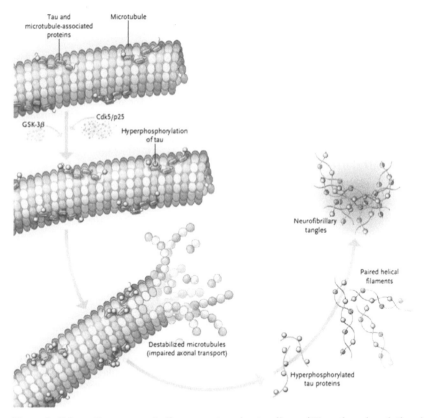

Figure 9. Schematic represents the current understanding of Tau phosphorylation in neurodegenerative diseases. When microtubule-bound Tau gets hyperphosphorylated, Tau detaches from microtubules. Subsequently, microtubules disassemble and Tau aggregates. GSK-3β refers to glycogen synthase kinase-3 beta and Cdk5/p25 refers to a cyclin dependent kinase 5 that is hyperactivated. Originally published in (Querfurth and LaFerla 2010).

Chapter 3

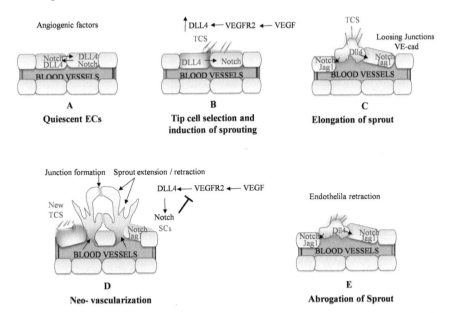

A
Quiescent ECs

B
Tip cell selection and
induction of sprouting

C
Elongation of sprout

D
Neo- vascularization

E
Abrogation of Sprout

Figure 1. Sprouting and lumen formation. (A-C): Sprouting Quiescent ECs:DLL4 and Notch signaling are balanced in endothelial cells (pink). **(B)** *Tip cells (TCs) selection and induction of sprouting*: VEGF -mediated signals induce DLL4 expression in TCS (blue). TCS are selected by DLL4 expression. **(C)**. *Elongation of sprout*: Dll4-expressing tip cells react strongest to VEGF signaling and acquire a motile, invasive and sprouting phenotype by loosening the contact to adjacent endothelial stalk cells (SCs, pink) by endocytosis of VE cadherin. DLL4 activates Notch proteins in SCs. DLL4-NOTCH signaling is suppressed in Jag1-expressing, SCs, which form the base of the emerging sprout.

(D, E): Lumen formation (D) *Neo-vascularization*: VE-cadherin and other other junction molecules are expressed on filopodia of tip cells (green) and might promote the formation of new inter-endothelial connections between bridging sprouts and the establishment of the newly forming junction. VE-cadherin seal the contact between adjacent endothelial cells, antagonize pro-angiogenic signals by stabilizing cell–cell contacts. At the same time, new sprouts (blue) are induced, which involves stalk-to-tip conversions. **(E)** *Abrogation of sprout*: When conditions do not favor the formation of new endothelial tubules, like the presence of repulsive signals or inhibition of VEGF, vessels aborted.

Chapter 6

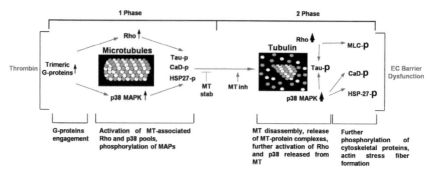

Figure 1. Schematic representation of thrombin-induced MT-mediated EC barrier compromise. Thrombin can induce cytoskeletal reorganization and EC barrier dysfunction in two stages. On the phase 1 thrombin-induced engagement of trimeric G-proteins activates Rho and p38 MAPK pools associated with microtubules leading to phosphorylation of microtubule-associated cytoskeletal targets and microtubule disassembly. On the second phase microtubule dissolution releases microtubule-associated protein complexes, further activates Rho and p38 MAPK pathways, increases phosphorylation of cytoskeletal targets and leads to stress fiber formation and EC barrier compromise. Abbreviations: MT stab: MT stabilization; MT inh: MT inhibitors; MAPs: MT–associated proteins.

T - #0421 - 071024 - C252 - 234/156/11 - PB - 9780367379513 - Gloss Lamination